Hong Kong Takes Flight

HARVARD EAST ASIAN MONOGRAPHS 454

Hong Kong Takes Flight

*Commercial Aviation and the Making
of a Global Hub, 1930s–1998*

John D. Wong

Published by the Harvard University Asia Center
Distributed by Harvard University Press
Cambridge (Massachusetts) and London 2022

© 2022 by The President and Fellows of Harvard College
Printed in the United States of America

The Harvard University Asia Center publishes a monograph series and, in coordination with the Fairbank Center for Chinese Studies, the Korea Institute, the Reischauer Institute of Japanese Studies, and other facilities and institutes, administers research projects designed to further scholarly understanding of China, Japan, Korea, Vietnam, and other Asian countries. The Center also sponsors projects addressing multidisciplinary, transnational, and regional issues in Asia.

Cataloging-in-Publication Data is on file at the Library of Congress.

ISBN 9780674278264 (cloth)

Index by Alexander Trotter

♾ Printed on acid-free paper

Last figure below indicates year of this printing
31 30 29 28 27 26 25 24 23 22

For Gregory and Ian

Contents

List of Figures and Tables	ix
Acknowledgments	xiii
List of Abbreviations	xvii
Introduction	1
1 Mapping Hong Kong: The Making of a Place in an Evolving Air Space	21
2 Reorienting Hong Kong: Resizing and Conforming to Emerging Geopolitics after World War II	60
3 Branding Hong Kong: Fashioning Cathay's Pacific	102
4 Upgrading Hong Kong: The Colony Takes Flight	148
5 Catapulting Hong Kong: Economic Liberalization and Geopolitical Transformations	193
6 Recasting Hong Kong: The Making of a "Hong Kong" Airline	238
Conclusion: What Is Next for Hong Kong?	287
Bibliography	311
Index	329

Figures and Tables

Figures

I.1	Crowds at Kai Tak Airport during its last flights	3
I.2	Growth of air traffic in Hong Kong, 1947–98	5
1.1	British imperial "horseshoe" service, 1939–44	28
2.1	Schematic representation of a map labeled "Composite Plan of Kowloon"	73
2.2	Schematic representation of Cathay Pacific's position in the regional pattern of Southeast Asia, 1950	94
2.3	Annual aircraft movements and passenger counts at Kai Tak, 1947–65	95
2.4	Passenger counts at Kai Tak Airport and Cathay Pacific's share	96
3.1	Different generations of Pan Am logos	106
3.2 a and b	*Betsy* and *Niki*, Cathay Pacific's first planes	107
3.3	Cathay Pacific coaster from the 1950s	107
3.4	Cathay Pacific brochure, 1960	108
3.5	Incoming visitors to Hong Kong by nationality, 1958–69	113

3.6 a and b	Cartoons boasting the cosmopolitan female cabin crew and the speed of the jets	118
3.7 a and b	The wide-ranging looks of Cathay Pacific female cabin crew, 1940s through 1962	124
3.8 a, b, and c	Cathay Pacific's 1962 uniforms for male and female cabin crew	130
3.9	Cathay Pacific's multiethnic cabin crew in national costumes	132
3.10 a and b	Cathay Pacific's 1969 uniform redesign	135
3.11 a and b	Cathay Pacific's 1974 uniform	139
3.12	"Discover the many faces of the Orient on Cathay Pacific Airways."	142
4.1	Number of passengers handled at Kai Tak Airport, 1962–76	150
4.2	Air cargo tonnage handled in Hong Kong, 1962–77	150
4.3	Kai Tak airport terminal	155
5.1	Schematic representation of Cathay Pacific's expansion beyond its regional network	199
5.2 a and b	"Our non-stop flights Hong Kong–Sydney"	201
5.3 a and b	"Today non-stop to Vancouver"	208
5.4 a and b	"Thank you for waiting. Today is a great day. We fly to London."	221
6.1	"People scramble for Cathay shares" at the headquarters of HSBC	251
C.1	Hong Kong exports and aircraft movements, 1947–98	292
C.2	Hong Kong newspapers reporting on the tandem visits of the leaders of the PRC and US on the inaugural day of the new airport at Chek Lap Kok	295
C.3 a and b	Air traffic statistics at the Hong Kong International Airport, 1947–2020	298

Tables

2.1   Cathay Pacific's percentage shareholding after the merger with Hong Kong Airways in 1959   92

4.1   Peer comparison of existing and projected runway lengths, 1968   159

6.1   Percentage shareholding of Cathay Pacific in 1978   246

6.2   Percentage ownership of Cathay Pacific before and after the initial public offering, April 22, 1986   253

Acknowledgments

In the five years during which this project grew from its embryonic formulation to the current book form, I have benefited from the generous support of various individuals. Elisabeth Köll, Angela Leung, Kam Louie, and Helen Siu offered critical help in framing this intellectual endeavor. I tried out my early ideas on my colleagues at the University of Hong Kong, as well as on friends at the Hong Kong History Project at the University of Bristol. Subsequent workshops and conferences took me to Australia, Europe, North America, and various Asian cities where I presented different parts of the project. From these interactions, my inchoate ideas began to take shape, thanks to the comments of colleagues including Prudence Black, Bram Bouwens, Maggie Cao, John Carroll, Carolyn Cartier, Rey Chow, Stephen Chu, Oscar De la Torre, Louise Edwards, Jane Ferguson, Linling Gao-Miles, Max Hirsh, Peter Hobbins, Jessica Hurley, Kendall Johnson, Su Lin Lewis, Lui Tai Lok, Elizabeth Manley, Laikwan Pang, Philip Scranton, Paul Sutter, Andrew Toland, Paul Ugor, Christine Yano, and Wen-hsin Yeh. As I prepared the manuscript, David Edgerton, Po-Shek Fu, and Christopher Munn provided indispensable input at crucial moments. Parts of chapters 5 and 6 were published in *Enterprise & Society* and the *International Journal of Asian Studies*. Comments from the reviewers and editors broadened and deepened this project's scholarly engagement. During the many rounds of revision, members of my reading group—Koji Hirata, Ghassan Moazzin, and Michael Ng—reviewed most if not all of the chapters and made

xiv *Acknowledgments*

insightful suggestions. In the final phase, Robert Bickers combed through my manuscript, recommended additional archival materials, and provided invaluable counsel.

The archival materials that form the basis of this study have come from numerous repositories. For access to government records and materials in the public domain, I am grateful to the Hong Kong Public Libraries, the Hong Kong Public Records Office, the Legislative Council of Hong Kong, the National Archives of Singapore, the National Archives of the UK, and the University of Hong Kong Libraries. The Special Collections Department at the University of Miami Libraries facilitated access to the corporate records of Pan American World Airways. The British Airways Heritage collection hosted my visit to the airline's archival collection, and I thank Tommy Tse for making the introduction. Qantas also allowed me access to their collection. Last, but certainly not least, I would like to express my gratitude to Matthew Edmondson and his colleagues who patiently guided my exploration of the Swire Archives in Hong Kong and London. Swire has also kindly allowed the use of many visuals from the company's collection in the production of this book.

This project would not have been possible without institutional support. A grant from the Hong Kong General Research Fund (project number 17605420), a Sin Wai-Kin fellowship, and an award from the Hsu Long-sing Research Fund provided the necessary financial backing. Apart from the archival visits, these grants allowed me to enlist support from a team of capable researchers that included Kelvin Chan, Jason Chu, Victor Fong, Edward Man, and Leo Shum. A Henry Luce Foundation grant enabled me to finish the book in the idyllic setting of the National Humanities Center in North Carolina. At the Harvard University Asia Center, Bob Graham shepherded the entire publication process from review to production. Jessica Ling professionally and meticulously copy-edited the manuscript.

This intellectual endeavor is inherently intertwined with my personal journey of growing with Hong Kong, enriched by the experience of friends and family. Gus Ng and friends in the airline industry offered their professional perspectives over many rounds of food and drinks. My brother Andrew provided me intellectual and spiritual support throughout the process. My wife Linda accompanied me on this long journey

Acknowledgments XV

of exploration, metaphorically and literally. Gregory and Ian, budding teenagers at the inception of this project, have blossomed into confident young adults as I finished the book. Without their geographical insistence and perseverance, this investigation of global networks in the making might not have been so deeply rooted in Hong Kong.

To Gregory and Ian, I dedicate this book. May they and our fellow Hongkongers spread their wings wide. And may global Hong Kong continue to flourish and reach new heights.

Abbreviations

ANA	Australian National Airways
ATLA	Air Transport Licensing Authority
BOAC	British Overseas Airways Corporation
CAA	Civil Aviation Authority
CAAC	Civil Aviation Administration of China
CAT	Civil Air Transport (of China)
CITIC	China International Trust and Investment Corporation
CMU	confidential memorandum of understanding
CNAC	China National Aviation Corporation
CTS	China Travel Service
EEC	European Economic Community
GDP	gross domestic product
HSBC	Hongkong and Shanghai Banking Corporation
LegCo	Legislative Council
MP	member of Parliament
Pan Am	Pan American
P&O	Peninsular & Oriental Steam Navigation Company
PRC	People's Republic of China
RAF	Royal Air Force
SAR	Special Administrative Region

INTRODUCTION

July 5, 1998. Dragonair's flight KA841 from Chongqing, China, landed at Hong Kong's Kai Tak Airport at 11:38 in the evening. Hong Kong was abuzz with anticipation as its hardy airport welcomed one last flight. Until the very last minute, its crew and passengers did not know that they would have the honor of being the last flight to land at Kai Tak. Cheers shook the cabin when the aircraft passed over Kowloon City at such a close range that passengers could peer into the apartments below.[1]

July 6, 1998. Two minutes past midnight, Cathay Pacific's CX251 took off for London, "one of the first cities to be served from Kai Tak and also the last." As gates 31 and 32 closed their doors on the last flight to depart from Kai Tak, the airport that had served Hong Kong for seventy-three years closed for the evening, and for good.[2] At 1:16 in the morning, Director of Civil Aviation Richard Siegel bade farewell to Kai Tak, which had become a civic icon for Hong Kong.

Kai Tak's story is Hong Kong's story.[3] Hong Kong's growth as a nexus of commercial aviation was no less breathtaking, and at times treacherous,

---

1. Swire HK Archive, *Dragonnews* No. 72, July 1998.

2. *SCMP*, July 6, 1998, 1; *Wen Wei Po*, July 6, 1998, A1, A3; *Ming Pao*, July 6, 1998, A1.

3. Many have written about Kai Tak itself (see, for example, Pigott, *Kai Tak*), and several industry participants have reminisced about it (e.g., Eather, *Airport of the Nine Dragons*). In the last couple of decades, Kai Tak has commanded widespread popular attention in and outside Hong Kong (see, for example, Chung, Kanazawa, and Wong, *Good Bye Kai Tak*; Ng and Ho, *Cong Qide chufa*; Ng, *Zaikan Qide*; Sekine, *Keitoku kaisō*).

2   *Introduction*

than the takeoffs and landings at Kai Tak.[4] Although Kai Tak's experience manifested the development of commercial aviation in Hong Kong and its story aptly symbolizes the trials and tribulations of the aviation industry, the history that this book tells goes far beyond the confines of the expanding footprint of Kai Tak Airport. Hong Kong's history of commercial aviation is also a story of the larger regional and global forces at work.

From its humble beginnings as a grass field to its position as the world's third-busiest airport, Kai Tak "has made a vital contribution to our economy and has also been a gateway for the hopes and dreams of millions of passengers," said Siegel.[5] Kai Tak's history "reflected the dynamic growth" of Hong Kong. The old airport was to have a special place in the hearts of people "not only here in Hong Kong but also . . . around the world." As the last passengers arrived and the last flight departed, Siegel turned off the lights in the air traffic control tower. "Goodbye Kai Tak, and thank you," he said. "Kai Tak has truly been one of the world's great airports, but tonight we must say goodbye to our old friend." Chief Secretary Anson Chan and Financial Secretary Donald Tsang watched as the runway, which had facilitated Hong Kong's takeoff to the world, went dark.[6] A new airport was to illuminate the way to the city's future.

In addition to these and other dignitaries, thousands gathered in the airport's vicinity to say goodbye (fig. I.1). The airport had caused severe noise pollution, and their lives would change with the relocation. No longer would schoolchildren and other residents of the adjoining neighborhood of Kowloon City hear the thundering noise of takeoffs and landings. Nor would airplane enthusiasts or benumbed Hongkongers see the underbellies of aircraft hovering overhead from the low-rise buildings that packed the area.

---

4. A news article published on Kai Tak's final day of operations spoke fondly of "the famed white-knuckle final approach into Kai Tak" that was considered "too difficult to leave to an autopilot" (*SCMP*, July 5, 1998, 64).

5. Unless otherwise specified, all quoted matter has been cited in sources noted at the end of each paragraph.

6. *SCMP*, July 6, 1998, 1; Hong Kong Civil Aviation Department, "Speech Delivered by Director of Civil Aviation."

FIGURE 1.1. Crowds at Kai Tak Airport during its last flights. Photo credit: Edward Wong. Source: *South China Morning Post*.

4   *Introduction*

Kai Tak had completed its mission. Overnight, the baton was passed to Chek Lap Kok, the new airport located in recently landfilled space on Lantau Island, removed from the urban center of Hong Kong, which had grown into a densely populated metropolis during Kai Tak's epoch.

Although local residents should have been eager to enjoy their new-found peace and quiet, something seemed to be missing. Madam Chan, who had lived in the Kowloon City area for decades, lamented the fact that the neighborhood would be less lively with the closure of Kai Tak. Although she was already "missing it dearly," she was ready to accept reality and was planning to check out the new airport right away.[7]

The airport's move to Chek Lap Kok was far from seamless.[8] Two decades later, however, the new airport's birthing pains have receded into memory. Skyscrapers have sprouted amid the tightly packed low-rises that cover much of Kowloon City. After decades of enduring deafening criss-crossing aircraft movements over their heads, local residents should be breathing a sigh of relief that the treacherous traffic did not produce any major disasters. Yet many in Hong Kong continue to wax nostalgic about the turbulent years that Kai Tak witnessed.

"In the early years, Kai Tak was not so busy but, by the late 1980s, there was a plane landing or taking off every couple of minutes from early morning until midnight," noted an interviewee twenty years after the airport's closure. "We had to stop talking to each other at home when aircraft were overhead, and pause phone calls for about thirty seconds, but you got used to it," he added, describing the impact air traffic had on daily lives in Hong Kong. "Strangely enough, I missed Kai Tak once it closed."[9]

One could actually clock the growth of Hong Kong by tallying the interruptions that airplane movements at Kai Tak entailed. The growth of commercial aviation in Hong Kong was indeed phenomenal. In the year ending in March 1948, aviation was a budding industry in Hong Kong, registering 3,662 aircraft landings and 3,647 takeoffs, or ten of each per day.[10] Half a century later, the bustling city was handling over 82,000

---

7. Li Xiaobing 李笑冰, "Zaijian! Qide Jichang 再見！啓德機場," *Wen Wei Po*, July 6, 1998.

8. To alleviate the chaos in the ensuing months, Kai Tak's cargo terminal 2 was temporarily reactivated (Hong Kong Memory, "Kai Tak.").

9. Peters, "Remembering Kai Tak."

10. *HKDCA*, 1955–1956, 35.

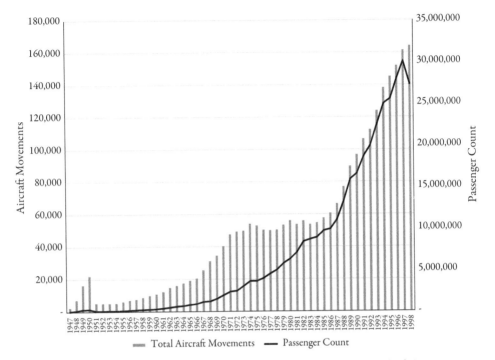

FIGURE I.2. Growth of air traffic in Hong Kong. Source: *Hong Kong Civil Aviation Department Annual Reports*, 1947–98.

aircraft movements in both directions, or some 450 landings and take-offs per day, on average, in Kai Tak's final year. Passenger counts exceeded thirteen million in both directions, more than double the city's population (see fig. I.2).[11]

Kai Tak was not always considered an appropriate location for what ultimately became one of the world's busiest airports. When the returning British colonial regime was planning in earnest for the start of commercial aviation through Hong Kong in the aftermath of World War II, a 1947 government survey reported that Kai Tak was "incapable of anything but very minor development. Kai Tak aerodrome lies in a horseshoe of high hills. . . . It is only safe for the operation of medium and light

---

11. *HKDCA*, 1997–1998, 66.

aircraft in conditions of good visibility."[12] Despite being condemned for its topographical drawbacks and incompatibility with the large-scale expansion needed for modern aviation, Kai Tak persisted for another four decades in its service of a thriving city. In the late 1940s, when the returning British presence and expansion of the United States in the Far East rendered Hong Kong a choice location for the confluence of air traffic, the transformation of the geopolitical landscape in China ironically derailed the relocation of Hong Kong's airport and saved Kai Tak from oblivion. In spite of numerous obstacles, the ensuing decades saw Kai Tak's expansion to accommodate resumed air traffic growth in a reconfigured global economic network in which Hong Kong assumed a larger role. Tensions among various interest groups impeded Kai Tak's development, but local interests prevailed to build the airport into an operation capable of handling the jumbo jets that passed through Hong Kong at escalating frequencies. As Hong Kong and its airport crafted their positions in the world, yet another major geopolitical shift threatened their survival. Delicate negotiations between Britain and China produced an uneventful transition for Hong Kong in 1997 before Kai Tak yielded to a new era under the auspices of the purpose-built Chek Lap Kok, part of a mega reclamation project launched in anticipation of Hong Kong's continuing role as a regional and global hub.

The expanding flight network that Kai Tak serviced was but the physical manifestation of Hong Kong's expanding regional and global connections. Hong Kong's expansion into a global hub certainly did not proceed without a hitch. Hong Kong was not simply drawn into the vortex of globalization. Rather, the city fashioned itself into an aviation hub by first mapping itself onto an evolving aviation network, and then asserting its place within that network through erratic but intensifying traffic flows that extended Hong Kong's reach to regional destinations and distant connection points. Historical developments fostered a conducive environment for the city's growth into a hub of commercial aviation. Yet, without the persistent efforts of resourceful participants, Hong Kong could easily have devolved into a peripheral location on a global map reconfigured by new technologies and geopolitical events. The city's evolution into an aviation hub parallels the making of Hong Kong, which was

12. TNA, T 225/597.

Introduction 7

rife with challenges that were overcome only through a combination of geopolitical happenstance and enterprising efforts. The turbulence of air traffic into Kai Tak, as well as its triumphant takeoffs, epitomizes the historical continuities and disruptions that punctuated Hong Kong's growth into a metropolis.

This study examines the globalization process with a focus on the role of commercial aviation in the budding cosmopolitan nexus of Hong Kong. Commercial aviation took shape as the city developed into a powerful economy. As I examine how commercial aviation and Hong Kong's historical trajectory influenced each other's developmental paths, I seek to understand how this infrastructural elaboration connected Hong Kong to the global economy and how the evolution of Hong Kong into an aviation hub reflected the city's position in the expanding network of a global industry. Rather than understanding air travel as an inevitable outcome of Hong Kong's arrival in the era of global mobility, I argue that Hong Kong's development into a regional and global hub was not preordained. By underscoring the shifting process that produced the hub of Hong Kong, this study aims to describe globalization and global networks in the making, a making that grew out of an inter-Asian, regional configuration. During the second wave of globalization, Hong Kong weathered the Cold War geopolitical shifts, the closing and subsequent reopening of the skyways over mainland China, and the economic liberalization of its major aerial connections to extend the reach of its commercial aviation network. Viewing the globalization of Hong Kong through the prism of its airline industry, this book examines how policymakers and businesses in Hong Kong asserted themselves alongside their international partners and competitors in a bid to accrue socioeconomic benefits, negotiated their interests in the city's economic success, and articulated their expressions of modernity.

The history of commercial aviation in Hong Kong is the story of how a new transportation technology facilitated Hong Kong's growth into a metropolis and remapped the city onto the modern world. Starting with aviation's humble foray into Hong Kong in the 1930s and concluding with the move in airports from Kai Tak to Chek Lap Kok in 1998, the book explores how commercial aviation fashioned regional and global connections against the backdrop of shifting geopolitics and in response to the exigencies of

Hong Kong's economic development. Initially developed to the exclusion of most local residents, air traffic drove Hongkongers to seek inclusion in a worldwide network that embraced Hong Kong. Subsequent economic growth enhanced access to the skies for the city's aspiring travelers. Privatization and deregulation in Western markets opened international airways to colonial Hong Kong just as the looming 1997 handover prompted a shift in business strategies among local operators to accommodate the incoming People's Republic of China (PRC) regime. At the same time, the transformed airways allowed certain anxious Hongkongers to envision livelihoods beyond the territory, fueling the city's diasporic impulse and enabling the expression of transnationalism. This book investigates how Hong Kong's political and economic development affected the structure and institutions of its commercial aviation industry, and how, in turn, the industry influenced and facilitated the city's development into a hub against the backdrop of a reconfiguring global network.

Aviation facilitated intensifying flows of people, goods, and capital. The industry's development allowed Hong Kong to merge onto a common international platform while accommodating articulations of local variations. More than merely enhancing physical connections and economic flows, commercial aviation shaped the people, spaces, and institutions that it connected. Focusing on policymakers in Hong Kong and London, and on Cathay Pacific and other local carriers that operated alongside such industry titans as British Airways (and its predecessors) and Pan American (Pan Am), this analysis combines the business history of aviation with an examination of Hong Kong's economic and political transformations.

What follows is not a history of technological progress but the story of Hong Kong from its initial struggles to establish a foothold in the skyways to its subsequent development of the financial wherewithal to rival technologically superior players despite its inherent topographical disadvantage. Nor is this book a cultural analysis of global connectedness. The rapid development of the airline industry in Hong Kong involved a multilayered and interactive process in which differences were (re)defined and meanings (re)negotiated in the context of Hong Kong's economic dynamism. Rather than involving myself in conceptual elaborations of global mobility, I subject these notions to empirical evidence and examine how

Introduction   9

the airline network evolved and mapped Hong Kong onto the world of commercial aviation. This analysis focuses on key historical junctures at which various interested parties wired Hong Kong into the international network, geopolitical upheavals transformed the city into an aviation hub, and participants in Hong Kong wove the city into a transnational infrastructure and in the process advanced their interests. It examines how Cold War politics reconfigured flight routes and remapped Hong Kong, how the infrastructures of air mobility shaped the city's development, and how air travel fueled aspirations in the high-growth economy of Hong Kong.

This book reveals various parties' efforts to contribute to and profit from Hong Kong becoming a hub of global flows through commercial aviation. The process was rife with challenges. In the 1930s, resourceful British business interests, alongside their American counterparts, put Hong Kong on the world's flight map. Yet, conflicting priorities would pit the colony against the metropole and frustrate progress in later periods. The city's proximity to and connections with the markets of mainland China proved instrumental to its establishment as an aviation hub. In the following decade, the Communist takeover of the mainland and the severance of China from commercial aviation with the capitalistic bloc threatened Hong Kong's survival as a hub. The PRC's reemergence in the late 1970s reinvigorated aspirations to connect with the mainland through the airways of Hong Kong while at the same time introducing political risks for an industry dominated by British players.

Commercial aviation extended the reach of Hong Kong in stages. In the aftermath of World War II, enterprising efforts conspired with geopolitical advantage to put Hong Kong on the aviation map just as seismic shifts in the international landscape thwarted sustained development. As the Cold War broke out, transformation in the global environment afforded Hong Kong another opportunity, albeit on a scale initially smaller than originally anticipated. In time, expanding tourist traffic and escalating air cargo volume fueled the growth of commercial aviation in Hong Kong. Through the insistence of local players, Hong Kong claimed a niche in the expanding industry and expanded its infrastructure at Kai Tak. Additional headwinds threatened the industry's survival in this city on borrowed time, and, yet, thanks to pragmatism, both corporate and individual, commercial aviation prevailed and reached new heights in the closing decades of the twentieth century. Empirical evidence on this trying process sheds light on how individuals and institutions in Hong Kong shaped and were shaped by commercial aviation during an era of spectacular economic growth, both physically and conceptually. A major theme throughout this study is the pragmatic attitude that various parties embraced in sustaining the development of commercial aviation in Hong Kong. Leveraging macro developments on the world stage, the industry made do with the resources available at each phase of its development. Throughout the 1970s, the commercial aviation industry created a regional nexus in Hong Kong that European and American titans connected to far-flung locations. In the subsequent period, the regional outpost grew into its own, maturing into a global hub that commanded flight routes to destinations far and near. The findings remain instructive today as shifting geopolitics reconfigure Hong Kong's centrality as a hub and Hongkongers reimagine their place in a world of rerouting traffic.

Commercial Aviation of Hong Kong in Regional and Global History

To date, scholarly studies have examined the expansion of commercial aviation as a global phenomenon. Existing scholarship has demonstrated the upsurge in global mobility that aviation generated in the twentieth century and Western control of the airways. Studies of the industry's development in Asia have celebrated business successes. Through archival research in Hong Kong, Britain, the United States, Australia, Taiwan, and Singapore, this study rejects the treatment of commercial aviation's development in Hong Kong as merely the outcome of the city's globalization. Instead, it reconceptualizes the city's growth into an aviation hub as a dynamic and interconnected process whereby participants, in an era of intensifying globalization and Hong Kong's own economic expansion, leveraged regional and global dynamics in shaping local developments.

Local Focus of Global Processes. Scholarly studies analyzing the beginnings of aviation often highlight its origins in military use.[13] In contrast,

---

13.   Edgerton, *England and the Aeroplane*; Melzer, *Wings for the Rising Sun.*

Introduction

this study focuses on commercial aviation alone. Previous work examining the expansion of commercial aviation as a global phenomenon has focused largely on the Western powers' assertion of their global influence through dominance of the airways.[14] Nonetheless, even as Western powers commanded such dominance, local and regional players also strove to establish a footing. The phenomenal economic growth of Hong Kong and other Asian cities afforded these locations mounting control of the skies and grew their carriers into formidable airlines. By focusing on the making of an aviation hub in Hong Kong and the development of commercial aviation in the city during its economic takeoff, this project underscores the reconfiguration of connections with Hong Kong via airways against the backdrop of international forces.[15]

Beyond the history of commercial aviation, the story this book tells involves the development of Hong Kong in the midst of broader global transformations. Contrary to the prevailing trend of decolonization in the post–World War II era, Hong Kong returned to British colonial administration. As mainland China came under Communist rule, Hong Kong became a foothold of the "free world" at the tip of southern China. In time, the little colony grew into its own, alongside other Asian economies, and sought increasing autonomy from the metropole. Yet, just as the city was developing its own sense of identity, it learned of its eventual return to Chinese sovereignty. The growth of commercial aviation in Hong Kong accentuated the peculiar local manifestations of these regional and global forces.

The development of commercial aviation into a global industry is illustrative of, and was instrumental in, the second wave of globalization that business historians consider to have started in the 1950s.[16] Just as steamships and railroads reduced distance in the first wave (circa the 1880s to 1920s), the rise of airways was a technological breakthrough that networked

---

14.   Van Vleck, *Empire of the Air*.

15.   Philip Chen, a veteran of the aviation industry, highlights the importance of aviation in the making of Hong Kong into one of the "greatest cities of the world." See Chen, *Greatest Cities of the World*.

16.   Jones, "Globalization." Rather than comparing and contrasting capital flows in different periods of globalization (see, for example, Schularick, "A Tale of Two 'Globalizations'"), this book primarily emphasizes the physical networks that commercial aviation facilitated for Hong Kong with a view toward the other types of connections engendered.

Introduction

regional and global systems.[17] Compared with the first wave of globalization, the second wave followed a fractured pattern in which the Cold War divided the world into two competing camps, recalibrating the political distance between locales according to their allegiance to one or the other camp. The development of commercial aviation in Hong Kong adhered to these political contours as flight routes remained confined to the capitalistic bloc until tensions eased in the late 1970s, generating new opportunities and challenges along with the shift in geopolitics. Commercial aviation connections that started with the regional soon extended to far-flung locations, providing the physical network that facilitated the second wave of globalization.

Wiring Globalization. This study espouses a regional and global framework that emphasizes the connections made possible by global resources while being mindful of the local elements at work.[18] Building on the literature on inter-Asian flows,[19] it investigates the concrete undertakings by various parties that together fashioned Hong Kong into an aviation hub. Aviation creates a liminal space that divides and connects at the same time. Similar to seaports and maritime routes, airports structure a boundary, both physical and mental, that airways transcend. For residents in a city undergoing economic transformation, this boundary produces anxiety, but its crossing (or at least the hope thereof) arouses aspiration. For Hong Kong, the connections provided by commercial aviation served both as facilitators and emblems of economic development. Yet, those connections were not preordained. Nor were they new. Seismic geopolitical shifts and technological advancements threatened the centrality of nexuses in fluid networks. This book explores how participants in and beyond Hong Kong wired and rewired Hong Kong into a regional and global hub.[20]

Mapping Hong Kong onto the airspace dovetailed with the city's earlier role as a shipping center. Not surprisingly, many local participants,

---

17. On steam globalization, see Darwin, *Unlocking the World*.

18. Tagliacozzo, Siu, and Perdue, *Asia Inside Out: Changing Times*; Tagliacozzo, Siu, and Perdue, *Asia Inside Out: Connected Places*.

19. Hamashita, "Tribute and Treaties"; Sugihara, *Japan, China, and the Growth*.

20. This study parallels Wang Gungwu's observations about the changing role of Hong Kong against the shifting geopolitics of the twentieth century ("Hong Kong's Twentieth Century").

Introduction

among them the companies Swire and Jardine Matheson,[21] were prominent maritime shipping enterprises. Although technological enhancements would progressively extend the reach of flight from Hong Kong, in the early decades, the commercial aviation routes radiating from this hub at the southern tip of China were largely regional in scope, extending to East and Southeast Asia, but conspicuously leaving out mainland China. Rather than focusing primarily on metropole-periphery dynamics, this study also considers Hong Kong in the regional context of decolonization and simmering Cold War tensions. The resulting inter-Asian flows engendered not only a special brand of cultural exchange but also a set of commercial connections specific to that era. Shifting political dynamics and technological progress stretched the reach of Hong Kong–centered airways just as mainland China was reentering the system. From the perspective of Hong Kong's airways, the creation of the global was the continued extension, and reconfiguration, of the regional. The proliferating flight routes to and from Hong Kong thus provide a concrete visual representation of the city's expanding regional and global network against the dynamic transformations of the second half of the twentieth century.

That vibrant era witnessed the development of a complex set of interdependent Asian economies, woven together by flows of capital, goods, people, and ideas, as well as reconstituted ties to mainland China.[22] Hubs emerged to facilitate inter-Asian flows and to serve as gateways for North America and Europe. The development of airline networks that sprang up to support these business linkages mirrors the enduring making and remaking of these regional and global networks.[23] By focusing on the development of commercial aviation during Hong Kong's eventful and dramatic economic rise, this book explores both the breadth and depth of the links between air transport and economic development in Hong Kong's transformation in the context of regional and global dynamics.

Ties to Economic Development. Despite significant public interest, scholarly analysis of air travel in Hong Kong has yet to produce a study of the development of the aviation industry in the context of the city's

---

21. John Swire & Sons (and its affiliate Butterfield and Swire, hereafter referred to as Swire) and Jardine Matheson & Co. (hereafter referred to as Jardine Matheson) are two prominent British conglomerates operating in post–World War II Hong Kong.

22. Urata, Yue, and Kimura, *Multinationals and Economic Growth.*

23. Taylor and Derudder, *World City Network.*

14   *Introduction*

transformation in the second half of the twentieth century. In contrast to academic studies that approach the airline industry from the perspective of politics and finance,[24] this book seeks to underscore the geopolitical context of the development of commercial aviation in Hong Kong against the backdrop of its economic growth. Considering the skies as a "space of flows,"[25] the book examines how commercial aviation shaped and reflected the economic development of Hong Kong and crafted the city's unique position in a rapidly changing global landscape.

The book provides a specific lens through which to understand the economic growth of Hong Kong. Whereas existing studies of the history of Hong Kong provide a comprehensive overview of the city's past,[26] this book focuses on an industry both critical to and reflective of Hong Kong's economic development, thereby providing an in-depth exploration of the crucial actors involved, both corporate and individual.[27] Building on the extant literature on the economic history of Hong Kong,[28] it offers empirical evidence on a pivotal industry and informs our understanding of the role of the state in facilitating global connections and economic growth.

Society and Technology. Focusing on commercial aviation as a technological breakthrough for mobility, this book responds to calls by scholars of the history of technology who draw our attention to the interaction between society and technology, and the consumption of technology.[29] Aviation contributed to an upsurge in global mobility in the twentieth century.[30] Like steam and rail travel in the preceding century, air travel inspired new understandings and imaginations of modernity, reshaped geopolitics, and transformed networks.[31] As the governance of aerial space

24.   See, for example, Rimmer, *Asian-Pacific Rim Logistics.*
25.   Castells, *Rise of the Network Society.*
26.   Carroll, *Concise History of Hong Kong*; Tsang, *Modern History of Hong Kong.*
27.   Compared with David R. Meyer's emphasis on social networks, this study underscores both the development of hardware infrastructure and the role of individual intermediaries. See *Hong Kong.*
28.   Schenk, "Negotiating Positive Non-interventionism"; Schenk, "The Empire Strikes Back."
29.   Wilkins, "Role of Private Business"; Edgerton, "From Innovation to Use"; Oudshoorn and Pinch, *How Users Matter.*
30.   Cwerner, Kesselring, and Urry, *Aeromobilities.*
31.   Köll, *Railroads and the Transformation.*

Introduction   15

evolved legally and politically,[32] historical routes conditioned the development of airways but the redefinition of space remained fluid. At the same time that this new form of transportation transformed Hong Kong, local manifestations of geopolitical shifts and the city's own trajectory of economic growth also shaped the manner in which commercial aviation developed in Hong Kong. Emblematic of modernity and economic progress, air travel also enabled certain developments in the postwar expression of gender.[33] I explore these gender issues in the context of the operators serving Hong Kong and further contribute to the discussion of local identities in the airline industry, a subject that has received only sporadic scholarly coverage.[34]

Transnational Hong Kong. As the airlines extended their aerial reach for the expanding middle class, they profited from, and operated with, aspiring worldviews rooted in the transnational particularities of their cities.[35] This study provides concrete manifestations of this process in Hong Kong as individuals and corporations interacted in a bid to profit from the city's development into a hub of global flows. As in other Asian contexts, when participants in Hong Kong developed their special brand of cosmopolitanism and transnationalism,[36] they drew energy from various local, regional, and global forces. Through its examination of one industry crucial to this development, this analysis explores in concrete terms the process that engendered cosmopolitan attitudes and reveals unevenness across elastic distinctions of ethnicity, social class, and gender constructs.

Although this study builds on theoretical frameworks on global flows, mobility, and connectivity, its primary contribution lies in its analysis of

---

32. Banner, *Who Owns the Sky?*; Goedhuis, "Sovereignty and Freedom"; Little, "Control of International Air Transport."

33. McCarthy, Budd, and Ison, "Gender on the Flightdeck"; Hynes and Puckett, "Feminine Leadership"; Yano, *Airborne Dreams*; Barry, *Femininity in Flight*; Mitchell, Kristovics, and Vermeulen, "Gender Issues in Aviation"; Rietsema, "Case Study of Gender."

34. Foss, *Food in the Air*; Hickson, *Mr. SIA*; Levine, *Dragon Takes Flight*; Ng, *Xianggang hangkong 125 nian*; Lau, Wong, and Chin, *Tiankongxia de chuanqi*; Dunnaway, *Hong Kong High*; Pirie, *Cultures and Caricatures*; Heracleous, Wirtz, and Pangarkar, *Flying High*.

35. Appiah, "Cosmopolitan Patriots."

36. Ong, *Flexible Citizenship*.

16 *Introduction*

empirical evidence on Hong Kong's place in the rapidly expanding world of commercial aviation. The analysis here underscores the backdrop of the city's economic takeoff, the efforts of specific parties at various points in the process, and the lived experience of local participants.

In this first academic study of an infrastructure development that proved crucial to Hong Kong in both economic and sociopolitical terms, I seek to reveal the process by which the airline industry, the city's phenomenal economic growth, and the expansion of its regional and global footprint unfolded in conjunction with one another. The book aims to understand how policymakers, businesspeople, and the public sought to derive benefits from commercial aviation during Hong Kong's dynamic transformation into a hub of global flows. It constitutes a study that foregrounds an industry and its associated infrastructure, corporations, and key players to shed light on the story of mid- to late twentieth-century Hong Kong and the globalizing context from which it emerged.

Structure of the Book

By examining how airlines reconfigured networks in Hong Kong from the humble beginnings of commercial aviation in the 1930s to the decommissioning of Kai Tak in 1998, the book explores the regional and global connections that flight routes facilitated in the city's late colonial era. These connections enabled the economic growth of Hong Kong on the one hand and, on the other, reflected the city's political and economic exigencies at various stages of its development. Against the backdrop of dynamic geopolitics and its own phenomenal economic growth, Hong Kong participated in the expansion of commercial aviation worldwide and grew into a formidable global player in the second half of the twentieth century, as regional and local developments punctuated the process at certain critical junctures.

Hong Kong developed into an aviation center in the context of regional and global history. Technological advancements promised enhanced connections, but geopolitical turmoil severed and reconstituted linkages. Hong Kong's emergence as a commercial aviation hub was not

Introduction 17

predestined. Chapter 1 explores how Hong Kong mapped onto the network of the budding aviation industry. In the 1930s, technological requirements such as those of the seaborne landing of early American clippers extended the city's seaport functions to aerial connections.[37] British hubris, ironically coupled with fears of being reduced to second-tier status, propelled policymakers to connect Hong Kong with aerial traffic operated by American and Chinese companies. The centrality of Hong Kong to international aviation was constructed at key historical moments. At the same time, this centrality was planned and negotiated, resulting from pragmatic adjustments as the geopolitical situation unfolded.

Chapter 2 examines the process by which shifting geopolitical dynamics influenced the pace and scale of infrastructure development for commercial aviation in Hong Kong. After the Communist takeover of mainland China in 1949, the British authorities made the expedient decision not to construct a new airport but rather to expand Kai Tak. Dimmed prospects of connections with mainland China prompted British efforts to construct a network of commercial aviation to link Hong Kong up with colonial ports in British Malaya and Singapore, as well as other regional hubs in Southeast Asian countries such as Thailand. Hong Kong then served to connect this region to a Cold War corridor that routed air traffic to avoid mainland China and provided a passage to the United States through the Philippines, Taiwan, Japan, and South Korea. Long-haul connections also extended these regional routes to Europe, and via a mid-Pacific passageway, to North America. This configuration formed the blueprint for the development of commercial aviation in Hong Kong through the early 1970s, as industry titans dominated long-distance traffic and burgeoning local carriers took up the regional connections with Hong Kong. The combination of such long-haul and regional aviation businesses underwrote the development of Kai Tak, which unfolded in incremental segments and assumed a more practical scale. The resulting infrastructure at Kai Tak ushered Hong Kong into the jet age and proved itself capable of coping with subsequent traffic growth and technological advancements through continuous improvement.

---

37. Unlike land planes which required runways, these flying boats, which Pan Am called "clippers," could operate in any city with a sheltered harbor.

18 *Introduction*

Against this backdrop, Hong Kong came into its own economically and politically. In the corporate arena, Cathay Pacific emerged as Hong Kong's iconic carrier. The construction of a brand identity for this Hong Kong carrier was a discursive process that involved issues of race, class, and gender with regard to both the airline's target clientele and its management team and workforce. Through the lens of the changing designs of its flight attendants' uniforms, chapter 3 explores the development of Cathay Pacific, which reflects the demilitarization of postwar reconstruction, the search for an identity in greater China and a decolonizing Asia, and intensifying industrialization at the height of the Cold War.

The expansion of flight networks required more than just the ingenuity of corporate participants. Chapter 4 studies the process of updating Kai Tak for technological progress and the difficulties it entailed. At the political level, the colonial government endeavored to develop aviation infrastructure in Hong Kong and to petition London for control over traffic rights. At the same time, London had to contend with its waning colonial power and shrinking influence on the global stage. In the process, the colonial administration diverged further from the metropole's policy and sought increasing autonomy in its effort to develop commercial aviation in the interests of Hong Kong.

The worldwide trends of privatization and deregulation, coupled with technological breakthroughs, fueled the growth of aspiring carriers. Chapter 5 investigates the process by which Cathay Pacific emerged as a powerful commercial aviation operator with flights radiating from the commanding base of Hong Kong's robust economy. The development of Hong Kong's aviation networks from the late 1970s to the 1980s reflected global economic trends, the availability of advanced technologies, and the period's shifting geopolitics.

As Cathay Pacific grew to rival the industry titans as Hong Kong was proving itself to be a formidable economic power, a new wave of political developments arose to redirect the trajectory of both the city and its airline. With the 1997 handover looming, the sustainability of British interests in Hong Kong became questionable. Chapter 6 discusses a corporate project to recast Cathay Pacific's identity in view of such changing political contours. In its effort to claim authenticity for its local status in Hong Kong, especially in the face of budding competition, Cathay Pacific altered the profile of its workforce to accommodate a local Chinese

Introduction 19

presence in the cockpit and accentuate Hong Kong representation in the flight crew. Complementing these changes in human resource policy, the airline modified its capital base. To rebrand the nationality of its ownership, Cathay Pacific raised the local profile of its investor base in the 1980s. In a bid to structure a working relationship with the incoming political regime for the benefit of not only the airline but also its parent company Swire, Cathay Pacific embraced "red capital" from mainland-controlled enterprises with connections to Beijing. Throughout this process, the carrier demonstrated considerable agility and resilience against a rapidly shifting geopolitical backdrop.

Rather than viewing the dawning of the jet age in Hong Kong as its incorporation into the global stage, I study how geopolitical contingencies and individual undertakings influenced the manner in which airways situated Hong Kong on the global map at different points in time. Focusing on the period from the industry's emergence in the 1930s through the city's economic takeoff, this book explains the reconfigured flows of people, goods, capital, and ideas in the evolving air space that connected Hong Kong, a space that air travel both compressed and enlarged at the same time. Like many other technological advances, aviation made and diminished localities and identities, and delineated spaces and bridged them. Combining the business history of aviation in Hong Kong with a study of the city's economic growth and expanding geographical reach, this analysis shows how people, places, and institutions shaped and came to be shaped by the development of commercial aviation through the city's growth from postwar destruction to a manufacturing powerhouse, and subsequently its transformation into a hub in the global economy. Instead of being passive partakers in globalization, industry participants— government officials, corporations, and individuals—took an active role in leveraging regional and global dynamics in shaping the local developments of commercial aviation in Hong Kong.

Far from being a predestined conclusion, the making of Hong Kong into a regional and then global metropolis was rife with challenges.[38] The trials and tribulations of its evolving commercial aviation network

---

38. Although I echo David R. Meyer's characterization of Hong Kong as a "global metropolis" as calibrated by the flows it facilitates, I see its development, as reflected in commercial aviation, as a protracted process punctuated by historical contingencies

20   *Introduction*

underscore the obstacles Hong Kong faced at the crossroads of traffic to and from mainland China and East and Southeast Asia, as well as farther afield to connections in Europe, North America, Australia, and beyond. Its position as a British colonial outpost and function as a gateway to mainland China secured for Hong Kong its earliest aerial connections, which dovetailed with its seaborne networks. In the subsequent development of its commercial aviation industry, geopolitical happenstance, some of it long in the making and some abrupt, threatened to derail the construction of airways through Hong Kong at various points. Practical minds and resourceful enterprisers prevailed and, through fits and starts, fashioned Hong Kong into a regional and global hub. In the face of foreboding geopolitics, financial hurdles, and identity crises, entrepreneurial players in Hong Kong leveraged the city's peripherality and liminality to map and remap Hong Kong time and again at the crossroads of traffic.[39] Such agility underwrote Hong Kong's growth as a metropolis and remains ever crucial as the city charts its way forward and upward.

---

during the geopolitically tumultuous era from World War II to the end of the century. See Meyer, *Hong Kong.*

39.   Prasenjit Duara has observed that Hong Kong has taken advantage of both its geographically peripheral location and "strategic liminality" for survival and prosperity ("Hong Kong," 228).

CHAPTER I

Mapping Hong Kong

The Making of a Place in an
Evolving Air Space

> In the not-so-distant future, civil aircraft will gather in Hong Kong. Hong Kong will become a nexus of civil aviation. The city has long dominated transportation by sea. It is possible that civil aviation will take its place in the future.
>
> —*Chinese Mail* 華字日報, October 13, 1936 [1]

Aviation held transformative potential in the remapping of regional and global networks. Yet, this mode of transportation, relatively new in the 1930s, grew out of an existing framework of connections. That Hong Kong was a seaport at the southern tip of China made it a natural candidate for an aviation hub. However, geographical location and topographical considerations were not the only factors that animated the development of commercial aviation. In the case of Hong Kong, the city's appeal lay primarily in its potential as both a point of connection with mainland China's cities and a point of confluence for traffic emanating from fledgling aviation hubs in North America and Europe.

In those early days, technological limitations, in particular the short distances that flights could cover, posed formidable challenges to the charting of flight routes against existing geopolitical formations. As Pan Am extended its reach across the Pacific, the airline plotted ways to link its flight network to the eastern side of the ocean. With its affiliate China National Aviation Corporation (CNAC),[2] Pan Am endeavored to

---

1. *CM*, October 13, 1936, 8.
2. Hoover Institution Archives, W. Langhorne Bond Papers, 1930–1998, Box 2. For a detailed history of CNAC, see Leary, *Dragon's Wings*.

construct a network that funneled traffic through the Pearl River Delta area at the southern end of the Chinese land mass. In Southeast Asia, colonial regimes were striving to connect their colonies with the metropoles. Although the British Empire may have lost some of its luster by the 1930s, the British airline Imperial Airways staked out its claim with aerial connections to Hong Kong, the empire's foothold in the region. In Hong Kong, the US and British aviation titans would meet, along with the flights that collected traffic from mainland China. Their convergence in Hong Kong as flight routes began to populate the region fashioned the British colony into an inchoate hub of commercial aviation.

The making of Hong Kong into a commercial aviation hub is part of both the political history of aviation and the business story of enterprising airlines operating in an emerging industry. Scholars focusing on political history have expounded on the relations between Hong Kong's interests and developments in China, producing studies that consider Hong Kong's interests as subsumed under British interests in China.[3] From a military standpoint, Hong Kong was an exposed outpost of the British Empire, a bargaining chip for different British military services in the fight for resources.[4] Business historians point to the agency of Hong Kong entrepreneurs who leveraged the city's special status as peripheral to both China and Britain to develop "transnational economic citizenship."[5] Hong Kong's success stemmed largely from its position at the edge of the Chinese and British empires and its special location as "not being part of China."[6] In this configuration, a complex network of relationships propelled colonial developments in Hong Kong.[7] From this context of geopolitical strategies and commercial interests grew a blueprint of commercial aviation that afforded Hong Kong a place on the map.

This chapter explores how commercial enterprises, alongside political powers, built Hong Kong into a promising hub that could assume control over the air space of the region. The mounting importance of this hub necessitated the development of an airport primed for the anticipated

3. Chan Lau, *China, Britain, and Hong Kong.*
4. Kwong and Tsoi, *Eastern Fortress.*
5. Chung, *Chinese Business Groups*; Kuo, "Chinese Bourgeois Nationalism."
6. Carroll, *Edge of Empires*, 57.
7. Law, *Collaborative Colonial Power.*

traffic growth of this budding industry. Functioning as an in-between place that connected China to the region and the world,[8] Hong Kong thrived on providing maritime connections before the age of aviation. Its evolution into a center of commercial aviation from the 1930s to the 1950s depended on many contingencies that arose from Hong Kong's status as a British colony, its attraction to American interests as a focal point in the region, and its role as China's outlet to the Western world even before the Communist takeover of 1949. The development of Hong Kong into a commercial aviation hub was much more than the physical expansion of the Kai Tak airfield into an international airport—it was the process of infusing significance into a place.

Vying for Connections: The Making of a Prewar Hub

Hong Kong emerged on the commercial aviation map as American and European carriers took control of their domestic markets and vied for connections in international airspace. Across the vast Pacific Ocean came Pan Am. Founded in 1927, Pan Am served as the exclusive international airline of the United States until the end of World War II. In its early years, generous postal subsidies and the diplomatic support of the US government allowed the fledgling carrier to expand out from its network with Central and South America to offer inaugural flights across the Pacific to the Philippines by 1935.[9] In competition with the long distances covered by Pan Am flight routes, along with the vast domestic market other US carriers commanded, European carriers focused on short distances to complement the services of Europe's steamship and rail networks while building their reach to colonies on the other side of the globe.[10] Among the European colonial powers, the British Empire was the largest. After a brief period of competition among British airlines, Imperial Airways was created in 1924 as the empire's "chosen instrument" and

---

8. Sinn, *Pacific Crossing.*
9. Van Vleck, *Empire of the Air,* 2; Davies, *History of the World's Airlines,* chap. 9.
10. Lyth, "Empire's Airway," 865; Davies, *History of the World's Airlines,* chap. 11.

operated as a private monopoly with a public subsidy.[11] By the mid-1930s, Imperial Airways had developed a network that had the longest routes in the world, with mail delivery considered the most important task of the airline, taking precedence over passenger transport.[12] Linking up with the Australian carrier Qantas (Queensland and Northern Territory Aerial Services), Imperial Airways in 1934 opened a thirteen-thousand-mile, twelve-day marathon route between London and Brisbane.[13] A branch to this network would eventually extend to Hong Kong and connect with Pan Am's expanded routes, albeit not until the various parties had finished negotiating their interests in the connection.

Despite earlier attempts by the government and private authorities,[14] the development of commercial aviation in Hong Kong began in earnest only in the 1930s. The rise of commercial aviation was not an isolated event in the British colony but part of a fervor that enveloped the region. Local governments and business enterprises, as well as foreign concerns, exerted control over the development of this burgeoning industry. In the business circles of Hong Kong, discussions emerged concerning how the city could become an aviation center. "That Hongkong has an assured future as an airport, despite its slow development, is the generally accepted local opinion," the local English-language newspaper *South China Morning Post* reported in 1933 from a meeting of the Hong Kong Rotary Club. The president of the United States Aircraft Export Corporation envisioned for the Rotarians a Hong Kong airport that was to serve "as a junction of the world's air lines, as well as a terminal base—the landing field of trans-Pacific planes and the centre for many 'feeder' lines." The news report highlighted Imperial Airways' plans for service from England through India and Singapore to Australia. It was "only to be expected" that this British aviation powerhouse would "endeavour to exploit the passenger and freight potentialities of South China." For the business community, connections from Hong Kong to various Chinese cities appeared particularly promising. "To-day the Colony ranks high among the shipping ports

---

11. Lyth, "Empire's Airway," 869; Lyth, "Chosen Instruments"; Davies, *History of the World's Airlines*, 34.

12. Lyth, "Empire's Airway," 873–74.

13. Lyth, "Empire's Airway," 877.

14. Ng, *Xianggang hangkong 125 nian*, 5–16.

of the world," noted the article. Foreseeing Hong Kong's transition from a seaport into an air hub, the article concluded, "There is no reason why in the future it should not rank equally high among the world's airports."[15]

British shipping powerhouse Swire explored the possibility of making the company an agent for Imperial Airways as early as 1933. In its application for Imperial Airways' agency in 1934, Swire promoted its shipping experience and trading network: "We have been trading in the Far East for over sixty years, during which time we have established a position as the leading British shipping organisation in the China trade." On the basis of its "experience as shipping agents, and the wide-spread nature of [its] organisation," Swire expressed confidence in providing Imperial Airways agency service in the new industry of commercial aviation.[16]

In addition to local interests in Hong Kong, mainland Chinese concerns, British commercial and political parties, and overseas airlines such as those from France and the Netherlands recognized the opportunities in Hong Kong and the region. Among these overseas ventures, Pan Am was particularly eager to capitalize on opportunities to link Hong Kong to its air routes.[17]

While Imperial Airways and Pan Am were plotting long-haul connections through Hong Kong, regional traffic also began to look promising, especially linkages to cities in mainland China. Excitement mounted in the early 1930s not only for air traffic within the mainland but also for connecting the mainland to Hong Kong, the United States, and Europe.[18] In line with its policy to attract foreign capital for cooperative ventures under strict Chinese control, the Republican government in the mainland had founded one Sino-American and one Sino-German corporation to pursue aviation business.[19] The Sino-American CNAC, functioning as an airline, began trial runs to Hong Kong from Shanghai.[20] However,

15. *SCMP*, June 9, 1933, 10.

16. JSS, 13/8/4/3 Imperial Airways, Correspondence (incomplete), 1933–1941.

17. *SCMP*, August 23, 1933, 15.

18. Hope, "Developing Airways in China."

19. Kirby, "Traditions of Centrality," 26; Davies, *History of the World's Airlines*, 188–91.

20. *Hong Kong Sunday Herald*, April 2, 1933, 1; *SCMP*, April 3, 1933, 14; *KSEN*, August 14, 1933, 1; *TiKP*, August 17, 1933, 3.

despite its attempt to work with the authorities in Hong Kong,[21] its requests to set up a station in Hong Kong did not readily win the approval of the British colonial government.[22] Nor did the Americans succeed when they sought to establish a base in Hong Kong for Pan Am's service to the Philippines.[23] The Americans' objective of using this connection as a foothold for their budding network in China was not lost on the British. Through their territorial control and political power, these two giants of the early era of commercial aviation began competing intensely on network expansion, with Hong Kong becoming the pawn in this game of chess.

The colonial authorities were indeed striving to make Hong Kong a hub of civil aviation. In 1934, Hong Kong legislators showed their impatience. Legislator William H. Bell, an oil executive and vice chair of the Hong Kong General Chamber of Commerce, lamented that he saw "no immediate prospects of Hong Kong being linked up with any of the main airways of the world." The city could ill afford "being left out" of the great race that before long would link up "most of the great commercial centres of the world . . . by air."[24] "Hong Kong is located at the choke point of southern China as well as the nexus of traffic between Europe and Asia," observed a local Chinese newspaper in March 1935. "The city occupies an important position in communication, both at sea and in the air. Anyone with foresight could predict Hong Kong becoming a hub of civil aviation in the Far East." The article reported the active participation of the Hong Kong government in this development. Understandably, British officials gave priority to the main British airline, Imperial Airways.[25] In 1935, the Hong Kong colonial secretary expressed hope that by the end of 1936, the colony would be "linked up with Imperial Airways or other air lines."[26]

Imperial Airways was hard at work to make that hope a reality. In October 1935, the airline sent a captain, a first officer, an engineer, and a

---

21. *KSMN*, October 25, 1933, 9.
22. *TiKP*, July 27, 1934, 3.
23. *SCMP*, July 26, 1934, 10.
24. LegCo, September 27, 1934, 172.
25. *TiKP*, March 8, 1935, 3; *KSMN*, March 8, 1935, 9.
26. LegCo, September 12, 1935, 162.

wireless operator to conduct a survey flight between Hong Kong and the British Straits Settlement of Penang. Immediately upon landing at Kai Tak, Captain W. Armstrong told reporters that the establishment of an air link between the two British outposts was "not only probable but practicable."[27] The first visit to Hong Kong by an Imperial Airways aircraft came with a letter from the managing director of the airline to the editor of the *South China Morning Post*. In the letter, the airline expressed its managing director's understanding of "the importance of an air service linking Hong Kong with the England-Australia route" and their decision to conduct test flights between Hong Kong and Penang, "the point of contact with the main route," with connections to "the through services operating in both directions."[28] In the pre–World War II world of aviation, the "horseshoe route" that formed the main trunk service of Imperial Airways sought to connect the British metropole, especially its mail services, with the far-flung corners of the British Empire. One line ran from London to the southern tip of Africa, while the other flew through continental Europe, the Middle East, India, Burma, and Malaya to Australia and New Zealand. The branch line to Hong Kong did not form an integral part of the trunk but connected the colony to the metropole and prepared the British network for northbound possibilities from this Far Eastern outpost (fig. 1.1).[29]

Because of its interests in protecting Imperial Airways, the British colonial government was not forthcoming with support for American efforts. Despite the incessant overtures of Pan Am, the Hong Kong government withheld permission for the airline's planes to land in the colony.[30] Pan Am hedged its bets by indicating its intention to establish

27. *SCMP*, October 5, 1935, 12; *KSMN*, October 6, 1935, 9; *TiKP*, October 6, 1935, 3.

28. *SCMP*, October 5, 1935, 12.

29. British Airways Archives, "O Series," Services, 6258; British Library, IOR/L/E/9/92, Collection 2/9A Civil Aviation —British Air Mail Service Flights across Siam to Hong Kong, Singapore and Australia; Davies, *History of the World's Airlines*, 325; Rimmer, "Australia through the Prism." Trunk routes are established air routes that connect the larger cities of a country or empire. The political entity that specifies such routes considers the connections that these routes facilitate to be of strategic importance, militarily and commercially, to the state.

30. *KSMN*, March 26, 1935, 10; *KSMN*, June 4, 1935, 10; *SCMP*, October 30, 1935, 15.

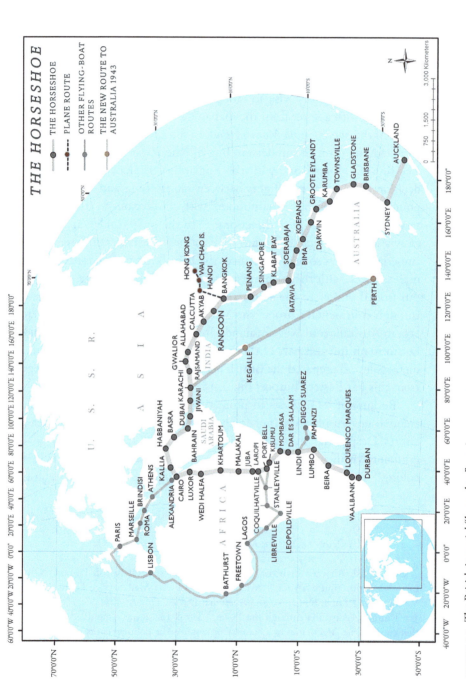

FIGURE I.I. The British imperial "horseshoe" service, 1939–44.

a base for its trans-Pacific route in Macao.[31] The Lisbon government was delighted with Pan Am's proposal to make the Portuguese enclave the terminus for its services to the Far East.[32] However, Pan Am officials confirmed only that surveys indicated Macao's suitability for the services the airline was considering offering. It soon became evident that as the terminus of Pan Am's trans-Pacific service, Macao "would not necessarily be the only one."[33] As late as October 1935, Pan Am representatives informed reporters of continued negotiations for permission to land in Hong Kong. The airline's president, Juan Trippe, refused to disclose the airline's definitive decision,[34] and Pan Am and the US post office would not disclose any details on the proposed location of their eastern air terminus, admitting only that they were negotiating with governments that might be any one of three: "Chinese, Portuguese or British."[35] Reporting on the difficulties that Pan Am encountered in dealing with the British authorities, a local Chinese newspaper in Hong Kong asked, "Is it really impossible to resolve political issues?"[36]

Pan Am's exploration of basing their operations in Macao succeeded in arousing fear among British officials. Writing to the secretary of state for the colonies on November 29, 1935, the officer administering the government of Hong Kong said that he was "much perturbed by [Pan Am's] . . . move to abandon Hong Kong as [an] *air-junction* for China" (emphasis added). His anxiety mounted when he learned that the French had granted Chinese air services landing rights in Hanoi, Vietnam, without demanding reciprocity. As competition was intensifying, he urged the British government to relax its insistent stance in its negotiations with various parties over aviation rights. In response, authorities in London issued instructions to communicate with the Portuguese ambassador, informing him of the talks between Imperial Airways and Pan Am to connect their services in Hong Kong. The ambassador was to be assured that Britain "would consider favourably" proposals for the establishment of

31. *SCMP*, August 29, 1935, 13.
32. *SCMP*, September 3, 1935, 10; *KSMN*, September 3, 1935, 9; *Hong Kong Telegraph*, September 3, 1935, 12.
33. *SCMP*, September 5, 1935, 11.
34. *SCMP*, October 23, 1935, 14.
35. *SCMP*, October 24, 1935, 14.
36. *KSMN*, November 20, 1935, 9.

air service between Macao and Hong Kong, which "would connect the Portuguese colony by air with the main trunk routes to Europe as well as to the United States."[37]

In the midst of escalating tension over the choice of an air terminus in the region, Imperial Airways rushed to extend its service to Hong Kong so as to buttress the colony's potential as an aviation hub. In the year leading up to the Imperial Airways inaugural flight to Hong Kong, the media reported on how Pan Am and Imperial Airways had prepared logistically and technologically for their entry into Hong Kong. Residents in the colony had hoped that Pan Am would deliver its first batch of trans-Pacific mail.[38] Although that anticipated batch did not arrive in 1935, Imperial Airways in that year took just twelve hours to make its first delivery of mail from Penang.[39] Hong Kong also witnessed the numerous trial runs that Imperial Airways conducted.[40] "Nine days from Hong Kong to London," read a headline that projected service to begin by February 1936.[41] That prediction, too, proved premature, with the long-awaited flight landing in Hong Kong on March 24, 1936.

The colony was abuzz with anticipation of the flight's arrival, which was expected "next week on the 23rd or the 24th." The media also reported the charges for the weekly mail and passenger service. The passenger fare from Hong Kong to London would be £175, whereas that from Hong Kong to Singapore would be £35.[42] At that time, passenger fares from Hong Kong to London and Singapore on a steamship were £60 and £9, respectively, approximately a third of the corresponding airfares.[43] To put these figures into perspective, the fare from Hong Kong to London translated into thousands of Hong Kong dollars at a time when the average

---

37. British Library, IOR/L/E/9/129, Collection 2/22 Civil Aviation - Pan-American Airways trans-Pacific service; British Library, IOR/L/E/9/129, Collection 2/22 Civil Aviation - Air Services over China; British Library, IOR/L/E/9/129, Collection 2/22 Civil Aviation - Use of Hong Kong by Foreign Airlines.

38. *KSMN*, October 23, 1935, 3.

39. *TiKP*, October 25, 1935, 3; *KSMN*, October 25, 1935, 9.

40. *TiKP*, November 13, 1935, 3; *KSMN*, November 13, 1935, 9; *KSMN*, November 20, 1935, 9.

41. *KSMN*, December 22, 1935, 9.

42. *KSMN*, March 17, 1936, 9.

43. *SCMP*, March 24, 1936, 21.

wage in Hong Kong was less than HK$2 a day.[44] A (half-ounce) letter to London, Malaya, and Australia cost 50 cents, 20 cents, and 80 cents, respectively. The highest postage, HK$1.50, was for destinations such as Sudan and Kenya.[45]

When the flight's arrival was confirmed on March 23, Imperial Airways indicated that it would welcome public viewing of the landing.[46] The plane arrived on schedule the following day. Nine Royal Air Force (RAF) aircraft guided its entry into Hong Kong, which was experiencing light rain.[47] Hong Kong Governor Andrew Caldecott, the colony's Director of Air Services Commander G. F. Hole, Superintendent of Kai Tak E. Nelson, Postmaster General H. R. Butters, Director of the Royal Observatory C. W. Jeffries, and other dignitaries gathered at the "aerodrome" to witness the arrival of *Dorado*, the Imperial Airways liner, from Penang. A crowd numbering over one hundred welcomed the arrival, which "inaugurat[ed] a direct service from London to Hongkong." The celebrations continued into the evening, with a cocktail party held by the Hong Kong business community to fete the pilots.[48]

The flight had departed Penang at 6:00 a.m. on Monday, arriving in Saigon, Vietnam, at 11:30 a.m. It then took off again at 2:15 p.m. after minor repairs, arriving in Tourane, Vietnam (present-day Da Nang), at 5:30 p.m. After an overnight stay in Tourane, the plane took off at 6:30 a.m., landing in Hong Kong at 11:35 a.m. on Tuesday morning. Imperial Airways planned to offer a weekly schedule following a similar pattern, with flights departing Hong Kong on Fridays and arriving in Penang on Saturdays. From Penang, Imperial Airways offered connections to England and Australia.[49]

Onboard the inaugural flight on the *Dorado* were sixteen bags of mail weighing forty-seven kilograms and a single passenger. Fourteen of the sixteen bags had come from London, with the other two originating in Singapore and Penang. The passenger, the first to land in the colony on

44. *Administration Reports for the Year 1935*, 24–26, 36.
45. *KSMN*, March 22, 1936, 9.
46. *KSMN*, March 23, 1936, 7.
47. *KSMN*, March 25, 1936, 9.
48. *SCMP*, March 25, 1936, 11.
49. *SCMP*, March 25, 1936, 11; *KSMN*, March 25, 1935, 9.

an Imperial Airways plane, was Ong Ee-lim of Kuala Lumpur, "an enthusiastic amateur flyer" who had traveled to Hong Kong to visit his brother.[50] For the return flight to Penang, the Hong Kong Post Office dispatched eighteen bags of mail that included 6,506 letters and 194 postcards to London and 3,648 letters and postcards to other places. This mail load did not represent the colony's usual demand, the superintendent of mails believed, as 75 percent of it, consisting mainly of philatelists' letters, would not be recurring.[51] Although the demand for airmail would stabilize to a regular level over time, the arrival of a single passenger alongside a heavy load of mail was revealing. The 1936 timetable for Imperial Airways listed "England, Egypt, Iraq, India, Malaya, Hong Kong, Australia" on its cover, with its map featuring Hong Kong as a destination on a branch route from Penang via Saigon and Tourane. Freight rates, separated into eastbound and westbound traffic, were accorded two pages of details arrayed in a grid configuration—equally important to the two-page coverage of passenger fares.[52] Air links facilitated connections not merely of the physical transport of people over long distances but equally importantly, if not more so, of information flows in the form of mail.

As the British Empire connected Hong Kong to the rest of its territories, the colony was reminded that the London–Hong Kong airmail service via Penang was "officially subsidised" and that the colony was expected to contribute to the cost of the service.[53] Similarly, the British secretary of state for air prepared detailed arrangements with Imperial Airways stipulating the terms of the subsidy for its operation of the Penang–Hong Kong route, with provisions for any directives that the government might deem appropriate for this "route connecting Hong Kong with Empire air routes."[54] Imperial Airways did not conceal its work for the British government. In his address to Hong Kong Rotarians, Captain J. H. Lock, who commanded the aircraft serving Hong Kong,

50. *SCMP*, March 25, 1936, 11; *KSMN*, March 25, 1935, 9.
51. *SCMP*, March 27, 1936, 12.
52. British Airways Archives, "Imperial Airways and Associated Companies / England, Egypt, Iraq, India, Malaya, Hong Kong, Australia / Timetable in force from 1 September 1936 until Further Notice."
53. *SCMP*, March 24, 1936, 10.
54. British Library, IOR/L/E/9/96, Collection 2/9E Civil Aviation - Air Service from Hong Kong to Penang.

remarked that his airline would continue to function as the British government's instrument "for the development of Empire air routes." Its ultimate objective was "the linking up of every scrap of our far flung Empire with the home country."[55]

Once Imperial Airways had finally connected Hong Kong to its network, the race was on. Its public reassurance about setting up a Macao base notwithstanding,[56] Pan Am continued to petition for a presence in Hong Kong. If Pan Am was to complete its trans-Pacific service in Macao, a link would remain missing for round-the-world air travel, one news article noted.[57] Rumors emerged that Pan Am would call at Macao and connect at Hanoi for a round-the-world service.[58] In the meantime, Imperial Airways bragged of plans to connect Hong Kong to England with just five days of flying.[59] The media even reported that nine hundred aspiring flyers were fighting for the eighteen seats on Pan Am's rumored inaugural flight to Macao.[60] Hype notwithstanding, reports also emerged that British resistance to Pan Am's entry into the colony had begun to abate. Within a month of Imperial Airways' inaugural flight to Hong Kong, news began to circulate in the media that the British and Hong Kong governments had offered Pan Am the facilities to extend its trans-Pacific service, which had previously ended in Manila, to Hong Kong. CNAC was also to be allowed into Hong Kong, connecting with Imperial Airways' European schedules. An article titled "Mutual Aid Permits" also noted that the British had secured a quid pro quo for British services to use certain US airports.[61]

That Imperial Airways had established its service in Hong Kong made the British colony a more desirable location for Pan Am and for CNAC. Highlighting the tremendous traffic potential from China, a newspaper article urged for a connection in Hong Kong. "Right moment to invite China's co-operation," said the article. "Few people in Hong Kong realise

55. *SCMP*, May 20, 1936, 18.
56. *SCMP*, January 13, 1936, 15; *SCMP*, March 13, 1936, 10; *TiKP*, May 29, 1936, 4; *CM*, May 30, 1936, 7; *Shenbao*, May 30, 1936, 3.
57. *TiKP*, May 17, 1936, 3.
58. *TiKP*, May 30, 1936, 4.
59. *CM*, June 17, 1936, 7; *TiKP*, June 17, 1936, 4.
60. *TiKP*, June 20, 1936, 3; *KSMN*, June 20, 1936, 11.
61. *SCMP*, April 22, 1936, 16.

34 *Mapping Hong Kong*

that the total mileage of the airlines in China exceeds that of Imperial Airways." It was in the best interests of Hong Kong to secure air service connections not just to the British network and China but also to North America. These connections "would definitely place Hong Kong on the aerial map of the world."[62]

Reports surfaced that the British and Chinese governments were negotiating the services of Chinese carriers to Hong Kong. The choice of Macao as "the Oriental terminus" would disrupt the route and add a forty-mile boat ride to any connection. As one news report speculated in August 1936, once the United States granted reciprocal landing rights to the British carrier for a trans-Atlantic service, Imperial Airways and Pan Am would link Europe and North America while Pan Am connected North America and Asia. With these services in operation and the British colony "becoming [the] terminus," Hong Kong travelers would be able to circumnavigate the globe entirely by air.[63]

After a month of intense negotiations,[64] the colonial government confirmed not only Pan Am's choice of Hong Kong for its "Asiatic terminal" for trans-Pacific service but also that CNAC would be welcomed at Kai Tak. It was hardly surprising that the various parties reached the two agreements at the same time. Pan Am's investment in CNAC likely facilitated both the diplomatic discussions and logistical arrangements. News also reached the colony that Pan Am and Imperial Airways had agreed to allow transfer of their passengers in Hong Kong. "H.K. Air Centre," one headline celebrated, "Clippers due soon."[65] "New Move Makes Hongkong Air Hub of Pacific," announced the *Hong Kong Telegraph* in an article calibrating the length of journeys that together would circumnavigate the globe—"Hongkong to London (Imperial Airways) 5 days; London to New York (Joint Service) 2 days; New York to San Francisco (Internal) 4 days; San Francisco to Hong Kong 4 days."[66] Hong Kong

62. *Hong Kong Sunday Herald*, March 8, 1936, 6.

63. *KSEN*, August 8, 1936, 4; *Hong Kong Telegraph*, August 8, 1936, 9; *TiKP*, August 9, 1936, 3; *KSMN*, August 9, 1936, 9; *SCMP*, August 10, 1936, 14.

64. *TiKP*, August 15, 1936, 3; *CM*, August 15, 1936, 8; *KSMN*, August 15, 1936, 9; *KSMN*, August 18, 1936, 12.

65. *KSMN*, September 13, 1936, 5; *CM*, September 13, 1936, 7; *SCMP*, September 14, 1936, 14; *CM*, September 15, 1936, 2.

66. *Hong Kong Telegraph*, September 12, 1936, 9.

was to become as critical a link as London, New York, and San Francisco. "Hong Kong will become a nexus of civil aviation worldwide," echoed a local Chinese newspaper, projecting the further expansion of routes not only to Britain, the United States, and China but also to the Netherlands, France, the Soviet Union, and Japan.[67]

Celebrations had begun, but the battles continued. A local Chinese newspaper in Hong Kong spoke of the enthusiastic development of civil aviation among Chinese businesses.[68] Pan Am set up an office in Hong Kong.[69] As the colony prepared for the arrival of "the giant Pan American Clipper," Kai Tak Airport boasted the "most modern installation in the Far East."[70] When the giant clipper arrived on October 23, 1936, representatives of the Hong Kong government, aviation specialists, and Chinese Legislative Council Chair Sun Fo (son of Sun Yat-sen, founder of the Republic of China; former minister of railways and onetime chief of CNAC) were among those in the welcoming party. Although it was a trial flight, the clipper carried nineteen people onboard.[71] The trial went smoothly.

The giant clipper was a "flying boat" that landed on water, a feature that allowed Pan Am to circumvent the rough state of early airfields. An English news article emphasized that in addition to its modern airport, Hong Kong also offered a "well defined" British law, which enforced the strict restrictions of ships when a flying boat was expected, eliminating the danger of collision experienced "in other parts of the world, notably in America."[72] The British had compromised with the Americans over air routes through Hong Kong, but that did not stop the British from asserting superiority in their colony.

In terms of the grand entrance of scheduled services to Hong Kong, Pan Am could not claim second place after Imperial Airways. That honor went to CNAC. The flying boat that inaugurated CNAC's regular service

---

67. *CM*, October 13, 1936, 8.

68. *KSMN*, October 20, 1936, 9.

69. *TiKP*, October 20, 1936, 3.

70. *SCMP*, October 22, 1936, 15.

71. *TiKP*, October 24, 1936, 3. This event is one of the "famous firsts" featured in a pictorial celebration of Pan Am's landmark occasions (San Francisco Airport Commission, *Famous Firsts*, 16–17).

72. *SCMP*, October 22, 1936, 15.

between Shanghai, Hong Kong, and Canton carried a full load of six passengers, arriving in Hong Kong on November 5, 1936. Governor Caldecott and his entourage, along with "a fairly large crowd . . . of over 50 people" welcomed "the grey and yellow Douglas Dolphin flying boat." Among those on board were CNAC's manager and the Far East representative of Pan Am. The former hosted a cocktail reception at Kai Tak.[73]

A few months passed before Pan Am's regular service connected Hong Kong to its network. Its introduction, long anticipated in Hong Kong,[74] constituted a component of the airline's expansion strategy in the Far East.[75] Operating from a Hong Kong office that it shared with CNAC, which also served as its agent, Pan Am was to fly its clipper from San Francisco to Manila (via Honolulu, Wake and Midway Islands, and Guam), then from Manila to Macao, from Macao to Hong Kong, and from Hong Kong back to Manila. This service reduced the twenty-eight-day steamer journey from San Francisco to Macao to a five-and-a-half-day service.[76] From San Francisco, the service would feed into the various connections in the continental United States, while in Hong Kong, it would connect with Chinese carriers such as CNAC, Imperial Airways, and carriers to Southeast Asia and Europe.[77] Postage was to be HK$0.35 to the Philippines, HK$0.80 to Guam, HK$1.80 to Hawaii, and HK$2.80 to San Francisco.[78] Passenger fare from San Francisco to Hong Kong or Macao was to be US$950, and from Manila to Hong Kong US$80. A 10 percent discount applied to roundtrip tickets.[79]

Taking pains to not sideline Macao, Pan Am called the Portuguese enclave "the Chinese base for the new venture." The company justified that choice by stating that their clippers could not land in the Canton river and that Hong Kong was "too open for basing operations." How-

---

73. *SCMP*, November 6, 1936, 15; *KSMN*, November 6, 1936, 10; *TiKP*, November 6, 1936, 3; British Airways Archives, "O Series," China, 3130.

74. *TiKP*, December 4, 1936, 3; *KSMN*, December 6, 1936, 3; *TiKP*, February 18, 1937, 3; *KSMN*, February 18, 1937, 11.

75. *KSMN*, December 12, 1936, 11; *KSMN*, December 24, 1936, 10.

76. *SCMP*, April 7, 1937, 15; *SCMP*, April 20, 1937, 13.

77. *TiKP*, March 1, 1937, 3.

78. *TiKP*, March 11, 1937, 4; *KSEN*, April 19, 1937, 4.

79. *KSMN*, April 7, 1937, 3; *TiKP*, April 7, 1937, 1; *SCMP*, April 7, 1937, 15; Pan Am, Series 5, Sub-Series 1, Sub-Series 2, Folder 1.

The *Making of a Place in an Evolving Air Space*   37

ever, Hong Kong would become a "port of call" and "a connecting point with Shanghai, interior Chinese points, Indo-China, the Straits Settlements and India."[80] Whereas Macao would serve as the official terminus, Hong Kong would function as the actual connection—the real pivot in the evolving network of civil aviation.

The *China Clipper* began its first trip from San Francisco on April 21, 1937. The destination was Hong Kong, where it was "to connect up with the Imperial Airways route, which has its Far East terminus at that Colony." To underscore the intended connection this Pan Am clipper was to initiate, the British consul general in San Francisco was at hand to bid the trans-Pacific passengers farewell at the send-off ceremony.[81] At Manila, the Hong Kong–bound portion would then transfer to another Pan Am clipper that had formerly served the east coast of the United States.[82] The shuttle clipper from Manila to Hong Kong had a capacity of thirty-two passengers. However, there was to be no passenger on this first trip "owing to the heavy demands on space for mail." It was scheduled to leave Manila at 5:00 a.m. and arrive in Macao at 11:00 a.m. Taking off again from Macao at 11:30 a.m., the clipper was expected to arrive in Hong Kong shortly before noon. After an overnight stop at Kai Tak, the clipper would skip Macao on the return flight and proceed directly back to Manila.[83] The itinerary provided a clear indication of the centrality of Hong Kong at the expense of Macao.

The British colony orchestrated a detailed "Official Programme of Welcome" for the arrival of Pan Am's inaugural service on Wednesday, April 28, 1937. Published in the *South China Morning Post*, the program indicated that about four hundred guests representing "every phase of the Colony's life" had been invited. The general public, which was expected to arrive "in considerable numbers," was to watch the ceremony from outside the enclosure reserved for dignitaries.[84] In Macao, reports of the enthusiastic multitude assembled for the plane's arrival suggested the

---

80. *SCMP*, April 7, 1937, 15.
81. *SCMP*, April 23, 1937, 12; TNA, CO323/1457/40.
82. *SCMP*, April 20, 1937, 13.
83. *SCMP*, April 26, 1937, 16.
84. *SCMP*, April 27, 1937, 9; *TiKP*, April 23, 1937, 3; *CM*, April 23, 1937, 2; *CM*, April 28, 1937, 2.

deference that the Portuguese colonial regime accorded it.[85] However, the scene in Macao paled in comparison to that in Hong Kong, where "over 4,000 persons witnessed one of the momentous events in the history of commercial aviation and of Hongkong."[86]

As the plane approached, spectators rushed to the waterfront and to Kai Tak, "where they had been inspecting Imperial Airways['] Dorado . . . also a CNAC Sikorsky, and the various smaller planes owned by the Far Eastern Aviation School." After circling over Hong Kong, the Pan Am clipper landed in Kowloon Bay, "taxying to her mooring ballard." After its punctual arrival at 11:55 a.m., His Excellency the Officer Administering the Government Hon. Mr. N. L. Smith welcomed the crew. "To-day we celebrate the final welding of perhaps the most important link in the chain of world communication," Smith said to the crowd. "Hongkong is only a tiny place but our magnificent harbour has been on the map for quite a long time. . . . And now it is our hope, that Hongkong will be equally on the air map, with London in one direction only nine days away and with New York in the other direction only six and a half days away." Pan Am's representative, H. M. Bixby, responded, "It is significant . . . that to-day at this magnificent *air-port* in the most beautiful of all harbours you have witnessed the first direct connection between the services of Pan-American and your great Imperial Airways" (emphasis added). Bixby's remarks revealed the sense of continuity between the functions of a sea *port* that connected water traffic and an aviation center (as reflected in our modern parlance of "air*port*") that connected traffic in the skies. Chu Chang-sing, director general of posts of the Chinese Government, called it "an historic event," as the flight established "the first direct air mail between China and America, *via Hongkong*" (emphasis added). That Hong Kong was not the ultimate destination did not matter; the city's value lay in its role as a nexus that connected traffic flows. The local radio station ZBW broadcast the proceedings, and a "worldwide 'hook-up'" connected the event with "listeners in America and Australia."[87] The welcoming party then proceeded to the clipper, which the daughter of the presiding British official christened the *Hongkong Clipper*. The Union

85. *SCMP*, April 29, 1937, 13.
86. *SCMP*, April 29, 1937, 12.
87. *SCMP*, April 29, 1937, 12.

The Making of a Place in an Evolving Air Space   39

Jack was raised, and the flags of the United States, Britain, and China were draped on the clipper.[88]

While the *Dorado* carried a single passenger on its inaugural flight to Hong Kong, the *Hongkong Clipper* carried none. The media calibrated the significance of this connection by the amount of mail dispatched instead of passenger count.[89] The Manila-bound flight from Hong Kong on April 29, 1937, carried "55 bags and five packets of mail, weighing 369,938 kilos, easily a record for the Colony." Similar to the load on the Imperial Airways inaugural outbound flight, much of the cargo consisted of commemorative materials. Notable among the Pan Am cargo was "a consignment of margarine rushed here from Bangkok by the Imperial Airways' liner."[90] Passenger traffic had to wait for the departure from San Francisco the following week. Twenty-four passengers were booked on that flight, which left San Francisco on April 28 and arrived in Hong Kong on May 5 after a connection in Manila. The majority of these passengers also booked the return journey from Hong Kong on May 6, which was scheduled to reach San Francisco on May 13.[91] The spectacle of this inaugural flight, as with its Imperial Airways and CNAC predecessors, was immensely exciting. However, there were few passengers on these early flights. The focus was on the novelty of the connections they facilitated (as well as the delivery of such perishable but hardly indispensable products as margarine) and the more practical aspects of timely information transmission in the form of mail.

A colonial report in 1938 boasted of the city's airport at Kai Tak, with its "facilities for marine and land aircraft." In the year after Pan Am's inaugural flight, civil aviation continued to develop, with the number of arriving and departing passengers growing from 3,685 in 1937 to 9,969 in 1938.[92] In June 1937, the Sino-German venture Eurasia had inaugurated its Peiping (Beijing)–Canton–Hong Kong service, a move regarded as "a final step" before the launch of a service to Berlin in conjunction with

88. *SCMP*, April 29, 1937, 12.
89. *SCMP*, April 28, 1937, 14.
90. *SCMP*, April 30, 1937, 12.
91. *SCMP*, April 26, 1937, 16; *SCMP*, April 26, 1937, 1; *SCMP*, April 29, 1937, 12; Pan Am, Series 1, Sub-Series 6, Sub-Series 2, Folder 10; Pan Am, Series 5, Sub-Series 1, Sub-Series 2, Folder 1.
92. *Administration Reports for the Year 1938*, 41–42.

Lufthansa.[93] In August 1938, when Air France connected Hanoi with Hong Kong, Consul M. Dupuy called it symbolic of a "collaboration so much alive, so active and so cordial of the British Empire with France."[94] In the same month, Imperial Airways doubled its service from Bangkok to Hong Kong to twice weekly, "connecting with the England-Australia trunk route." Pan Am maintained its weekly schedule to San Francisco via Manila. The Japanese occupation of China spelled the end of operations to Hankou and Canton, but CNAC continued to offer services to Guilin and Chongqing.[95]

In September 1938, the mainland-based Chinese newspaper *Shenbao* noted the rapid development of civil aviation in Hong Kong, despite its short history. Thanks to its flight connections with "such companies as CNAC, Eurasia, Imperial Airways, Air France and Pan Am," as well as the pending arrival of KLM, "Hong Kong holds a rather important position in the world of aviation." However, the colony needed to update its airport facilities to respond to the demands of traffic at this "nexus of civil aviation in the Far East."[96]

Hong Kong did not lead the development of commercial aviation in the region. Pan Am had reached Manila, and Imperial Airways had reached Singapore, before Hong Kong entered the picture. Although initially a laggard, however, Hong Kong became the linchpin of global aviation infrastructure in 1937 when the pioneers of commercial aviation met in this British colony situated at the southern tip of China at one end of the vast Pacific Ocean. Physical landscape played an important role in the choice of landing sites, although such considerations played a secondary role in this early period. Technology had not matured into a common platform, as evidenced by the landing mechanism of Imperial Airways' *Dorado*, which differed from that of Pan Am's flying boat christened the *Hongkong Clipper*. The discrepancy between Pan Am's alleged reasons for its choice of sites (Hong Kong was "too open" compared with Macao) and its eventual flight schedule (overnight in Hong Kong versus

---

93. *SCMP*, June 30, 1937, 14. On the issue of Sino-Western ventures in aviation, see Kirby, "Traditions of Centrality," 26, and Kirby, *Germany and Republican China*, 76–77.

94. *SCMP*, August 11, 1938, 9; *Shenbao*, August 10, 1938, 4.

95. *Administration Reports for the Year 1938*, 41–42.

96. *Shenbao*, September 25, 1938, 4.

a short stop in Macao) underscores the overriding urgency of the construction of a traffic node.

The political landscape trumped the topographical landscape in the design of networks in the early days of commercial aviation. The confluence of flights from Imperial Airways and Pan Am, in addition to CNAC's services, made Hong Kong a nexus in this evolving network of global aviation. "Hongkong Now the Junction for Three National Lines," the *South China Morning Post* boasted the day after Pan Am's inaugural flight. Imperial Airways connected the colony to Penang and Europe. CNAC flew from this city just south of China proper to all parts of China. Pan Am provided a service linking Hong Kong to North America via Manila.[97] Although none of the three lines could boast of any local representation of Hong Kong, their convergence in Hong Kong pointed to the city's peculiar situation, both politically and geographically, as facilitator of global flows through this newly created nexus of commercial aviation.

Jules Verne's *Around the World in Eighty Days* (1873) provided fictional motivation for a circumnavigation that only global business enterprises could realize in practice. In the world of commercial aviation, airlines were national enterprises. In the case of Hong Kong, colonial status did not afford the city a local carrier in this era. The British Empire privileged Imperial Airways by giving it a head start in flying to this colonial outpost, not only to provide a connection to the colony and tighten its colonial grip but also to further the empire's interest in the Chinese market. Expanding its network from the other side of the Pacific, Pan Am not only benefited from US control of Manila but also penetrated the Chinese system through its commercial alliance. Although Macao had offered an alternative to Hong Kong as an aviation hub in the area, Portugal could not offer a network remotely comparable to the UK's global reach. Despite the trials and tribulations of diplomatic negotiations and the scheduling of "first" flights, the world connected in Hong Kong on April 28, 1937. "Round the World Route Now an Actuality," read a newspaper headline. The first regular air service between Manila and Hong Kong bridged the last remaining gap in the "dream of aviators, business men and travellers."[98] The foundation of Hong Kong as an aviation hub

97. *SCMP*, April 29, 1937, 12.
98. *SCMP*, April 28, 1937, 14; *SCMP*, April 29, 1937, 12.

42 *Mapping Hong Kong*

owed much to the liminality of Hong Kong as a British imperial outpost at the edge of China and to its allure as a connection at the far end of the Pacific for the extending reach of the United States.

A British Hub for Chinese Traffic, or Not

The competition over prewar aerial connections in Hong Kong manifested itself in the conflict over the Kai Tak "airfield." The area owes its name to two Chinese entrepreneurs who had planned in the 1910s to develop a garden city on land to be reclaimed from Kowloon Bay. By the mid-1920s, their investment company had run out of funds, and the Hong Kong government took over the reclamation work and redesignated the land for use as an airfield.[99] These are the humble origins of Kai Tak Airport, which the colonial government's Civil Aviation Department traces back to 1930. The confluence of air routes in Hong Kong resulted in the airfield's slight enlargement in 1938 to "a grass field, without runways, covering 171 acres."[100] The Japanese capture of Canton in late 1938 increased the hazards of flying close to the Chinese border. In December 1941, the outbreak of war with Japan further disrupted civil air service. Kai Tak was evacuated on December 11, 1941, after the British authorities had destroyed all equipment potentially of use to the Japanese. Civil aviation came to a grinding halt in Hong Kong and did not recommence for almost four years.[101]

As World War II escalated in Europe and Asia, the colonial authorities in Hong Kong explored expansion opportunities for its "aerodrome" months before the Japanese occupation of the city.[102] Ironically, despite Britain's wartime ambitions, it was the Japanese authorities who expanded Kai Tak. In a bid to extend their aerial reach, the Japanese occupation forces decimated "a great deal of the surrounding Chinese property" and

---

99. Chu, "Speculative Modern," 128–33; see also HKPRO, HKRS558-1-141-12.

100. *HKDCA*, 1946–1947, 1.

101. *HKDCA*, 1952–1953, 2–3. For a detailed account of wartime activities at Kai Tak, see Kwong and Tsoi, *Eastern Fortress*, chap. 9.

102. LegCo, February 20, 1941, 56.

reclaimed land to expand Kai Tak to 376 acres and laid two hard-surface concrete runways. Thus, when the British reoccupied Hong Kong in 1945, Kai Tak occupied twice the area it had in 1941.[103]

Heightened anxiety over aerial control continued to fuel the expansion of aviation infrastructure during the war. During the war, the maintenance of Britain's imperial footprint, even vis-à-vis its allies, took on renewed importance. Discussions of civil aviation in Hong Kong formed a component of British strategy in the Far East. After what various parties had billed as the grand connection in Hong Kong when the routes of Pan Am and Imperial Airways had met in the colony, Britain continued to contend with other global powers, principally its wartime ally the United States, over control of aerial routes on the basis of territorial claims to aviation hubs. In 1943, the War Cabinet spelled out the British government's overriding claims to the internal traffic of the United Kingdom, as well as traffic within the "Colonial Empire" and along "the great Empire routes . . . in cooperation with the Dominions and India." Britain's aim was to "secure as much international traffic as may be obtained."[104] In April 1945, anticipating the British return to Hong Kong, Lord Knollys, Chairman of Imperial Airways, now reincarnated as the nationalized British Overseas Airways Corporation (BOAC), called on J. C. Sterndale Bennett, head of the Far Eastern Department of the British Foreign Office, stressing the importance of Hong Kong "not merely as the headquarters of some company to operate in China but also as an air route stage."[105]

In the world of commercial aviation, Britain's interests conflicted with those of its wartime ally the United States, which was seeking to expand its global influence by constructing an infrastructure of airways. Aviation precipitated a new global imaginary for the United States.[106] The "logic of the air" foreshadowed a new world order characterized by the decline of European empires and the emergence of the United States as the new global leader. American dominance of the world was to be predicated

---

103. *HKDCA*, 1946–1947, 1; 1952–53, 3; see also HKPRO, HKRS115-1-28; HKPRO, HKRS156-1-390-2; HKPRO, HKRS156-1-399.

104. TNA, CAB66/42/12.

105. TNA, FO 371/50297.

106. For more extensive discussions of the development of Anglo-American relations, see Dobson, *Anglo-American Relations*, and Woods, *Changing of the Guard*.

not on territorial control but on the access that airways would facilitate.[107] Adolf A. Berle, US assistant secretary of state and the architect of Roosevelt's postwar aviation strategy, set the plan in motion as he advocated the policy of "open skies"—an international aviation standard of unfettered access that would benefit all nations.[108] Such a policy would place the territory-rich empires at a disadvantage while granting an unassailable position to the United States, already a world leader that commanded some 70 percent of the world's aviation passenger miles by 1944.[109]

In 1944, the United States convened a conference to "make arrangements for the immediate establishment of provisional world air routes and services" and to "discuss the principles and methods to be followed in the adoption of a new aviation convention." From November 1 to December 7, some seven hundred delegates representing fifty-two nations attended the Convention on International Civil Aviation held in Chicago.[110] The Soviet Union was noticeably absent, having recalled its delegation at the last minute, foreshadowing the conflicts over the international skies in the decades to come. Despite the anticolonial sentiments that Berle marshalled to the advantage of the United States, Britain and the other European colonial powers refused to allow "freedom of the air" to their colonies, signaling the continuation of the imperial footprint in the skyways at the conclusion of World War II. While calling for the establishment of an international organization to ensure competitive practices elsewhere, Britain insisted on the franchise of BOAC over its imperial routes.[111] Although disappointed that US resistance prevented a configuration based on multilateral agreements, the British minister of civil aviation anticipated that this arrangement would allow Britain and its dominions to secure reciprocal facilities from the United States. Although its focus was not on Hong Kong, a 1945 policy established that the state-owned BOAC would be "responsible for Commonwealth services and for

107. Van Vleck, *Empire of the Air,* intro.
108. Engel, *Cold War,* 95–96.
109. Van Vleck, *Empire of the Air,* 168–70.
110. The Danish minister and Thai minister in Washington also attended in their personal capacities. See International Civil Aviation Conference, *Proceedings of the International Civil Aviation Conference,* 1, 113–19.
111. Van Vleck, *Empire of the Air,* chap. 5; Lyth, "Chosen Instruments," 50–51.

service to the United States, China and the Far East." Reaching an agreement with the United States became vital to British aviation interests.[112]

The United States had a different agenda. At the Chicago conference, the international aviation heavyweight advocated five "freedoms of the air":

1. Freedom of peaceful transit, granted by one state to another to fly across its territory without landing (First Freedom of the Air).
2. Freedom of non-traffic stop (to refuel, repair, or [take] refuge), granted by one state to another to land in its territory for non-traffic purposes (Second Freedom of the Air).
3. Freedom to take traffic from the homeland to any country (Third Freedom of the Air).
4. Freedom to bring traffic from any country to the homeland (Fourth Freedom of the Air).
5. Freedom to pick up and discharge traffic at intermediate points, granted by one state to another to put down and to take on, in the territory of the first state, traffic coming from or destined to a third state (Fifth Freedom of the Air).

Of these five freedoms, the fifth was paramount to the United States' aviation future because of its lack of overseas territories.[113]

Britain would not concede in Chicago, but in less than two years' time the struggling imperial power succumbed to its mounting economic vulnerability. At the Anglo-American conference on civil aviation in Bermuda, the two sides signed the Bermuda Agreement on February 11, 1946, after several British concessions for the sake of US assistance on other issues. Britain yielded to the United States, granting all five freedoms in an air rights agreement between the two countries.[114] The bilateral approach of the Bermuda Agreement, which specified the routes each country's carrier would fly, along with stops along the routes, became the blueprint of subsequent arrangements to govern international aviation. Although the significance of the geopolitical power of territories may have

112. TNA, CAB66/56/42; TNA, CAB129/3/21; TNA, CAB129/3/22.

113. International Civil Aviation Conference, *Proceedings of the International Civil Aviation*, 3; Van Vleck, *Empire of the Air*, chap. 5.

114. *New York Times*, February 12, 1946, 4. For more on this Anglo-American rivalry in civil aviation, see Dobson, *Peaceful Air Warfare*; Dobson, "Other Air Battle."

been muted by the introduction of the "Fifth Freedom," Hong Kong remained an important bargaining chip for the British at the far end of their shrinking imperial footprint. In the world of commercial aviation, the colony conferred negotiating value on Britain not merely for its attraction as an end point of aviation routes to a British-controlled city but also for its appeal as a connection point for global air traffic.

The Bermuda Agreement allowed British carriers to run a service from Singapore and Hong Kong to San Francisco via Manila, Guam, Wake and Midway, and Honolulu. In return, American operators could serve the route from San Francisco and Los Angeles to Hong Kong via the same intermediary points.[115] Shortly after the conclusion of the conference in Bermuda, a British official announced a civil aviation meeting for the governments of Australia, New Zealand, Britain, Canada, and the United States to map Pacific air services. The media highlighted Hong Kong as a subject for discussion, even reporting that the US Civil Aeronautics Board had recommended Pan Am and Northwest Airlines as carriers for new routes to and through Hong Kong.[116]

Hong Kong's involvement in the postwar reconstruction of air routes began as part of US-British competition in the region. In its attempt at a "second colonial occupation,"[117] Britain justifiably viewed aviation as a crucial tool for reviving its imperial footprint and did not always consider its interests to be aligned with those of its major wartime ally. Compared with ports and railroads, aviation infrastructure could be constructed more economically and swiftly.[118] Within months of the British regime's return, Lawrence Kadoorie, a prominent businessman in Hong Kong, was already in communication with colonial officials. On October 11, 1945, he wrote to the Civil Affairs Office about Macao being an alternative to Hong Kong for Pan Am. Encouraging the colonial govern-

---

115.   TNA, CAB 128/5/11; TNA, CAB 129/6/37; TNA, Treaty Series No. 3 (1946), Agreement between the Government of the United Kingdom and the Government of the United States of America relating to Air Services between their Respective Territories [With Annex] Bermuda, 11th February, 1946.

116.   *SCMP*, February 20, 1946, 2.

117.   Coined by Low and Lonsdale to characterize British activities in Africa in the aftermath of World War II ("East Africa"), the term "second colonial occupation" has been extended to other contexts (see Darwin, *Empire Project*, 559).

118.   Bickers, *China Bound*, 306.

The Making of a Place in an Evolving Air Space   47

ment to grant aviation facility access to American and other aircraft operators, Kadoorie advised the British regime that "every opportunity should be taken to show that it is the Government's intention to make the Colony as important an airport as it is a seaport." As the colonial government endeavored to reestablish itself in Hong Kong, Kadoorie expressed his belief that "we shall gain much more than we shall lose by encouraging visitors and re-establishing Hong Kong on the map." On August 21, 1946, he communicated with the colonial secretary, urging the British regime to upgrade its infrastructure. Reporting on Pan Am's alleged choice of Canton over Hong Kong as a port of call once trans-Pacific services resumed, Kadoorie warned that "in spite of the desire of the air transport companies to use Hong Kong as a base, unless the necessary facilities are provided, the Colony will be by-passed."[119]

Originally preoccupied with the Commonwealth routes that it saw as an instrument to secure an aviation order free of US competition, the British focus came to rest on Hong Kong after the 1946 Bermuda Agreement, which granted both British and American carriers rights to operate air services to Hong Kong. In a 1946 report on the colony, British officials called Hong Kong "a most important link in the net-work of postwar aviation." Within easy reach from Hong Kong were Saigon, Singapore, the Republican Chinese capital of Nanjing, Manila, Tokyo, and other locations in Japan. BOAC launched a weekly flying boat service in August 1946, connecting Hong Kong to the United Kingdom with a six-day route.[120]

This renewed enthusiasm for building Hong Kong into a connecting hub propelled the British authorities to step up surveying efforts in identifying an appropriate site for the construction of greater aviation infrastructure in Hong Kong. Despite the Japanese expansion of the airfield, Kai Tak remained a primitive site. Upon their return to Hong Kong, the British authorities surveyed the two existing runways. The northwest–southeast runway was 4,580 feet long and 330 feet wide, whereas the northeast–southwest runway was 4,730 feet long and 225 feet wide. "The approaches except from [the] south east are very bad," observed the colonial administration in 1947, highlighting the need for "a first-rate

---

119. The Hong Kong Heritage Project, SEK-3A-064 B02/18.
120. *Annual Report on Hong Kong for the Year 1946*, 86–87.

48 *Mapping Hong Kong*

modern aerodrome."[121] Despite the "considerable extension" undertaken by the Japanese that had doubled its size, Kai Tak "remained inadequate for heavy aircraft." Wartime improvements in aeronautics accentuated Kai Tak's deficiencies. The authorities continued to search for a suitable airfield site, noting that, fortunately, "in spite of the sometimes hazardous approach to Kaitak," few accidents had occurred.[122] In March 1947, a technical team representing the Ministry of Civil Aviation, Air Ministry, and BOAC arrived to examine aerodromes not just in Hong Kong but throughout the Far East.[123]

The colonial government, in conjunction with London, had considered Ping Shan in the northwestern corner of the colony as a possible site of development. However, the Civil Aviation Ministry officials and BOAC found its southern approach to be too weather dependent. The mission dispatched by the Air Ministry to Hong Kong suggested as an alternative Deep Bay, an area two miles northwest of Ping Shan that bordered China. Construction at Deep Bay would cost some £4 million and take thirty months. In the meantime, Kai Tak needed to be rehabilitated to ensure the operations of Chinese airlines, the RAF, and BOAC.[124]

It was not only the British authorities but also the media that expressed the urgency of constructing a new airfield. In November 1946, the *South China Morning Post* relayed an article from *The Times*. Thanks to its geographical location and operation as a free port, Hong Kong enjoyed "pre-eminence as the distributor centre for South China before the war," the article said, and deemed the city destined to become "the junction or terminus for most of the air traffic in the Far East." Kai Tak was but a "small and, by modern standards, quite inadequate airfield" in "the Leased Territory." Reportedly, even before their return, the British knew that Kai Tak "could never be reconstructed or expanded in such a manner as to conform to modern aviation standards." The approach to the airfield, "almost [entirely] surrounded by hills, often enveloped in mist,"

---

121. *HKDCA*, 1946–1947, 1.

122. *Annual Report on Hong Kong for the Year 1946*, 86–87.

123. *SCMP*, March 8, 1947, 1.

124. LegCo, July 19, 1946, 56–57; TNA, T225/597. The Swire group learned of the government's plan at Deep Bay as early as June 1946 (JSS, 1/3/16 Director in the East Correspondence).

was "at times positively dangerous for modern high-speed aeroplanes." The article went so far as to say that "only the American CNAC pilots who flew in and out night after night when Chungking [Chongqing] was largely isolated" would contemplate landing at and taking off from Kai Tak with equanimity. It therefore questioned whether any airline operating modern aircraft would provide a regular service to Hong Kong if Kai Tak remained the only airfield. The article reported on the proposal to build at Deep Bay, which would offer virtually "a mist-free landing ground, with clear approaches from the sea and usable in practically any [weather] but typhoon weather." Demand for aviation facilities in Hong Kong was growing—Kai Tak was already in daily use, weather permitting, by the British military and by British and Chinese airlines, and American, Australian, French, and Dutch carriers were seeking permission for regular services. Lamenting the lack of any sense of urgency, the article expressed worries that Hong Kong could be "relegated to a secondary status . . . instead of being the heart of the network of Far Eastern air services." Hong Kong was projected to recover its leading position in shipping in the Far East, and its development into an aerial hub of Southeast Asia would only enhance its maritime status.[125]

In 1947, the Ministry of Civil Aviation technical survey party continued to report that Kai Tak was "incapable of anything but very minor development" and that its location "in a horseshoe of high hills" made it safe only for operations of "medium and light aircraft in conditions of good visibility." Reports to the Colonial Office argued for the construction of the Deep Bay airport before the Chinese completion of "any major work at Canton," assuring the office that "an aerodrome in British territory would be administered and run so much more efficiently than an airport in China that the traffic would gravitate to Hong Kong."[126]

The British had to fend off the Americans, who sought to extend their control over the routes in the skies. However, they competed for influence over Hong Kong not necessarily for the sake of Hong Kong but for the sake of traffic through the city, especially its connections with China. Impressive as the British network might be, British aviation represented only 17 percent of passenger traffic in Hong Kong from 1946 to 1947.

125. *SCMP*, November 6, 1946, 4.
126. TNA, T225/597.

China became the predominant user of Kai Tak, with Chinese-registered C-46 and C-47 aircraft bound for the interior accounting for half of all departures. Chinese carriers provided air transport to Shanghai, Nanjing, Chongqing, Kunming, Hainan, and Canton. In May 1946, CNAC was the only operating civil airline, bringing twenty-four arrivals to Hong Kong with 475 passengers onboard.[127] Traffic was poised to increase after the Anglo-Chinese air agreement of 1947, which granted Britain rights to fly routes from Hong Kong to Canton and Shanghai, along with other routes from London or Singapore. In return, Chinese carriers could fly from Shanghai or Canton to Hong Kong and Singapore.[128] In its annual report for 1947, the colonial government's tally listed thirteen airlines (four British, two Chinese, three Philippine, and one each from the United States, France, Norway, and Siam [Thailand]) that provided regular services to Kai Tak. During the year ending March 31, 1948, the airport handled 7,309 aircraft movements (a more than 200 percent increase from the previous year) and 113,326 passengers (a nearly fourfold increase compared to the previous year). As in the previous year, Chinese carriers dominated the traffic, accounting for half of the passengers carried from 1947 to 1948.[129] Britain enjoyed technological and operational advantages over the Chinese, but British officials were certainly aware of the opportunities, as well as the potential rivalry, that could arise across the British colony's northern border.

By early 1947, the lack of government action had made local Hong Kong political circles impatient. In a March 27 meeting of the Legislative Council (LegCo), Ronald D. Gillespie, chair of the Hong Kong General Chamber of Commerce and its representative on the council, complained that "I see no signs of progress, and I hope Government are really awake to the urgency of the matter [the need for a modern airport in Hong Kong]."[130] Local politicians understood that Hong Kong would need to contribute substantially toward the cost.[131] However, the relative

127. *Annual Report on Hong Kong for the Year 1946*, 86–87; *HKDCA*, 1946–1947, 2, 14, 15.
128. TNA, CO937/104/2; *The Times*, July 24, 1947, 3.
129. *Annual Report on Hong Kong for the Year 1947*, 108–9; *HKDCA*, 1947–1948, 1, 8, 9.
130. LegCo, March 27, 1947, 81.
131. LegCo, April 24, 1947, 127.

The Making of a Place in an Evolving Air Space   51

sharing of the financial burden remained undecided, and politicians in London were hesitant. In October 1947, the Labor government was questioned again on its commitment to modernize Kai Tak. The under-secretary of state for the colonies replied that although he was aware of Hong Kong's significance for the British Empire, he was reluctant to commit the British government to modernizing Kai Tak Airport because of the "very large capital expenditure" that the Hong Kong government might not be able to bear. Finding it imprudent to ask British taxpayers to bear the cost, the under-secretary did not reply directly to the question of whether his government had put the Kai Tak project "into cold storage indefinitely."[132]

The British government finally decided on the construction at Deep Bay in February 1948. The decision, based partly on reasons of defense, also stemmed from three civil considerations. First, unlike Kai Tak, Deep Bay would allow for an expeditious and economical expansion for future types of trunk-route aircraft. Second, Hong Kong's value as an entrepôt was deemed important not only to the colony but also to British interests in the Far East. Third, internationally, the need for a trunk-route airport in southern China was unquestionable. The British government needed to take action before the Chinese government established competitive facilities in Canton.[133] Hence, the British government announced that it stood ready to extend a £3 million loan for the construction of an airport at Deep Bay, which by September 1947 was expected to cost £4 to 5 million and take three to five years.[134]

The British government remained cautious about the reaction of the Chinese government in Nanjing. Although the British embassy in China did not believe that the announcement of the loan would cause any embarrassing reactions from China, the Hong Kong governor thought it best that the Chinese government be informed of the British decision to proceed with Deep Bay and that the Chinese minister for foreign affairs and the governor of Guangdong be approached personally. In the meantime, the Civil War in China had escalated. British documents indicate that

132. House of Commons debate, October 29, 1947, vol. 443, col. 862–63.
133. TNA, FO371/69580.
134. TNA, FCO141/16981; *HKDCA*, 1948–1949, 6.

Mapping Hong Kong

Deep Bay construction was underway in late 1948.[135] By February 1949, British officials agreed that the Chinese government should be given advance notice of any announcement. At the same time, they had also become uncertain about the changing political situation in China.[136]

Political uncertainties in China notwithstanding, Hong Kong remained confident of its role in communication and transportation in early 1949. In his address to the Legislative Council on March 16, 1949, Governor Alexander Grantham boasted about the growth of cargo and passenger traffic through the colony in 1948. With 25,000 passengers arriving and departing every month, Kai Tak was "handling almost as much traffic," claimed Grantham, "as any airport in the Empire."[137] In response, a legislator complained about Kai Tak in light of the colony's promise, calling the existing aerodrome "not suitable as a first class and major air traffic center."[138] Similarly, the governor reported good aviation growth to the secretary of state for the colonies in early 1949. Between 1938 and 1948, aircraft arrivals increased fourteen-fold to 9,144 per year, the passenger count (entering and leaving Hong Kong) increased twenty-three-fold to 232,558, and freight and mail grew five-fold to 1,875 metric tons.[139] The year ending in March 1949 alone registered a doubling in the number of passengers, with 20,043 passengers passing through Kai Tak in March of that year, 75 percent of them traveling to or from China. Reiterating the inadequacies of Kai Tak and the role of Hong Kong as a hub, the colonial authorities reported that the city's "only airfield, at Kai Tak" had "on the whole successfully handled the large number of aircraft and passengers using it." However, the difficult air approaches had inevitably resulted in delays and diversions in bad weather. As the city awaited the decision on a new airport, Kai Tak's "temporary passenger building" was "at times grossly over-crowded."[140]

Grantham pleaded for a commitment to the new airport at Deep Bay, which would become "one of the busiest in East Asia . . . a boon both to

---

135. TNA, DEFE5/8/136.
136. TNA, FO371/69581; TNA, FO371/69582B; TNA, FO371/75923.
137. LegCo, March 16, 1949, 60.
138. LegCo, March 30, 1949, 107.
139. TNA, FO371/75923.
140. *HKDCA*, 1948–1949, 5–6.

Hong Kong and to China." However, the surge in traffic from 1948 to mid-1949 proved to be a blip, as the geopolitical landscape underwent a dramatic transformation. British officials had seemed sufficiently optimistic about the project to suggest on March 5, 1949, an "announcement in Hong Kong, China and London." However, they soon retracted the decision, stating in April that they should "in present circumstances do everything possible to avoid stimulating political reactions, particularly in China" and were thus "anxious to avoid making any public announcement here [London] or in China at this stage," noting further that "publicity should be restricted to [a] minimum local announcement . . . necessary to enable [a] survey and other preliminary work to start."[141]

Although the British made no public statements about their plans for Deep Bay, they continued to pursue diplomatic exchanges with the Chinese government through the Chinese Ministry of Foreign Affairs. The issue had become an open secret, and a ministry representative offered unsolicited advice on the acquisition of land in May 1949, citing a special article governing the 1898 lease of the New Territories that the British had extracted from the Qing government.[142] In the meantime, the retreat of Nationalist forces to the south had fueled an exodus, and the Hong Kong airport was handling an average of one thousand three hundred passengers per day in mid-1949. The situation remained fluid throughout the year. In October 1949, Communist forces captured Canton and Kunming, completely disrupting air transport between Hong Kong and mainland China, which registered an immediate drop of some 60 percent.[143]

As 1949 drew to a close, the British Foreign Office expressed discomfort over the British government's "leisurely way" of handling the construction of the new airfield. Officials worried that with the increasing consolidation of the Communists' power, local landholders would likely react more violently to eviction. Communist advances to the border with Hong Kong did not put a stop to the Deep Bay project. Reports of the

141.   TNA, FO371/75923.

142.   TNA, FO371/75923.

143.   *HKDCA*, 1952–1953, 5–6. The British government, along with the colonial authorities in Hong Kong, became embroiled in an ownership dispute about Chinese aircraft left at Kai Tak at the time of the Communist takeover of the mainland. For a detailed discussion of that controversy, see Kaufman, "United States, Britain."

cessation of commercial service operations to Shanghai notwithstanding, an Air Ministry survey party was supposedly finally "getting down to the initial job of surveying the site" in February 1950, causing concerns that such a dilatory approach would end in the project's abandonment or result in "a first class example of officially inspired Chinese agitation." The Foreign Office warned that unless construction was expedited, the new PRC government could actively oppose the project, and no diplomatic channel existed for any discussion. The Colonial Office could only respond by saying that "it would be a great pity if any unavoidable delay caused us to have to abandon the project at a later stage."[144]

In June 1950, British officials began to capitulate on the Deep Bay project. Governor Grantham had come to the conclusion that "Deep Bay was not a sound proposition" not simply because of its high cost but, more importantly, because of its risky location within "small arms range of Communist held territory." Although not as technically satisfactory as building at Deep Bay, pivoting the main runway at Kai Tak would be a superior solution, Grantham believed. Talks thus began on the extension of the existing runways, although British officials considered it prudent to cap the costs of any such extension. It was also generally agreed that Hong Kong needed more than one airfield for defense purposes. Having ruled out Deep Bay, it was decided that two airfields be developed at Kai Tak and Shek Kong. By early July, British officials had met in London to discuss the costs of the two projects.[145]

Thus ended the discussions on constructing a new airport at Deep Bay. British officials reverted to Kai Tak, once considered topologically inferior and unsafe. The technical experts were reluctant, but the government's rationale was irrefutable.[146] Deep Bay's proximity to Communist territory evidently played an important role in its demise as an airport location. However, the dramatic reduction in air traffic volume in Hong Kong also gave the British pause in making any significant investment in the budding aviation hub. The cessation of aerial connections between Hong Kong and mainland China eliminated the largest component of the city's aviation flow. The immense impact was felt by early 1950, when

144. TNA, FO371/83447; TNA, BT245/991.
145. TNA, CO537/5619.
146. TNA, CO537/5620.

The Making of a Place in an Evolving Air Space 55

Chinese operations ceased entirely.[147] In July 1949, the Chinese operator Central Air Transport Corporation (previously Eurasia) had operated twenty-eight weekly services to Canton and twenty-nine to other points in China, and CNAC had offered thirty-five weekly services to Canton, fifty-eight to other Chinese destinations, one each to India and Thailand, and two monthly services to the United States. By March 1950, the services of both operators had ceased. Also eliminated were the twenty-eight weekly services to Canton that Hong Kong Airways Limited had offered in July 1949. Air traffic in March 1950 was one-sixth the level that it had been in August 1949.[148]

Aircraft arrivals to and departures from Hong Kong totaled 11,057 and 11,016, respectively, in the twelve months ending March 31, 1950.[149] The year ending March 31, 1951 registered 2,640 arrivals and 2,650 departures. Thus, total aircraft movements plummeted 76 percent compared with the previous year. The loss of 17,091 landings and takeoffs associated with mainland Chinese traffic was only marginally offset by the 1,425 landings and takeoffs connecting Hong Kong to the new Nationalist base in Taiwan. Similarly, the total passenger count plunged 74 percent to 73,064 in that year, as the 4,109 inbound and 5,403 outbound Taiwanese passengers could hardly compensate for the elimination of some 100,000 mainland Chinese passengers traveling in either direction in the previous year.[150] Aircraft movements remained at these low levels until the mid-1950s, and the colonial government saw little indication of a resumption of scheduled air services to mainland China.[151] Although British officials differed on the suitability of an expanded Kai Tak for future aircraft types, an air advisory board recommended in February 1951 the deferral of the Deep Bay project and pressed the Hong Kong government to approve the extension at Kai Tak.[152] Debates ensued on the applicability of the loan that London had extended to Hong Kong for a new airport. Unofficial members of the Executive Council in Hong Kong argued that an expanded Kai Tak would benefit the United Kingdom and that it would

147. LegCo, March 8, 1950, 36.
148. *HKDCA*, 1940–1950, 3–4; TNA, CO537/5619; LegCo, March 8, 1950, 48. Chapter 2 will provide details on the history of Hong Kong Airways Limited.
149. *HKDCA*, 1949–1950, 12; TNA, CO537/5619.
150. *HKDCA*, 1949–1950, 12; *HKDCA*, 1952–1953, 60; TNA, CO937/274.
151. *HKDCA*, 1954–1955, 25; *HKDCA*, 1956–1957, 38.
152. TNA, BT245/991.

56 *Mapping Hong Kong*

therefore be reasonable for London to bear part of the cost.[153] The issue of cost sharing and technical challenges persisted in the following years, but by the mid-1950s officials in London and Hong Kong had focused their attention on Kai Tak.[154] It would take the rest of the 1950s to complete the construction.[155]

That Hong Kong was already an international center for sea traffic did not necessarily presage the making of Hong Kong into an aviation hub. Civil aviation represented a disruptive technology that held the potential to rechart the routes of global traffic. It was therefore understandable that the British authorities, along with the vested interests in British shipping that operated in Hong Kong, endeavored to develop the new routes in the skies according to the old footprint of sea traffic through the British colony. In 1939, Jock Swire, who was to spearhead Swire's efforts in developing commercial aviation in Hong Kong (chapter 3), suggested an "aeronautical wing" for the company's dockyard to safeguard "against the future."[156]

Technology evolves; it does not develop anew. In fact, as aviation technology was maturing into a commercial form, the colonial authorities listed Kai Tak in official documents in the early 1950s as both a "Land Aerodrome" and "Water Aerodrome" with the same "Co-ordinate Location" but different specifications for "Elev[ation]," "Landing Area," and "Lighting."[157] This amphibious configuration was necessary in the early days of commercial aviation, as operators such as Imperial Airways opted for flying boats to circumvent the difficulty of constructing and maintaining airstrips all along its extensive network.[158] While injecting inventiveness into the system, the nascent technology for commercial aviation also had to contend with the inertia that made it natural for air routes to conform to the old traffic of hopping from port to port. Emerging powers (most notably the United States) and the opening of new

153. TNA, T255/599; LegCo, March 27, 1952, 141–42.
154. LegCo, March 6, 1958, 45; LegCo, March 26, 1958, 84–85, 101–2.
155. LegCo, March 4, 1953, 28–29; LegCo, March 17, 1954, 67, 78–80; LegCo, June 16, 1954, 216–20, 225–27; LegCo, January 12, 1955, 8–9; TNA, FO 371/141231.
156. JSS, 13/8/4/3 Imperial Airways, Correspondence (incomplete), 1933–1941.
157. *HKDCA*, 1952–1953, 12, 29; *HKDCA*, 1954–55, 14.
158. Lyth, "Empire's Airway," 881; see also HKPRO, HKRS156-1-409.

space (in the mid-Pacific) introduced new elements to the evolving aviation industry. In Hong Kong, established interests, represented by Britain's Imperial Airways, connected with burgeoning forces, led by Pan Am and its Chinese joint venture. Its role as the region's connector was not preordained, as evidenced by the prolonged discussions on the possibility of establishing Macao as an aviation base.

Thanks to its geographical location at the southern tip of mainland China and its political configuration as a British colonial outpost in the Far East, Hong Kong was an appealing point of connection for enterprising Chinese, British, and other business interests from around the world seeking to establish a foothold in the region. For access to this entry point to the alluring Chinese market, the weakened British wrestled with ascending American influence to negotiate their broader interests. Hong Kong's development into an aviation hub may have seemed a natural extension of its maritime development. In reality, the city became a hub only after overcoming numerous obstacles, not only in terms of technological contingencies but also in the context of regional and global dynamics.

Hong Kong's appeal as a commercial aviation center took surprising twists in the 1940s. Wartime events produced further entrenchment, as the Japanese occupation expanded the infrastructure at Kai Tak, primarily to meet its own military needs. Wartime deliberations over the rules of commercial aviation cemented Hong Kong's crucial role as a British foothold at this far end of its flagging imperial reach. Plans for postwar reconstruction began even before the conclusion of World War II, as the winning powers vied for control over air space. As commercial traffic resumed at the conclusion of the war, the city required earnest development of its physical infrastructure to serve connections in the skies. Policymakers proceeded cautiously owing to their anxiety over the mainland Chinese reaction.

In the brief interlude of peacetime development, the intense traffic into and out of Hong Kong from China generated the financial interests that underwrote its commercial aviation hub. In Hong Kong, Chinese air traffic connected with routes to Europe and Southeast Asia and enjoyed additional linkages to the United States. Such traffic held tremendous promise for British interests not only in the colony but also in London. After the Communist takeover of mainland China, the enthralling

58 *Mapping Hong Kong*

opportunities of China's emerging aerial market disintegrated. British negotiations between the colony and the metropole over the cost sharing of infrastructure development in Hong Kong came to an abrupt halt. Diminished business prospects of commercial aviation in Hong Kong led the British authorities to consider infrastructure construction on a reduced scale. Commercial calculations and geopolitical considerations resulted in the expedient decision to return to the old Kai Tak, which ambitious planners had considered inadequate for Hong Kong's commercial aviation potential just a few short years before.

David Edgerton has convincingly argued against the narrative of British decline in the twentieth century.[159] He contends that Britain remained technologically innovative.[160] In particular, his study of Britain's aircraft industry shows that the country continued to be a "warfare" state that prioritized technological, industrial, and military developments.[161] From the perspective of developments in Hong Kong in the aftermath of World War II, British ambitions endured. However, British investment in aircraft development followed a path that diverged from the country's resolve to preserve its infrastructure footprint. Although the application of aircraft technologies is not limited geographically, airport infrastructure is spatially specific. Britain lost its interest in constructing a new airport to match its place in Hong Kong precisely as the Communist takeover of mainland China heightened the geopolitical risks for its colony. The truncated airport project then followed a mostly financial logic and proceeded only through the perseverance of British officials in the outpost of Hong Kong.

Infrastructure development, even on a reduced scale, marked Hong Kong's place on the global aviation map. Its prominence, albeit erratically projected during these early days, owed much to its location both geographically and politically. The British presence in Hong Kong connected the colony to Britain's imperial network, then being refashioned into the Commonwealth. American interests met this British network in Hong Kong, completing the long-awaited route of circumnavigation in the skies. Notwithstanding its internal and external military conflicts, China had

159. Edgerton, *Rise and Fall*; Edgerton, "Decline of Declinism."
160. Edgerton, *Science, Technology*.
161. Edgerton, *England and the Aeroplane*.

until 1949 promised potentially enormous global aviation traffic flows, and Hong Kong was in a favorable position to capture those flows. The altered political landscape to the north of Hong Kong pared the city's network in one direction, but the commercial aviation network continued to proliferate around Hong Kong on the basis of the infrastructure already in place.

The physical manifestation of Hong Kong's aviation infrastructure, as well as the wiring of the city into the network of flight routes, unfolded against shifting geopolitical dynamics. In those early decades, Hong Kong's centrality in the world of commercial aviation was constructed through careful negotiation and pragmatic adjustments in a tumultuous historical period.

CHAPTER 2

Reorienting Hong Kong

Resizing and Conforming to Emerging Geopolitics after World War II

Kai Tak Airport Desolated Like Never Before

There is no news yet on the resumption of air traffic between Hong Kong and the Communist area [the PRC]. Under the current circumstances, the Kai Tak airport is unprecedentedly quiet. Hong Kong, which used to be the busiest air terminus in the Far East, has been drastically transformed. For the last few days, idled airport personnel at Kai Tak left before five in the afternoon.

—*Kung Sheung Morning News* 工商日報, January 9, 1950[1]

Initially established to provide the British Empire with a foothold for northern expansion, the aviation hub of Hong Kong had to change course after 1949 with the Communist takeover of mainland China. The hub had come into being through the confluence of air traffic for British, American, and Chinese carriers, among others. The retreat of the Nationalist forces from the mainland rerouted air traffic to Taiwan as Communist-controlled China cut the mainland off from the aviation hub of Hong Kong along with the rest of the non-Communist world. How did the commercial aviation industry in Hong Kong persevere in the face of the loss of its largest flying partner?

In the 1950s, the aviation hub downsized its expansion plans significantly. Yet infrastructure development continued and its evolution paralleled the geopolitical transformations in the subsequent period. The loss

---

1. *KSMN*, January 9, 1950, 5.

of the connections to mainland China was but one of the factors reshaping the development of commercial aviation in Hong Kong. Infrastructure investment in Hong Kong contributed to the legitimacy of the British regime in the age of decolonization, albeit at a reduced level. World War II had exposed the vulnerability of colonial powers, epitomized by their swift defeat at the hands of Japanese forces in Hong Kong and other parts of Southeast Asia. In the aftermath of the war, imperial powers, among which the British occupied a premier position, had to reformulate their policy to cope with rising nationalism and escalating economic competition.[2] In the case of Hong Kong, independence was out of the question due to British disputes with the Chinese regimes over sovereignty.[3] Britain came to view Hong Kong less as its imperial symbol than as a critical component to its strategy to remain a global power.

To execute this strategy successfully, Britain needed to invest in infrastructure in its colony to strengthen Hong Kong and mobilize the city in the global competition for power and wealth.[4] However, the British machinery did not march to a single drumbeat. Different forces in the British system sought to profit from this process. In Hong Kong, the development of commercial aviation demonstrated that in the immediate postwar period, the industry grew to conform to a hierarchy of British power. The rivalry among British interests in London and Hong Kong underscores the contentious process by which business enterprises vied for control and influence in the economic reconstruction and development of the colony.

Compounding the global dynamics of decolonization, the Cold War was taking shape, and the rival camps turned Hong Kong into a frontier, a contact zone between imperialism and nationalism. The principles of free trade in colonial Hong Kong confronted national and developmental ideas.[5] Local concerns animated Cold War dynamics,[6] which in Hong Kong's case precipitated popular support of the colonial government in

---

2. Duara, *Decolonization*, 7.

3. Gibraltar and the Falkland Islands were in similar situations (Ashton, "Keeping Change Within Bounds," 49).

4. Duara, *Decolonization*, 16.

5. Duara, "Hong Kong," 211–12.

6. Zheng, Liu, and Szonyi, *Cold War in Asia*.

the age of decolonization. In the geopolitical reconfiguration that ensued, Hong Kong reoriented its airspace. As commercial aviation connected Hong Kong with budding aviation hubs on its side of the Cold War divide, the industry not only redirected the proliferation of air routes to a different region but also fashioned a sphere of activities for the economic and cultural exchange that airways facilitated.

While continuing to serve to connect the traffic of Britain, the United States, and their allies, the air space around Hong Kong developed to reflect the reconfiguration of the Southeast Asian region and connected to a northern, island-hopping route that rimmed Communist China, linking Hong Kong to North America. This process took shape in the proliferation of flight routes, the construction at Kai Tak, and the commercial reorganization of the airline industry based in Hong Kong.

Linchpin of a Reconfigured Air Space

After the Communist takeover of mainland China, the euphoria over Hong Kong's extraordinary growth in commercial aviation proved unfounded. The explosion of traffic, particularly in 1948 and 1949, reflected not sustainable flows but unrelenting flight from the collapsing Nationalist regime via Hong Kong. A new baseline on which the government needed to craft its policy on aviation development began to emerge in 1950.

By March 1951, regular flights into Hong Kong had reduced to those of ten airlines offering twenty-nine weekly services in total. Pan Am offered the most flights with eight weekly flights: five to the United States "via Pacific," and three to London via Bangkok, Thailand. BOAC offered six a week: two each to London (via Bangkok), Tokyo, and Singapore. Hong Kong-registered Cathay Pacific offered five per week: three to Singapore (two via Bangkok and another via Saigon), one to Hanoi via Haiphong, and one to Labuan via Manila. These leading players were followed by Air France (with three weekly services to Paris via Saigon) and Australia's Qantas Empire Airways (with two per week: one each to Tokyo and Sydney). The rest of the carriers offered one weekly flight each (Norway's Braathens SAFE Airtransport A/S to Oslo; Canadian Pacific to Vancouver; Hong Kong Airways Ltd. to Taipei and connecting with Northwest Airlines to Minneapolis in the United States; Philippine Air Lines to Manila and thence to the United States and Europe; and Thai Airways to Bangkok).[7]

BOAC lamented that British aviation was missing the opportunity to establish "predominance in the regional traffic of the Far East . . . for lack of wholehearted support from the United Kingdom Government." The company viewed Hong Kong as the "natural centre of gravity" for the region, situated along "the routes between Japan and South East Asia, between Japan and Burma, E. Pakistan and India, with various branches." In the prevailing geopolitical environment, even "ignoring China traffic as outside practical politics for the time being," BOAC listed potential routes "radially from Hong Kong" to Korea, Japan, Taipei, Manila, Borneo, Indonesia, Singapore, Bangkok, Saigon, Rangoon, Chittagong, Calcutta, Haiphong, and Hanoi. "All of these can be served," the airline claimed, "by an aircraft which returns to the base (in Hong Kong) within 36 or at most 48 hours." The route between Hong Kong and Australia would also be "a natural and legitimate expansion," noted BOAC. The airline complained that, in these aspects of "air politics," it was not sufficient to rely on the restrictive power the United Kingdom exercised through bilateral treaties. Restricting air traffic would only add to "the rapid wane of British influence." Instead, the British government should leverage its colonial possessions in the region and fill the vacuum of air traffic with British services so that the UK government could "with reason and strength" oppose intruding forces, especially from the United States.[8]

Accordingly, in the year ending March 1952, BOAC overtook Pan Am as it increased its weekly frequency to nine flights with two extra services to London via Bangkok and an additional service to Tokyo. Philippine Air Lines also doubled its frequency to two. Two more operators entered the fray: Air Vietnam took over two of the three services from Air France, offering two weekly services to Saigon, one each via Hanoi and Haiphong. Nationalist China's Civil Air Transport (CAT) entered the picture with

---

7. *HKDCA*, 1952–1953, 39–40.
8. British Airways Archives, "O Series," Geographical, 3316.

one weekly service to Bangkok, two to Taipei, and two to Tokyo via Taipei.[9]

By March 1953, British and colonial concerns had expanded their shares further. BOAC stepped up its activities to eleven services a week (five to London via Bangkok, four to Tokyo, and two to Singapore). Cathay Pacific added one flight to Labuan via Manila, and Hong Kong Airways tripled its service to Taipei.[10] In spite of the lackluster recovery of aircraft movements through Hong Kong, passenger count showed healthier growth. After an uninspiring showing of 5 percent growth for the year ending March 1952, the following two years registered 16 and 12 percent increases, respectively. Thailand dominated both inbound and outbound passenger traffic in the 1950s, not surprisingly as Bangkok served as the connection to Europe. While passengers traveling to and from Indo-China remained a significant share of the traffic, the Philippines soon became the second most important contributor to passenger count as Manila served as a gateway to the United States and points south. During these years in the early 1950s, mail volume registered decent growth while commercial cargo struggled at low levels.[11]

Through 1953, in spite of scant resources, the British colonial government continued to position Hong Kong as a global hub that welcomed connections with "neighboring countries." "Hong Kong is on the trunk routes of a number of major airlines of the world," boasted Hong Kong's director of civil aviation in his 1953 report. BOAC "operated trunk route services from the United Kingdom through Hong Kong to Tokyo, some by way of Bangkok and other *via* Singapore." Canadian Pacific Air Lines linked Hong Kong with Canada on a northern route by way of Japan. In terms of connections with the southern hemisphere, Hong Kong served as the connector for a regular service of Qantas Empire Airways, which linked Australia to Japan. Beyond the British Commonwealth, friendly colonial partners and other allies also called at Hong Kong. Air France and "its subsidiary regional Air Vietnam" connected Hong Kong with Indo-China and Paris. More expansive than the French network was of course the American one. Pan Am World Airways' trunk route emanated

9.   *HKDCA*, 1952–1953, 39–40.
10. *HKDCA*, 1952–1953, 39–40.
11. *HKDCA*, 1952–1953, 58, 60; *HKDCA*, 1954–1955, 42.

"from the United States of America across the Pacific, through Asia to Europe," passing through Hong Kong. Associated with the Pan Am trunk route was Philippine Air Lines, which linked the West Coast of the United States "through Manila to Hong Kong and to Europe and the United Kingdom." In conjunction with its position on these major thoroughfares, Hong Kong also entertained "frequent services" to nearby destinations "such as the Philippine Islands, Formosa [Taiwan], Thailand, North Borneo and Singapore." Missing from the picture was the vast mainland Chinese market, and the director took caution to note that "air communications with the China mainland and Hainan Island remained suspended."[12] In the early 1950s, the British Commonwealth took a leading role in civil aviation in Hong Kong, buttressed by commercial interests of the United States and other allies in the region.

Escalating Cold War conflicts invigorated the transformation of commercial aviation through Hong Kong. The loss of French control over northern Vietnam introduced new dynamics to commercial aviation flows through Hong Kong. The Geneva Accords of 1954 stipulated the creation of a military demarcation line at the 17th parallel.[13] Although Air France retained its weekly service from Hong Kong to Paris via Saigon, Air Vietnam, which had operated under the French colonial regime, discontinued its service from Hong Kong to Saigon via Hanoi in September 1954. Similarly, Cathay Pacific discontinued its fly-through service via Hanoi in March 1954.[14] It should come as no surprise that in January 1955 British authorities considered it "inopportune for any talks with the Chinese about a resumption of service" between Hong Kong and the mainland.[15]

In the meantime, other connections through Hong Kong came to be established and a new regional configuration emerged. In addition to its two weekly connections from Hong Kong to Bangkok, with onward service to Calcutta, Thai Airways added two weekly connections between Hong Kong and Tokyo in 1954. Nationalist China's CAT extended all its Taipei-bound flights from Hong Kong to Tokyo, resulting in a total of

---

12. *HKDCA*, 1952–1953, 39; Pan Am, Series 2, Sub-Series 1, Sub-Series 3, Box 2, Folder 1.

13. Matejova and Munton, "Western Intelligence Cooperation."

14. *HKDCA*, 1954–1955, 26–27.

15. TNA, FO371/115397.

66   *Reorienting Hong Kong*

four weekly services to Tokyo by March 1955. By March 1955, Air India had also introduced a weekly service from Hong Kong to Bombay via Bangkok. Adding to "the long list of foreign Airline Companies operating Regional services to and through Hong Kong," Japan Air Lines started a weekly service between Hong Kong and Tokyo. Upon the conclusion of the Korean War, Korean National Airlines began a weekly service to Seoul via Tokyo. The two Hong Kong–registered carriers continued to report satisfactory performance in "the regional sphere." In particular, Cathay Pacific expanded its fleet and promised "a substantial increase" in its service in 1956.[16] Hong Kong became a hub in Southeast Asia and connected regional traffic not only to Europe but also to the trunk routes of North America through the corridor of the Philippines, Taiwan, Japan, and Korea.

The government reported that Hong Kong was "amply served by major Airlines from all parts of the World, notably the British Overseas Airways Corporation and Pan American World Airways." Between 1953 and 1954, BOAC had retrenched from its leading position in Hong Kong, cutting two of its five weekly connections between Hong Kong and London via Bangkok, and one of its four weekly services between Hong Kong and Tokyo. Reclaiming the top position in the number of connections with Hong Kong, Pan Am added two weekly flights to the United States on its Pacific route and one weekly connection with London via Bangkok, bringing its total weekly connections to eleven, compared with BOAC's eight in March 1955. In addition to flying with the highest frequency, Pan Am also carried "the greatest number of passengers to and from Hong Kong," and used "the largest types of aircraft" at Kai Tak.[17]

All told, these reconfigurations of traffic flows saw further increases in flights flying the flags of the United States and its friendly partners. For the year ending March 1955, Tokyo topped the chart in passenger count origin (distributed primarily among Nationalist China's CAT, BOAC, and Thai Airways), followed by Manila (dominated by Philippine Air Lines). Ranked behind these two regional destinations were Los Angeles and New York, both served by Pan Am. Cathay Pacific held its own with the leading share in the popular market of Singapore (shared with

16.   *HKDCA*, 1954–1955, 25–27.
17.   *HKDCA*, 1954–1955, 26–27, 31; TNA, BT245/860.

BOAC) and its reach to other regional destinations such as Labuan and Saigon.[18] In terms of aggregate flows, the year registered a 20 percent surge in civil aircraft movements, 22 percent increase in passenger count, 19 percent growth in mail volume, and 8 percent increase in commercial cargo.[19] It should be noted that despite such phenomenal expansion in traffic volume, air travel remained beyond the reach of most Hong Kong residents. A flight from Hong Kong to Manila, the cheapest ticket for travel from Hong Kong, cost HK$300 for a one-way flight on Cathay Pacific in 1953. That fare was more than half a month's salary for an accountant working at the University of Hong Kong, or two months' pay for a semi-skilled male wage-earner. A one-way trip from Hong Kong to Tokyo cost HK$992 on BOAC, approximately the monthly salary of an assistant lecturer.[20] Local Hong Kong residents were evidently not the prime target customer for airlines; foreign tourists were. Toward the end of the decade, Cathay Pacific published newsletters carrying comments from certain passengers: "a 9-year-old American girl," an American couple, a Japanese businessman who traveled between Hong Kong and Singapore, a Chinese resident in Singapore who flew between Bangkok and Singapore, a German traveler, and an American family in Delhi that flew between Hong Kong and Bangkok.[21]

The intensified air traffic flows through Hong Kong unfolded not only in the context of escalating Cold War conflict but also in conjunction with global aviation growth. The International Civil Aviation Organization reported that in 1956 scheduled services carried 78 million passengers over an average distance of 575 miles, "almost the equivalent of airlifting the entire population of Ceylon from Colombo to Sydney." That passenger count reflected a 15 percent increase from 1955 and was nearly 4.5 times as large as it was a decade ago. The industry had made tremendous technological progress: while the average airplane carried 17 passengers at 177 mph in 1945, in 1955 it carried 28 passengers at 200 mph.[22]

18. *HKDCA*, 1954–1955, 50–51.

19. *HKDCA*, 1956–1957, 25.

20. *HKDCA*, 1952–1953, 44; *Report of the Salaries*, 20–21; Ma and Szczepanik, *National Income of Hong Kong*, 20.

21. Swire HK Archive, CPA/7/4/1/2/4–6 Newsletter [August and September 1959].

22. *HKDCA*, 1954–1955, 42.

68 *Reorienting Hong Kong*

Against this global backdrop, commercial aviation kept growing in Hong Kong. A local Chinese newspaper celebrated the growth of traffic through Kai Tak, even bragging that it was the fourth-busiest airport in the world. That Hong Kong had made up for lost ground and achieved such success (albeit exaggerated in the article) was the inevitable result of the efforts of the Hong Kong government, the article claimed.[23] In the expanded air traffic through the hub in Hong Kong, US representation became more noticeable. Pan Am deployed the DC-8, the "World's Largest, Fastest, Over-Ocean Jet Fleet," on the Los Angeles to Hong Kong route, shortening the trip to 23 hours and 45 minutes.[24] For the year ending March 1958, Pan Am represented 15 percent of all aircraft takeoffs and landings in Hong Kong, eclipsing BOAC's 11 percent. (The two Hong Kong–registered carriers were responsible for 14 and 10 percents.) In terms of passenger count, Pan Am's share was even larger at 19 percent, versus BOAC's 10 percent (and the local carriers' 14 and 11 percents).[25] British interests in commercial aviation were keenly aware of this development and monitored the situation closely.[26] In the late 1950s, BOAC and Pan Am continued to vie for market share, especially over the connection to Tokyo.[27] Although British interests had vowed in the early 1950s to seize control of the hub in a bid to arrest declining British influence in the age of decolonization, Cold War dynamics propelled a surge in American interest in the British colony by the late 1950s, at least in the critical industry of commercial aviation.

Airport Development

The timing of this Cold War–related surge in air traffic volume coincided with the colonial government's stepped-up commitment to developing aviation infrastructure in Hong Kong. After discussions of a new airport at Deep Bay had subsided, British authorities continued to consider

---

23.   *WKYP*, September 12, 1956, 10.
24.   Pan Am, Series 2, Sub-Series 1, Sub-Series 4, Box 1, Folder 1.
25.   *HKDCA*, 1957–1958, 45–58.
26.   TNA, BT245/873.
27.   *TKP*, April 7, 1959, 4; *TKP*, April 19, 1959, 4.

equipping Hong Kong with "an airport up to the international standards." The first tangible progress toward that goal was the Ministry of Civil Aviation's dispatch of a new technical mission in 1951 "to examine on site the various schemes which had been proposed for the provision of an international airport in the Colony." The mission produced in June 1951 the Broadbent Report, which decided on the development of Kai Tak. Accepted in principle by the Hong Kong government and the Ministry of Civil Aviation in 1952, the report proposed a development plan that included a new main runway and the extension of the existing runway. Survey work began that year.[28]

Although British officials had confined their discussions of airport construction to Kai Tak,[29] the government did not approve a plan for the substantial development of Kai Tak and make the necessary financial provisions until June 1954. The project approved in 1954 called for the construction of "a promontory, 795 feet wide and 8,300 feet in length, containing a 7,200 feet paved runway, with prepared over-runs of 300 feet at the south-east end, and 800 feet at the north-west end with a taxiway 60 feet wide parallel to the runway." This 150-acre promontory was to come from a consolidated reclamation. The authorities planned to flatten hills in the northwest approach to the main runway "to provide a one-in-forty clearance angle," with the demolished hillside providing landfill for the runway project. The project was estimated to cost HK$96.75 million, to be covered in part by an interest-free loan of HK$48 million from the UK government. On some 9 acres of additional reclaimed land, the government would build parking facilities, a maintenance area, and a new terminal building. The terminal building was designed to facilitate efficient passenger and baggage flow and would accommodate, if necessary, a 100 percent expansion of control handling units without interruption to the services. The runway and the associated facilities were to be "stressed to take aircraft with an all-up-weight of 250,000 lbs." (BOAC was flying the Britannia which had an all-up-weight of 150,000 pounds.) The plan was to finish all major work by the end of 1958, with the new terminal

28. *HKDCA*, 1952–1953, 33–35; LegCo, March 4, 1953, 28–29; Government of Hong Kong, *Papers on Development*, 1.

29. TNA, CO937/273; TNA, CO937/274; LegCo, March 17, 1954, 67, 78–80; LegCo, June 16, 1954, 216–20, 225–27; HKPRO, HKRS163-1-1324.

area ready for use by late 1959.[30] The airport development project was said to have "attract[ed] world interest." Construction progress received due attention from the colonial authorities as members of the Airport Progress Committee and the governor made inspection tours of the site and the hills being leveled.[31]

On September 12, 1958, the new runway was officially opened. The public had eagerly anticipated the opening for days as the colonial authorities announced that all were welcome at the ceremony.[32] Governor Robert Black and his family crossed the harbor in a helicopter to officiate the opening ceremony. Six hundred official guests and some fifty thousand spectators were treated to an air display after the ribbon cutting and a two-minute salvo of firecrackers. While certain news coverage mentioned the arrival of the jet age and the further development that the new airport infrastructure would facilitate,[33] excitement mounted primarily over the acrobatics.[34] Three RAF jet fighters flew in formation, followed by a fly-by of airplanes from Cathay Pacific (a DC-3 and a DC-6B), Pan Am, Air India, Qantas, Hong Kong Airways, Canadian Pacific, and BOAC. The eight civil aircraft showcased in the ceremony represented the types in operations at Kai Tak. The governor called the new runway "a very fine and unique engineering achievement," "a project of which the people of Hongkong can feel justly proud." The opening of the runway was "an important step in the development of civil aviation facilities on one of the world's major air routes." "Henceforth," said Black, "Hongkong will not only be noted for its fine natural harbour but also for its imaginatively planned runway. . . . As with our harbour in the past, so will this runway help us in this modern air age, to develop the trade on which Hongkong's prosperity was founded." With the rest of the facilities

30. *HKDCA*, 1954–1955, 18–19; *SCMP*, June 3, 1954, 1; LegCo, March 6, 1958, 45; HKPRO, HKRS163-1-1897; HKPRO, HKRS163-1-1329. The secretary of state for the colonies informed the governor of the fifteen-year, interest-free loan of £3 million (HK$48 million) on July 24, 1954 (HKPRO, HKRS1448-1-149; HKPRO, HKRS1764-1-5). The official rate for conversion to pound sterling was HK$16 = £1.

31. *HKDCA*, 1956–1957, 12; *SCMP*, January 27, 1955, 12; HKPRO, HKRS1764-3-18.

32. *KSMN*, September 9, 1958, 5; *TKP*, September 9, 1958, 6; *KSEN*, September 10, 1958, 6; *WKYP*, September 12, 1958, 5; HKPRO, HKRS70-1-1.

33. *KSEN*, September 10, 1958, 6; *WKYP*, September 12, 1958, 5.

34. *KSMN*, September 12, 1958, 5.

that the government promised to construct, Kai Tak was to be "one of the most modern and best equipped airports in the Far East, comparable with any major airport in the world," boasted the governor.[35]

Chinese newspapers reported the festivities with multipage coverage filled with illustrations of the spectacle. "Kai Tak turns the page in history," and "the new terminal building that will be completed in two years will be Asia's finest," read the headlines of one newspaper, which featured photographs of the new runway and the flyover of the jet planes.[36] Another newspaper arrayed pictures of the runway, plans for the terminal, the helicopter that carried the officiating party, alongside photographs of the Comet 4, which had newly entered the BOAC fleet, the Bristol Britannia 314 of Canadian Pacific Air Lines, and the Boeing double-decker plane of Pan Am. These photographs were followed on another page by a detailed report of the project; over the next two days, follow-up pieces in each issue detailed the growth of Kai Tak.[37] Even a leftist newspaper could not ignore the extravaganza. The newspaper published an oblique critique of the capitalistic venture as it reported that a female spectator lost her HK$1,300 diamond ring in the airport toilet.[38] Two days after the runway's opening, other local newspapers continued to carry congratulatory messages. "The newly opened runway will bring prosperity to the Pacific region," reported a Chinese newspaper that conveyed the well wishes from Pan Am and mayors of US cities.[39]

With some follow-up construction, Hong Kong was ushered into "the advent of the modern pure jet transport aircraft on scheduled services" by 1960, fulfilling the promise of the plan that was finalized in 1952. "The phasing and the timing of the completion of the various projects of the development plan could hardly have been bettered," noted the director of civil aviation in 1960, "for it was only at this time that the airlines could take delivery of and bring into operation their high performance aircraft, and the runway and other operational movement areas were ready for them." Colonial officials were proud of the timing and efficient

---

35. *HKDCA*, 1958–1959, 18; *SCMP*, September 13, 1958, 1; HKPRO, HKRS70-1-1.
36. *KSMN*, September 13, 1958, 5.
37. *WKYP*, September 13, 1958, 4, 7, 12; *WKYP*, September 14, 1958, 7.
38. *TKP*, September 13, 1958, 6.
39. *KSMN*, September 14, 1958, 5.

construction. The completion of the new facilities eliminated load penalties and extended the hours of operation, allowing more flexibility for the operations of "the major world routes."[40]

Ironically the completion of work at Kai Tak aroused even more concerns for the British authorities. In the midst of the celebrations of the new runway and airport facilities, British officials came to question whether they had constructed the new structures in areas China had ceded or leased to Britain. Kai Tak was situated to the east of Boundary Street, which marked the northern border of the ceded territories covered in the 1860 Convention of Peking.[41] The area to the north of Boundary Street, stretching to the south of the Sham Chun (Shenzhen) River, later known as the New Territories, was leased to Britain for ninety-nine years in the 1898 Convention for the Extension of Hong Kong Territory. In February 1959, the Colonial Office questioned "whether any part of Kai Tak Airport (including the area of its reconstruction) lies within the boundary of Kowloon, which was ceded by China in 1860, and if so, which part is in Kowloon and which in the adjoining New Territories, the latter leased by China to Great Britain in 1898 for 99 years."[42]

The governor of Hong Kong concluded that "the new Airport buildings would be regarded as being in the leased territories, while the runway might or might not be included in the British share of the waters of Kowloon Bay after the expiry of the lease in 1997, depending on the exact alignment of the northwest/southeast dividing line." Legal advisers suggested that it was "impossible to draw a precise line" unless they could secure "a more authoritative chart of the coastline of 1898" (fig. 2.1). Rejecting appeals to an article in the 1958 Convention on the Territorial Sea and the Contiguous Zone because it was doubtful that Communist China would become a party to the convention, the Colonial Office concluded tentatively that "there was an implied agreement in 1898 as to the dividing line between the ceded and the leased areas and consequently we are unable to alter the line to our advantage by subsequent constructions in

40. *HKDCA*, 1959–1960, 38.
41. The area was originally leased in March 1860.
42. TNA, FO371/141231.

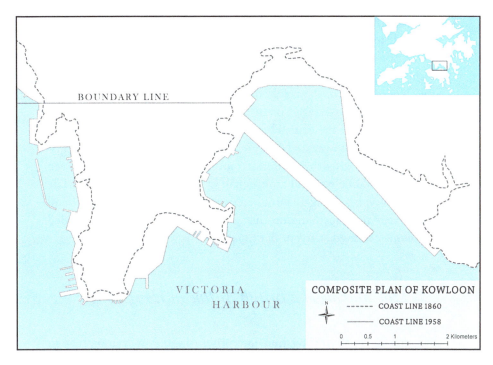

FIGURE 2.1. Schematic representation of a map labeled "Composite Plan of Kowloon showing Coast Line 1958 in Solid Line / Coast Line 1860 in Dotted Line. The map of 1860 is very inaccurate and the coast line as shown is approximate only." Source: TNA, FO371/141231.

the bay." In the end, the officials left the issue unresolved and found it not necessary at that time to discuss the problem further.[43]

Kai Tak did provide Britain with an airport location that was more militarily defensible than the originally suggested Deep Bay. However, technological advancements in reshaping landscape through reclamation led to a potential diplomatic problem that nineteenth-century negotiators of the Sino-British treaties could not have foreseen. The British authorities, in spite of years of deliberations, had overlooked an important difference between their leased and ceded territories. Ironically, in

43. TNA, FO371/141231.

specifying the boundaries in 1860, the maritime power that had secured additional territories for Britain paid attention only to land formations and overlooked the waters around them. As the lease was not to end for another thirty-eight years, any diplomatic issue would be decades away. Nonetheless, at the height of the Cold War, lines of demarcation, even within the boundaries of the colony, activated the alarms in the highest echelons of the British bureaucracy.

As the new runway and facilities began service, Hong Kong registered consistent growth in aircraft movements through the city even though the volume was still around half of the bloated levels in the late 1940s. However, passenger count had expanded by over 20 percent in all but one year since 1955, and on that score, by 1960 traffic had exceeded the volume achieved at the peak of 1949.[44]

The Making of a British Airline in Hong Kong

The competition among aviation enterprises to stake an aviation base in Hong Kong extended beyond business interests overseas. Besides the foreign businesses whose activities had fostered Hong Kong's development into an aviation hub, local interests sprouted to participate in the development of this new industry at this crossroads of regional and global traffic. After overseas interests had jumpstarted commercial aviation in the mid- to late 1930s, the onset of World War II put further development on hold. However, wartime development proved to have produced more than mere disruption. In addition to the military surplus that came to be repurposed for commercial use, flying mavericks who traversed the region during the period of conflict also played a significant role in the launching of civil aviation in the postwar years. Of the contenders vying for a home base in Hong Kong, the most prominent and earliest was Cathay Pacific.

Cathay Pacific owes its establishment to two enterprising pilots, one American and one Australian, who spotted the needs of air transport in post–World War II Asia. The two had served in CNAC, which operated in

44.   *HKDCA*, 1959–1960, 41.

Resizing and Conforming to Emerging Geopolitics 75

Republican China with American investment and managerial assistance. Australian Syd de Kantzow and American Roy Farrell were "reluctant to leave the attractions and entrepreneurial opportunities of the Orient after the rigours, hazards and adventures of the war."[45]

Roy Farrell was among the courageous pilots who flew "The Hump" over the Himalayas to transport supplies from India to the Chinese army. When the war ended, he dreamed of starting up a "bit of an airline" in the Far East. Within months of the conclusion of World War II, the enterprising veteran aviator bought in New York an aged C-47 completed in 1942, which had served the US military during the war years. With this first aircraft, Farrell carried a shipment of piece goods to Shanghai via South America, Africa, India, and Burma. Landing in Shanghai on New Year's Day in 1946, this service spelled the birth of the Roy Farrell Export-Import Company. The shipment turned into an economic disaster as most of the goods were stolen at a Chinese airport. However, Farrell persisted. With Syd de Kantzow, whom he had met during wartime, the maverick pressed on and continued the expansion in the region with the purchase of another C-47 in Australia, opening branch offices in Hong Kong, Manila, Singapore, and Sydney.[46]

Thus began the pair's flying operations. The war had deprived the shipping industry of ocean transport facilities and gave the incipient airline an opportunity in the air-cargo charter business. Responding to the great demand of goods everywhere in the southwest Pacific, the founders operated unscheduled services in the region. However, the opportunity was short lived as the airfreight business could not compete with ocean freight as shipping resumed.[47]

Fortunately for the freelancing mavericks, civilian air travel grew with the restoration of peace and stability. Farrell and de Kantzow moved their operations to Hong Kong. They announced plans to acquire "modern British Registered aircraft" to be piloted by "experienced Airline crews"

45. Swire HK Archive, CPA/7/4/1/1/170, Newsletter [November 1981]; Bickers, *China Bound*, 308.

46. Swire HK Archive, CPA/7/4/1/1/151, Newsletter [October 1976]; Swire HK Archive, CPA/7/4/1/1/170 Newsletter [November 1981]; Swire HK Archive, CPA/7/4/1/1/172 Newsletter [November 1983]; Swire HK Archive, *The Weekly*, Issue 67 [January 12, 1996], 4. See also Young, *Beyond Lion Rock*, part 1.

47. Swire HK Archive, CPA/7/4/1/1/170, Newsletter [November 1981].

that would be available for chartered services "within the British Empire."[48] On September 24, 1946, Cathay Pacific Airways Limited was incorporated under the 1932 Companies Ordinance in the British colony.[49]

In fact, the fledgling company extended its business beyond the imperial network. British records indicate that the first plane to be registered in Hong Kong after the war concluded belonged to the Roy Farrell Export-Import Company.[50] Seizing the business opportunity, the two founders refurbished the two C-47s up to DC-3 standard and offered regular unscheduled passenger and cargo flights between Hong Kong and Macao, Shanghai, Manila, and Sydney (via Morotai, Darwin, and Cloncurry, Australia). Its nonscheduled service between Manila and Hong Kong made it the first British airline to be authorized by the Philippine Department of Foreign Affairs for commercial air traffic after the war.[51] Its service to Sydney became the first direct air link between Australia and China.[52]

Cathay Pacific also provided chartered flights for immigrants and students to Australia and Britain. In October, the company operated a charter to the United Kingdom, along with additional charters to Singapore and Bangkok. Their initial charter flight to the United Kingdom on October 23, 1946, was billed as an "expansion of the service of The Roy Farrell Export-Import Co. (Hongkong) Ltd., The First International AIRMERCHANDISE Service in the World, Australia—Manila—Hong Kong—China and now United Kingdom." By November 1946, the airline had secured coverage in a local Chinese newspaper. In the first three months of operations, the new airline carried about three thousand passengers and fifteen thousand kilos of cargo. By the end of 1947, the fledgling airline carried its growing traffic, most of which remained charter, with seven DC-3s and two Catalina flying boats. Traffic demand continued to grow in the following year, so much so that Cathay Pacific began its scheduled services. By March 1948, the airline flew from Hong Kong

48. *SCMP*, September 6, 1946, 13.
49. HKPRO, HKRS 163-1-700.
50. TNA, FO 371/53649.
51. *SCMP*, November 26, 1946, 5.
52. *SCMP*, December 26, 1946, 1.

Resizing and Conforming to Emerging Geopolitics   77

to Macao twice a day, to Manila four times a week, to Singapore via Bangkok twice a week, and to Rangoon and Bangkok once a week.[53]

Business growth notwithstanding, emerging competition and tighter government control made the founders reorganize Cathay Pacific in 1948. Government supervision over air traffic concluded the era of laissez-faire traffic rights and spelled the end to freelance airline operators. The establishment of a new airline in the British colony, Hong Kong Airways, also heightened the competition in the airline business in Hong Kong.[54]

Hong Kong Airways, Cathay Pacific's archrival in Hong Kong, could trace its beginnings to a conceptual framework that BOAC developed with Jardine Matheson for shuttle service between Hong Kong and Canton. BOAC had paid close attention to Cathay Pacific's embryonic formation in 1946. In particular, the British mammoth noted in its files Cathay Pacific's use of the Union Jack in its advertisement and its tagline of "Hong Kong's Own Airline" in 1947.[55] By the time Cathay Pacific launched its first charter flight in 1946, BOAC and Jardine Matheson had explored the commercial viability of providing such service through a venture, initially called the Oriental Airways Express Company. The interested parties assessed their proposal, particularly in the context of local competition with Cathay Pacific. Renamed Hong Kong Airways in 1947, the venture was both BOAC's proactive effort to expand the market to Canton and elsewhere, and its defensive move in response to Cathay Pacific's encroachment on its services to such places as Bangkok.[56]

Hong Kong Airways was formed and registered with the Registrar of Companies in Hong Kong on March 4, 1947. A subsidiary that BOAC had funded entirely, the airline operated initially with BOAC-appointed

---

53. Swire HK Archive, CPA/7/4/1/1/151 Newsletter [October 1976]; Swire HK Archive, CPA/7/4/1/1/170 Newsletter [November 1981]; *SCMP*, October 23, 1946, 1; *SCMP*, December 11, 1946, 1; *SCMP*, December 16, 1946, 1; *KSMN*, November 24, 1946, 3.

54. Swire HK Archive, CPA/7/4/1/1/170 Newsletter [November 1981]. With the infusion of new capital, the company requested a name change and on October 18, 1948, was incorporated as a private limited company called "Cathay Pacific Holdings Limited" (HKPRO, HKRS163-1-700; Swire HK Archive, Cathay Pacific Airways Limited Prospectus, April 22, 1986, 10, 68).

55. British Airways Archives, "O Series," Geographical, 3278. *KSMN*, October 23, 1935, 3.

56. British Airways Archives, "O Series" 6812.

directors at the helm. BOAC held almost all of the 399,998 shares directly or indirectly through local representatives, some of whom (including Lee Hsiao-Wo, a prominent Chinese resident in Hong Kong) received 1,000 each. When first registered, it was "a paper concern as far as Hong Kong knows"; while the avowed intention of the airline was to fly between Hong Kong and China, the United Kingdom and the Republic of China had not signed an aviation bilateral agreement yet.[57]

Time was of the essence as Hong Kong Airways and Cathay Pacific asserted their route claims. The two airlines appealed to the Ministry of Civil Aviation and the Colonial Office to confirm the definition of regional services as distinct from BOAC's trunk services, and to receive licenses to develop regional routes in their respective spheres. In a colonial civil air service conference in London in April 1947, the various parties reached an understanding. The director of air services in Hong Kong maintained that Cathay Pacific was considered "something in the nature of a chosen instrument" and BOAC capacity was deemed insufficient for "Hong Kong's growing trade for example to Manila and to Bangkok." Cathay Pacific was thus assigned the airline responsibilities "for the operation of the area south of Hong Kong." It was noted that Cathay Pacific was to be allowed service to Bangkok "but without prejudice to BOAC services."[58]

Hong Kong Airways was then granted the license for the development of the sphere to the north of Hong Kong. The BOAC subsidiary was to be charged with services between Hong Kong and China, as well as Macao, including the extension of the BOAC trunk route from the United Kingdom to Shanghai "until trunk route aircraft ran through to Shanghai." It was understood that BOAC had plans to operate services competing with Cathay Pacific's when aircraft became available, and that Hong Kong Airways might fly to Japan. BOAC also desired to fly to Chinese coastal ports and Taiwan. Although a formidable international power, BOAC maintained a policy at that time to keep their commitment in the Far East to a minimum. The company treasury had also placed a limit on the capital of Hong Kong Airways. As such, the divi-

57. British Airways Archives, "O Series" 169, 3318, 3321; *SCMP*, October 15, 1947, 1.
58. British Airways Archives, "O Series" 3132, 6812.

Resizing and Conforming to Emerging Geopolitics   79

sion between the two Hong Kong–based carriers and the restrictions placed on Hong Kong Airways "was not unwelcome."[59]

Against a backdrop of BOAC capacity inadequacy, their assessment of the potential of the northern market, and their confidence in recouping market share within Southeast Asia, the British authorities settled on an arrangement among their designated carriers. BOAC held firmly onto the trunk routes. Its subsidiary Hong Kong Airways received the routes between Hong Kong and China. Cathay Pacific enjoyed some latitude in expanding within the Southeast Asian market, essentially Route 3 of the Sino-British bilateral agreement, which covered "Hongkong via Manila, Penang, Singapore, Bangkok, French Indo China to Hong Kong (circular route)."[60] This arrangement would form the basis of Cathay Pacific's area of growth and guide the regional traffic flow emanating from Hong Kong.

On the basis of this understanding and with the execution of the bilateral agreement on July 23, 1947, Hong Kong Airways proposed to commence service based in Hong Kong on two routes, both into China and both covered under the new bilateral agreement. The two routes, one between Hong Kong and Shanghai and the other between Hong Kong and Canton, were subject to the approval of the British and Hong Kong governments, which had to designate the airline as the British operator on these routes. The first aircraft and the chief pilot arrived in Hong Kong on July 29, 1947, only to find that the company was still in negotiations on route designation. Apart from these two routes, the airline also applied for service between Hong Kong and Macao, and expressed its desire to extend the Shanghai route to Tianjin and to develop a new route to Taiwan. The airline's home staff were on secondment from BOAC, and the Hong Kong operations hired only a manager and a few locally engaged staff members. All four of the initial aircraft were chartered and their crews had come from BOAC. Hong Kong Airways finally received the route designation in October 1947. The company inaugurated the Shanghai service on December 2, 1947, with three weekly services in each direction. The Canton service began on January 10, 1948, with two

---

59. British Airways Archives, "O Series" 3132, 6812.
60. British Airways Archives, "O Series" 3132.

services daily at first, which increased to four in each direction in two months.[61]

In spite of their agreement with Hong Kong Airways, Cathay Pacific continued to complain about the forced entry of a BOAC subsidiary into the market in Hong Kong. In a letter to the *South China Morning Post* on October 3, 1947, Cathay Pacific, along with three other companies that had purported to represent local interests in Hong Kong, complained that the operation of Hong Kong Airways was "clearly imposed upon the Colony by the Home Government." The group of companies characterized Hong Kong Airways' invitation of local commercial participation as mere "white-wash to endeavor to give the company the appearance of being locally controlled," when in fact it would be "almost entirely owned and remotely controlled by BOAC." They feared that the establishment of Hong Kong Airways had begun the process of "Colonial Nationalisation" of certain industries in Hong Kong.[62]

Flaunting its status as "the Government's chosen instrument," BOAC complained about the treatment it received in Hong Kong.[63] The Foreign Office observed that three companies stood in the way of BOAC as it strove to assume control over civil aviation in Hong Kong. In its calculation, among the air operating companies already established in Hong Kong were Cathay Pacific Airways, Far Eastern Aviation, and Skyways. Far Eastern Aviation was deemed adequate to operate a flying training school, while Skyways was considered by the governor of Hong Kong to "have not yet established themselves there." Cathay Pacific was found to have violated "the normal principle that substantial ownership and effective control of a British-designated company should be held by persons belonging to the United Kingdom (or Hong Kong)."[64]

The colonial government in Hong Kong showed considerable sympathy and support toward the bold and enterprising Cathay Pacific venture, but "the fact that it was half owned by Americans" made it difficult to win official franchises against the vociferous opposition of BOAC and the British Ministry of Civil Aviation. As it mediated the dispute between

61. British Airways Archives, "O Series" 169, 2898, 3132, 3317, 6812.
62. *SCMP*, October 3, 1947, 5.
63. Bickers, *China Bound*, 309.
64. TNA, FO 676/357.

Resizing and Conforming to Emerging Geopolitics 81

the two competing local carriers, the Hong Kong government refused to grant the franchises to either Cathay Pacific or Hong Kong Airways and gave Cathay Pacific an opportunity to dispose of its American shareholding.[65] Echoing the prewar dispute over Pan Am's connection in the British colony (discussed in chapter 1), the rivalry between British and US interests over commercial aviation in Hong Kong persisted as development resumed after the war.

The sale of the American-Australian interests to a British concern was inevitable. As the British colonial infrastructure returned, it was decided that civil aviation routes should revert to British control. In 1947, the Foreign Office in London wrote to the British embassy in Nanking stating that as "a matter of policy," local and regional air services "operating from British territory should connect with and set as tributary lines to the main routes," and that "reliable companies under United Kingdom or Colonial control" should operate such services.[66]

The Cathay Pacific founders, aware of the weakness of the half American-owned airline, looked for a British buyer in early 1947. Skyways' Chairman Brigadier-General Alfred Cecil Critchley flew to Hong Kong to negotiate the purchase of Cathay Pacific, but the companies could not reach an agreement on price. Swire entered the fray in late autumn of 1947. The Hong Kong government gave Cathay Pacific an ultimatum in January 1948, making it clear to the company that it could only hope to secure the franchises if it reduced its American shareholding to a maximum of 10 percent.[67]

The founders of Cathay Pacific sold their majority stake to Swire, which had been active in Asia for almost a century.[68] The local newspaper called the sale, which took effect on July 1, 1948, an "airways merger." "CPA, Australian National Airways Pty. [Proprietary], Ltd. [ANA], and

---

65. Swire HK Archive, Cathay Pacific Airways Limited Board Minutes, June 25, 1951.

66. TNA, FO 676/357.

67. Swire HK Archive, CPA/7/4/1/1/170 Newsletter [November 1981]; Swire HK Archive, *The Weekly*, Issue 67 [January 12, 1996], 4; HKPRO, HKRS163-1-361.

68. Swire HK Archive, CPA/7/4/1/1/170 Newsletter [November 1981]; Swire HK Archive, *The Weekly*, Issue 67 [January 12, 1996], 4. Swire had monitored keenly the development of commercial aviation in the region since the conclusion of the war (JSS, 13/1/2). For a magisterial account of the long history of Swire, see Bickers, *China Bound*.

Butterfield & Swire" were the principal interests concerned. The merger was to "combine the experience of the ANA, which is recognized as one of the largest and most efficient airlines in the world, with the local experience already gained by Cathay Pacific Airways who are noted for their enterprise and good service." The sale was but the inevitable absorption of an intrepid startup into a solid British establishment in this geographic sphere: "Since the war, Cathay have been in the forefront of air pioneering in this part of the world, against the background of Butterfield & Swire's wide and old established interest throughout the Far East."[69]

This "airways merger" resulted from the orchestration of John Kidston (Jock) Swire, chairman of the Swire Group,[70] and ANA's Ivan Holyman.[71] The Swire affiliates, ANA, Skyways, and Far Eastern Aviation formed a new Hong Kong company which would take over all of Cathay Pacific's assets. De Kantzow accepted the valuation ANA placed on the assets, and the various parties agreed that de Kantzow would remain a 10 percent shareholder and serve as the manager of the company with a seat on the board, and that his American friends would also retain a 10 percent shareholding. When Skyways and Far Eastern Aviation dropped out of the negotiations, the new company started operations on July 1, 1948, under the new ownership structure.[72]

ANA took a 35 percent stake, but Swire, through the aggregate holdings of its affiliates, became the largest shareholder in Cathay Pacific. Upon the completion of the reorganization in 1948, Swire held 10 percent of the company's shares and the China Navigation Co. Ltd., a Swire group company, held 35 percent.[73] An initial agreement had allowed for the infusion of capital by either Malayan Airways or BOAC through an equal

---

69. *SCMP*, June 6, 1948, 14.

70. For extensive coverage of Jock Swire, see Bickers, *China Bound*.

71. JSS, 13/6/1/1, Establishment of Cathay Pacific Agreements, 1948–1949; JSS, 13/1/3 General Correspondence, July–December 1948; JSS, 13/10/1 copy of letter from J. K. Swire in Melbourne to C. C. Roberts, Hong Kong, dated June 3, 1948; JSS, 13/10/1 entry dated December 1950 / June 1951.

72. Swire HK Archive, Cathay Pacific Airways Limited Board Minutes, October 27, 1948; JSS, 13/6/1/1, Establishment of Cathay Pacific Agreements, 1948–1949; JSS, 13/1/3, General Correspondence, July–December 1948; *SCMP*, May 29, 1961, 23.

73. Swire HK Archive, Cathay Pacific Airways Limited Board Minutes, October 27, 1948.

reduction in the holdings of ANA and the China Navigation Co. Ltd.[74] That infusion did not materialize as BOAC continued to pursue its strategy with Hong Kong Airways.

In 1949, the airline entered into a general agents agreement with Swire, its majority shareholder.[75] Thus began Swire's role of spearheading Cathay Pacific's business developments. Through its equity stake in Cathay Pacific, ANA expanded its footprint after its failed campaign for rights to fly outside Australia. The shareholdings of Farrell and de Kantzow were reduced to 10 percent each. Through a holding company he created, Farrell would retain his ownership for a couple of years. De Kantzow stayed on as general manager.[76]

As de Kantzow had reduced the American stake to 10 percent, equal to his own, in the 1948 reorganization, the Hong Kong government officially confirmed the allotment of air service space between Cathay Pacific (the first to enter the field, established in the area, and now 90 percent owned by British and Australian interests) and Hong Kong Airways (a Hong Kong company that was in fact wholly owned by BOAC, with powerful connections in London). The various parties arrived at an arrangement in May 1949 when BOAC and Jock Swire signed an agreement for Hong Kong Airways and Cathay Pacific Airways, with the blessings of the Colonial Office and the Ministry of Civil Aviation, that assigned Cathay Pacific "the area South of Hong Kong" and Hong Kong Airways, "the area North of Hong Kong" with the exception of Macao, the franchise for which Hong Kong Airways already held, and Manila, which was made accessible to both. In conjunction with this agreement, BOAC relinquished their right to Route 3 in the China Treaty. The Hong Kong/Australia franchise remained unresolved as the Hong Kong government insisted on having its own airline on this service and the Australian government, under Qantas influence, refused any entrant but BOAC.

---

74. Swire HK Archive, Cathay Pacific Airways Limited Board Minutes, August 12, 1948.

75. JSS, 13/6/1/1, Establishment of Cathay Pacific Agreements, 1948–1949.

76. Swire HK Archive, CPA/7/4/1/1/170 Newsletter [November 1981]; Swire HK Archive, *The Weekly*, Issue 67 (January 12, 1996), 4; Swire HK Archive, CPA/7/4/6/34 *Cathay News* No. 34 [1965]; Swire HK Archive, Cathay Pacific Airways Limited Board Minutes [April 17, 1950, March 31, 1951, and June 25, 1951]; *SCMP*, November 22, 1957, 8. See also Young, *Beyond Lion Rock*, chap. 11.

84 *Reorienting Hong Kong*

Discussions over the Hong Kong/Singapore direct service also remained unsettled.[77]

When de Kantzow scaled back his holdings by half on April 17, 1950, Swire, ANA, and the China Navigation Co. Ltd. absorbed the shares according to their original percentage holdings. On June 2, 1951, Swire acquired the 10 percent American shareholding for HK$50 per share, or $150,000 in total. On the same day, de Kantzow severed his connection with Cathay Pacific and sold his remaining 5 percent shareholdings at the same price, half to Swire and half to ANA.[78] De Kantzow's position as manager of the company was absorbed into "the functions of B&S' [Butterfield & Swire's] Management." After the various sales which spanned almost three years, Cathay Pacific came to be 39.69 percent owned by ANA, 37.19 percent by China Navigation Co. Ltd., and 23.12 percent by Swire. Through the latter two, the Swire group commanded majority ownership.[79]

The Swire contingent would remain the dominant shareholding group of Cathay Pacific, but the airline did welcome other Commonwealth interests into the company from time to time. In 1954, in conjunction with a capital call of HK$5 million to finance the purchase of a new DC-6, the airline welcomed Peninsular & Oriental Steam Navigation Company (P&O), a British shipping company which had operated in the Far East since the nineteenth century, "through their association with Australian National Airways Pty. Ltd."[80] With P&O becoming a 31.25 percent shareholder by February 1955, ANA's share dropped to 14.88 percent. China Navigation's shares stood at 31.25 percent, on par with P&O, and John Swire and Sons 22.62 percent.[81] In other words, the Swire contingent remained the majority shareholder. In 1956, the Borneo

---

77. *SCMP*, September 28, 1948, 3; Swire HK Archive, Cathay Pacific Airways Limited Board Minutes, June 25, 1951; TNA, CO 937/69/6; Bickers, *China Bound*, 313–14.

78. Swire HK Archive, Cathay Pacific Airways Limited Board Minutes, April 17, 1950; Swire HK Archive, Cathay Pacific Airways Limited Board Minutes, June 25, 1951.

79. Swire HK Archive, Cathay Pacific Airways Limited Board Minutes, June 25, 1951.

80. Swire HK Archive, Cathay Pacific Airways Limited Board Minutes, December 8, 1954. P&O's vessels were classified as "British" in a study of "Scheduled Voyages, Europe-China, by Owner and by Registry arriving during 1953" (TNA, FO 371/110276).

81. Swire HK Archive, Cathay Pacific Airways Limited Board Minutes, February 16, 1955.

Company Limited, another British venture with a longstanding interest in the Far East,[82] bought a 4.19 percent interest in the airline for HK$500,000.[83]

Such was the transformation of Cathay Pacific from an Australian-American venture to a professionally managed British Commonwealth concern, one that would rival another group of British interests keen to direct the flow of commercial aviation through the colony.

Competing to Be Hong Kong's Airline

Through its Hong Kong Airways subsidiary, BOAC had hoped for opportunities for northern expansion from the British colony of Hong Kong. It was already serving parts of Southeast Asia on its trunk route, and the agreement with Cathay Pacific only allowed the regional carrier to share in some of the traffic. The first step of the northern expansion of Hong Kong Airways lay in the Chinese market. However, successive explorations of its two inaugural routes were hampered right from the start. The Shanghai route faced strong competition with the Chinese carrier CNAC carrying 80 percent of the total passengers with its superior equipment. The Canton service, where Hong Kong Airways competed with CNAC and Nationalist Chinese carrier Central Air Transport Corporation, was plagued by a rapidly depreciating Chinese dollar as the ruling regime lost control of the country both militarily and economically. The situation deteriorated further as the Chinese carriers scheduled services in excess of demand. These Chinese carriers had lost many of their internal northern routes due to the advances of the Communists. To maintain aircraft and crew utilization, they operated any profitable route, especially from the safe haven of Hong Kong. With full knowledge of the British authorities, these Chinese carriers flew routes through Hong Kong without official

82. The Borneo Company was already active in the region in the nineteenth century. See, for example, *North China Herald*, November 1, 1856, 55; *North China Herald*, December 19, 1857, 81. For an account of the history of the company, see Longhurst, *Borneo Story*.

83. Swire HK Archive, Cathay Pacific Airways Limited Board Minutes, April 25, 1956.

86 *Reorienting Hong Kong*

authorization or operating permits. British-designated carriers were not afforded any opportunities for flying the routes on a basis of reciprocity. What BOAC had hoped to be a winning proposition turned out to be a loss-incurring venture.[84]

When it started the Manila service in October 1948 upon being designated as one of the two British scheduled operators, Hong Kong Airways also found this route dominated by Pan Am as through load to the West Coast of the United States. Hong Kong Airways' traffic consisted primarily of "second and third class Chinese Traffic" between the Philippines and southern China, a market handled by Chinese brokers that competed on fare.[85] The following year only brought more disastrous developments for Hong Kong Airways. Communist advances first led to the suspension of services to Shanghai. Despite attempts to negotiate with the new regime, the airline recognized that "it would be some time before any regular service to Shanghai would again come into operation." When the Nationalist government proved incapable of defending Canton, to which it had retreated, operations to Canton also ceased in November 1949.[86] The north-south split was an untimely deal for Hong Kong Airways as the Communist takeover of the mainland eliminated routes and curtailed Hong Kong Airways' expansion. Gone were both of the routes that BOAC had hoped would underwrite the operations of Hong Kong Airways. BOAC moved quickly. It had already offered Hong Kong Airways to Jardine Matheson before Canton fell. Within days of the Communist seizure of Canton, the sale closed.[87]

Although BOAC had passed the ownership of Hong Kong Airways to Jardine Matheson, it stayed involved in the airline's operations and continued its effort in exerting regional influence.[88] The airline attempted to reroute its northbound traffic from Hong Kong after the Communist takeover of the mainland primarily by initiating "private" discussions with the regime in Taiwan for services to Taipei, with the possibility of extend-

84. British Airways Archives, "O Series" 169, 3317; TNA, FO 371/76337.
85. British Airways Archives, "O Series" 3317.
86. British Airways Archives, "O Series" 3319; Bickers, *China Bound*, 314.
87. British Airways Archives, "O Series" 169, 3132; TNA, FO 371/76337, CO 937/69/7; *SCMP*, November 28, 1949, 1.
88. British Airways Archives, "O Series" 68.

ing it to Okinawa and Tokyo.[89] Although Hong Kong Airways managed to open its Hong Kong–Taipei route in November 1949,[90] issues with travel documentation put a damper on the traffic between Hong Kong and Taipei. The financial performance of Hong Kong Airways continued to suffer as it awaited the results of its applications for flights to Japan, which required both the approval of the American authorities and a license from the Japanese government.[91]

By the end of 1950, Hong Kong Airways retained none of its aircraft and flew its remaining routes with planes chartered from Cathay Pacific.[92] This business initiative proved to be a thorny issue for BOAC. When Hong Kong Airways received the green light for services to Taipei and Japan, it chartered planes from the US carrier Northwest Airlines. That Hong Kong Airways flew as an operator sanctioned by the British authorities using equipment with the US flag was an embarrassment to the United Kingdom.[93] These Hong Kong Airways flights, granted by the British government, connected with Northwest Airlines' services to Okinawa, Tokyo, and the United States. The ministry in London did not welcome such an arrangement, which it considered to be playing into the Americans' hands.[94]

BOAC had planned to equip Hong Kong Airways with its own planes so as to eliminate Northwest Airlines' involvement. However, the British operator was also realistic that the profitable traffic rights at Okinawa were also "the gift of the United States Government" that entailed reciprocal exchange. The proposition to remove Northwest Airlines was expensive, but BOAC, in conjunction with the London authorities, considered it a worthwhile undertaking because Northwest Airlines' service to Hong Kong would open the floodgates to international competition over the Hong Kong/Japan sector.[95]

89. TNA, BT245/1073; TNA, FO371/84794.

90. *SCMP*, November 19, 1949, 14.

91. British Airways Archives, "O Series" 169; TNA, FO371/84794.

92. Swire HK Archive, Cathay Pacific Airways Limited Board Minutes, June 25, 1951.

93. *SCMP*, August 17, 1950, 11; TNA, FO371/84783; TNA, FO371/84793; TNA, FO371/84795; TNA, FO371/93126.

94. British Airways Archives, "O Series" 9953.

95. British Airways Archives, "O Series" 9954.

88   *Reorienting Hong Kong*

In the meantime, Cathay Pacific's backer Swire suffered numerous setbacks in their attempts to expand their operations. At a board meeting on June 25, 1951, the airline's chairman C. C. Roberts spoke of Swire's interests in air services in the Far East. Their application for the agency for BOAC was unsuccessful because of BOAC's close relationship with Jardine Matheson, with whom they had planned to operate jointly throughout China. Although their ambitions in China failed due to US opposition, BOAC granted Jardine Matheson its agency in 1946. In the meantime, ANA's attempt to fly to Hong Kong was rejected by the Australian Labor government, which was determined to keep air services in government hands. Swire then appealed to the Hong Kong government for a license to ANA for Hong Kong–Australian service. That too failed in the face of Australian government opposition in the Colonial Office, coupled with the influence of BOAC in the British Ministry of Civil Aviation, which had set aside that route for the Australian government–owned, BOAC-affiliated Qantas.[96]

The British authorities had expressly protected the trunk route monopoly of BOAC. In 1949, the Colonial Secretary informed Cathay Pacific that regional operators might only run "sector traffic which BOAC cannot take on its through services, or if there is a need for short-stage services or for services catering for differing classes of traffic, in addition to, and not competing with, the trunk services."[97] A regional company under the auspices of BOAC would ensure that such subsidiary operations remained subordinate to the backbone of the British network.

BOAC had hoped to materialize the objective of the British and Hong Kong governments to merge Hong Kong Airways and Cathay Pacific to form a regional company that would "serve the interests of Hong Kong and of the Corporation." However, BOAC could not agree with Swire on the assignment of a controlling interest in a merged entity. Cathay Pacific had developed within the southern region defined in the 1949 agree-

---

96. Swire HK Archive, Cathay Pacific Airways Limited Board Minutes, June 25, 1951. Early entries in the minutes of Cathay Pacific indicated Swire's belief, "for some time," that in the interests of Swire and its partners, "it was essential . . . to get into the Air somehow in the Far East" (JSS, 13/10/1 entry dated December 1950 / June 1951). See also Bickers, *China Bound*, chap. 12.

97. HKPRO, HKRS163-1-361.

ment. In addition to Bangkok, Singapore, and Manila, Cathay Pacific had added by 1953 service to Rangoon, Jesselton (Kota Kinabalu), Sandakan, Labuan, Saigon, Hanoi, and Haiphong, as well as Calcutta. An industry magazine lauded Cathay Pacific for its "enviable record of regularity and high utilization," calling the British independent airline's business "a solid achievement by one of the few private scheduled operators under the British flag." By comparison, the northern region remained undeveloped, leaving "a British vacuum." BOAC blamed it on the "inactivity" of Jardine Matheson which had further complicated the matter by "allow[ing] the Americans to infiltrate." A British concern must develop the northern region and not "leave a vacuum to be gradually filled by the Americans and Japanese," BOAC maintained. Having sold its ownership to Jardine Matheson in late 1949, BOAC once again took up an equity interest, this time at 50 percent in Hong Kong Airways in January 1956.[98]

Having stepped up its ownership in Hong Kong Airways again, BOAC took a firm stance against Northwest Airlines' entry into Hong Kong. Characterizing this move as being "in line with BOAC's worldwide policy on American and particularly PAA [Pan Am's] activities," the British operator saw no reason to agree, "even temporarily, to two American trunk operators across the Pacific flying into Hong Kong." Its deputy chairman went as far as to say that he would rather accept losses at Hong Kong Airways because of lack of rights to Okinawa than see Northwest Airlines flying to Hong Kong.[99]

BOAC's style appeared to suit Hong Kong Airways. The Hong Kong carrier even registered a surplus for a month in 1957. However, the carrier could not overcome the issues of currency and noncompetitive fares. In October 1957, Hong Kong Airways offered services from Hong Kong to Tokyo through Taipei, Hong Kong to Manila, Hong Kong to Seoul through Taipei, Hong Kong to Taipei, and even began the long-awaited service from Hong Kong to Tokyo through Okinawa. Few of these routes generated any gains, and the monthly tally showed continued losses. Its Manila services looked the most productive on paper, but the company had to incur significant costs in unremittable pesos. To break through

98. British Airways Archives, "O Series" 9953, 9954; British Airways Archives, "O Series" N584; TNA, BT 245/1060; *Flight & Aircraft Engineer*, January 22, 1954, 88–89.

99. British Airways Archives, "O Series" 9954.

such conundrums, BOAC discussed its desire to achieve a single Far East regional company by merging the two Hong Kong–based regional carriers and Malayan Airways. However, the rivalry between Swire and Jardine Matheson remained a formidable roadblock. Besides, not only had Swire refused to approach any such discussion on the basis of an even three-way ownership split with BOAC and Malayan Airways, but the political campaign for full Malayan autonomy had also made it unlikely that the Malayan government would agree to subordinating Malayan Airways' identity, even partially.[100]

Another year of an unpromising outlook compelled BOAC to cut losses. In December 1958, BOAC purchased the rest of Jardine Matheson's shares in Hong Kong Airways, setting the stage for its negotiation with Swire for establishing a regional company in which Swire would hold 85 percent, and BOAC, 15 percent. On June 30, 1959, Hong Kong Airways ceased operations. Ostensibly wishing "to cut down its commitments in Associated Companies in various parts of the world and to confine its interests to their own trunk route services," BOAC unwound its investments and operations in Hong Kong Airways and confined its role to consultation for Cathay Pacific on tariff issues. It would henceforth "promote and encourage" regional services, leaving the control and management whenever possible in local hands. "Being in agreement with the wisdom of this policy," Cathay Pacific consented to the formation of a new company to operate both the spheres covered by Cathay Pacific and Hong Kong Airways.[101]

The new entity was to be predominantly owned, managed, and operated by Cathay Pacific "in the same manner as [Cathay Pacific] has hitherto been managed." It was also to adopt Cathay Pacific's articles of association to the extent possible. Its board would consist of Cathay Pacific's board, and BOAC would continue to enjoy the right to nominate one director as long as it held no less than 15 percent of the new company's

---

100. TNA, FCO141/15126; TNA, FCO141/15127; British Airways Archives, "O Series" 9953; British Airways Archives, "O Series" 9954; National Archives of Singapore, ABHS 950 W4627 Box 2739, 110/3/8 Pt 1, 1949–1967. BOAC had long desired to form a regional company. An investigative team had prepared a report on this possibility in 1952 (TNA, CO937/236).

101. British Airways Archives, "O Series" 9953, 9954; British Airways Archives, "O Series" N584.

shares. BOAC agreed to confine its operations in the region to trunk services and pledged not to operate services originating from and terminating locally in the region. The new company agreed to strive to promote the interests of BOAC and British civil air transports. Similarly, BOAC would use its influence whenever possible to promote the interests of the new company in the region.[102]

BOAC had expressed discontent with the name Cathay, which "meaning China, is politically a poor word."[103] Accordingly, the new regional company was provisionally designated Far East Regional Air Services. However, BOAC eventually relented. The final agreement of June 18, 1959, referred to the holding company as "Cathay Holdings Limited" or "such other name as CPA may think fit." As for Jardine Matheson, the agreement assigned "a token shareholding" so as to "make the transfer of HKA from the Jardine Matheson & Co. Ltd. orbit to the Butterfield & Swire orbit as smooth as possible" and to assuage Jardine Matheson's "somewhat difficult problem."[104]

Explaining the agreement in the annual general meeting on April 7, 1959, the chairman of Cathay Pacific Airways attributed the merger to the realization that "for some time" it was "uneconomical to have two regional airlines operating out of Hong Kong." Upon the completion of the merger on July 1, 1959, BOAC received 15 percent of the shares in the combined entity, and Jardine Matheson, 0.5 percent. The shareholders in Cathay Pacific before the merger became owners of the combined entity with their holdings prorated to form the remaining 84.5 percent (table 2.1).[105] This basic structure, which continued to allow Swire's control of the airline, would continue until 1970.

---

102. British Airways Archives, "O Series" N584; JSS, 13/10/2 Special and Extraordinary Resolutions of Cathay Pacific Airways Limited passed on 16th June, 1959.

103. British Airways Archives, "O Series" 9953. Jardine Matheson had made similar complaints about the name "Cathay" in 1955 and called for a compromise "such as 'Hong Kong Pacific Airways'" (TNA, CO 937/439).

104. British Airways Archives, "O Series" N584; JSS, 13/6/1/1. Envisioning a more substantive participation of a parent company based in the metropole, an earlier investigative report prepared in 1952 had suggested the name "British Oriental Airlines Company" so as to share the initials BOAC (TNA, CO937/236).

105. Swire HK Archive, Cathay Pacific Airways Limited Annual General Meeting, April 7, 1959.

92   *Reorienting Hong Kong*

Table 2.1   Cathay Pacific's percentage shareholding after the merger
with Hong Kong Airways in 1959

	Shareholding
John Swire & Sons	18.31%
The China Navigation Co. Ltd.	25.30%
Australian National Airways	12.05%
Peninsular & Oriental Steam Navigation Company	25.30%
Borneo Company	3.54%
BOAC	15.00%
Jardine Matheson & Co. Ltd.	0.50%

Source: Swire HK Archive, Cathay Pacific Airways Limited Board Minutes, December 8, 1954;
Annual General Meeting, April 7, 1959.

A local Chinese newspaper had announced prematurely the demise
of Hong Kong Airways on January 7, 1959, expecting its operations to cease
in the following month and anticipating its absorption by Cathay Pacific.[106] A local English newspaper presented a more accurate account of the
"airline merger" on February 24, 1959, stating that Cathay Pacific had
formed a new company with BOAC "to absorb the administration and
complete share capital of Hongkong Airways Ltd," including Hong Kong
Airways' routes to Seoul, Tokyo, and Taipei. The news article also reported
that the combined entity would commence operations "on or about July 1"
and "continue to operate under the name of Cathay Pacific Airways
Ltd."[107] When the long-rumored merger was consummated on July 1, 1959,
the local Chinese newspaper sounded the death knell: "The reorganized
aviation business of Cathay Pacific officially begins its business today.
BOAC interest is injected into [the combined concern] which also ab-
sorbs the business of Hong Kong Airways."[108]

Initially viewing Cathay Pacific as a competitor and aiding Hong
Kong Airways in its route development, BOAC had to reverse course and
agreed to the merger of the two Hong Kong–based airlines into a regional
carrier. The insistence of Swire in keeping Cathay Pacific as the surviv-
ing entity displeased BOAC at first. However, as Hong Kong Airways

106.   *WKYP*, January 7, 1959, 12.
107.   *SCMP*, February 24, 1959, 6.
108.   *WKYP*, July 1, 1959, 15.

continued to struggle for expansion of northern routes and the defense of its base of northbound service from Hong Kong (in particular against North American encroachment), Cathay Pacific succeeded in its penetration of the southbound traffic from Hong Kong and strengthened its fleet through aircraft purchase. BOAC had shouldered the continued financial burden of Hong Kong Airways and underwrote the airline's losses. In the end, the British titan relented and consented to the merger on such otherwise undesirable terms.

Thus ended the decade-long rivalry between Swire-led Cathay Pacific and Hong Kong Airways, backed by Jardine Matheson and BOAC. BOAC managed to unwind its involvement, but the victory in the region belonged to Swire and Cathay Pacific. Yet, it is important to note that BOAC remained the operator of "trunk" routes as Cathay Pacific fed subsidiary regional traffic into the backbone of the ex-imperial network.

Aggregating the Traffic and Figuring Out the Local Share

Commercial air traffic flow through Hong Kong unfolded thus: after the disastrous loss of connections with mainland China in 1949, commercial aviation in Hong Kong languished in the early 1950s as the colonial government recalibrated its strategy to develop the city's potential as a regional air hub. With the mandate of connecting Hong Kong with air traffic to the south of the colony, Cathay Pacific became a key player in the emerging network of commercial aviation in Southeast Asia (fig. 2.2).

Cold War dynamics rejuvenated commercial air flow through Hong Kong by the mid-1950s, as evidenced by the steepening climb in both aircraft movements and passenger counts (fig. 2.3). The sharper increase in passenger count indicates the use of higher-capacity planes operating closer to full capacity. This trend intensified after the opening of the new runway at Kai Tak as passenger count growth consistently outpaced the increase in aircraft movements in the early to mid-1960s. The upgrading of Kai Tak to leverage technological improvements paid off: the runway and the associated facilities allowed the Hong Kong airport to accommodate larger and heavier planes, and the market demand responded favorably.

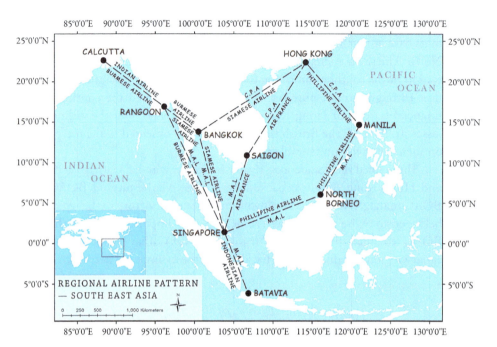

FIGURE 2.2. Schematic representation of Cathay Pacific's position ("C.P.A.") in the regional pattern of Southeast Asia (1950): Hong Kong/Bangkok route shared with "Siamese Air[ways]"; Hong Kong/Saigon route shared with Air France; Hong Kong/Manila route shared with "Phillipine [sic] Air [L]ine[s]." Source: Adapted from materials in British Airways Archives, "O Series" 3658.

Cathay Pacific, the locally registered carrier, grew with the industry. From its meager five weekly frequencies in March 1951 (17 percent of scheduled aircraft movements),[109] Cathay Pacific had almost quadrupled its service to nineteen weekly frequencies by March 1960, extending its reach to Calcutta via Bangkok, Vientiane, Phnom Penh, Tokyo via Taipei, Kuching via Brunei, and Sydney via Manila. Beyond increasing its share to 20 percent of scheduled aircraft movements, Cathay Pacific in 1960 had overtaken in number of services BOAC and Pan Am, each with sixteen weekly.[110] As BOAC and Pan Am continued to command the long-range services to London and the United States, Cathay Pacific's

109. HKDCA, 1952–1953, 39–40.
110. HKDCA, 1959–1960, 21–23.

Resizing and Conforming to Emerging Geopolitics 95

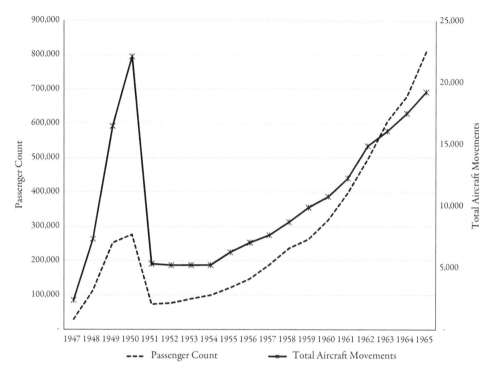

FIGURE 2.3. Annual aircraft movements and passenger counts at Kai Tak Airport, 1947–65. Source: Director of Civil Aviation, *Hong Kong Annual Departmental Reports,* 1947–65.

gains came partly from its assumption of Hong Kong Airways' services, in particular to Tokyo. The locally based carrier had established itself amid the formidable rivalry of the industry's mammoths. As passenger count continued to climb in the 1960s, Cathay Pacific consistently captured over a fifth of the traffic, rising to 30 percent of the total by the end of the decade (fig. 2.4).

Cathay Pacific's accomplishments are no mean feat. British officials were dismissive of the airline's potential, especially as it ran into competition with more established players such as Qantas, which possessed experience in long-haul international services and technological superiority. Indeed, Cathay Pacific would not even have been allocated aviation rights had it not been considered a British colonial concern. British officials did look out for Cathay Pacific's interests, so long as they did not

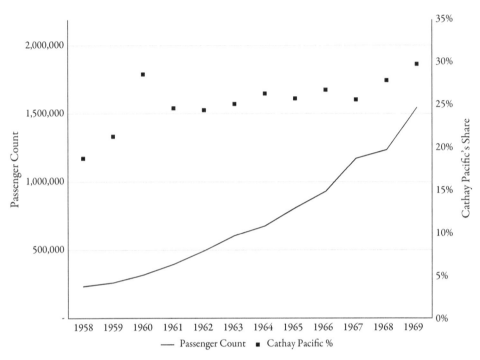

FIGURE 2.4. Passenger counts at Kai Tak Airport and Cathay Pacific's share. Sources: Director of Civil Aviation, *Hong Kong Annual Departmental Report*, 1958–69; *Cathay Pacific Annual Reports*.

conflict with those of BOAC.[111] However, its status as a "colonial airline" accorded it a lesser status in British officials' route negotiation even within the Commonwealth.[112] Disappointed in Cathay Pacific's refusal to set up a regional airline, officials in London remarked to the colonial government in 1959 that Cathay Pacific "would inevitably have to face competition from other airlines." They noted that it would be impossible for Britain to convince other Commonwealth members to adjust their policies "to suit the convenience of a *Hong Kong Airline* at the possible expense of their own airlines" (emphasis added).[113]

111. TNA, BT245/1510.
112. TNA, BT245/552.
113. TNA, BT245/1005.

Hong Kong's critical position was amply clear in Qantas's request in 1962 for rights at Hong Kong for a "Kangaroo service"—the air route between Australia and the United Kingdom. By one calculation, some 43 percent of Australian traffic to Europe traveled over Hong Kong, and so it was only fair that Qantas could share the business. London had long denied these rights to the Australians. British officials admitted, "Apart from London, Hong Kong was the only point of great value." They felt the need "to maintain BOAC's network which had been built up by the gift of traffic rights in the Colonies." This "gift" was becoming elusive as these colonies "were emerging as independent territories." London ferociously guarded traffic rights in Hong Kong because "with the rapid loss of other Colonial points, it is becoming an ever more important main source of negotiating strength in maintaining the United Kingdom's world-wide airline network." Appreciating that ease of access into Hong Kong would only depreciate the value of Hong Kong rights, British officials privileged BOAC in their cost-benefit analysis. "A large part of BOAC's rights in Europe and elsewhere is sustained by rights given in Hong Kong. Its value as a bargaining point must be kept high." Hong Kong's criticality notwithstanding, its governor was sidelined in London's negotiations with Australia over aviation rights through Hong Kong, a matter "of close concern both to Hong Kong and to its airline, Cathay Pacific Airways," British officials in London acknowledged.[114]

Hong Kong remained important to British interests in the aviation industry, but the benefits did not necessarily have to accrue to Cathay Pacific. In a June 1962 briefing of the minister of aviation ahead of his visit to Hong Kong, British authorities acknowledged, "Hong Kong is of outstanding and increasing importance in our international negotiations on civil aviation." Hong Kong was significant to Britain for its own connections: "The traffic that passes through it, and stop-off (at any rate for awhile) is growing." Equally importantly, Britain's footprint was shrinking in the age of decolonization: "The number of other colonial points we can dispose of in negotiations is reducing fast." The officials in London recognized the danger of playing Hong Kong "as a negotiating card" and not taking care of local interests. The colonial government in Hong Kong was quick to criticize London for subordinating the interest of "their

114. TNA, BT245/1187.

98   *Reorienting Hong Kong*

locally based airline, Cathay Pacific" to "BOAC's trunk operational interests, or the exigencies of the Commonwealth partnership with Qantas." Although the officials in London appreciated the strategic significance of Hong Kong in commercial aviation and recognized Cathay Pacific as an ally, albeit as a "local" British Airline, the minister was advised only to suggest that "trunk and local interests should be worked out by frequent consultations."[115] In other words, Hong Kong remained important to British interests as a strategic stronghold in a shrinking empire for route expansion and negotiations with aviation officials in the Commonwealth and beyond. However, the hierarchy was clear: priority was to be given to trunk traffic and Cathay Pacific, as an offshoot that refused to conform to the British grand plan, needed to survive as an ancillary concern.

Commercial aviation developed in Hong Kong in earnest in the aftermath of World War II. In certain respects, the industry's reinvigorated development followed the same pattern as the one that animated its embryonic growth in the 1930s. The prized value of Hong Kong continued to be its location at the crossroads of British, American, and Chinese air traffic. In fact, the wartime allies had strategized on the proliferation of flight routes in the air space around Hong Kong even before the war ended. As the British returned to Hong Kong, colonial and Commonwealth interests moved in swiftly to claim primacy in the city's aviation industry. US airlines followed, albeit at a slower pace. The surge in Chinese aviation connections with Hong Kong, driven by the rapid and dramatic transformation of the political landscape in the mainland, took the city by surprise. Although the principal actors in this confluence of air traffic through Hong Kong remained the same, the preponderance of Chinese connections introduced a new element to the developmental aspirations of commercial aviation in Hong Kong.

Although airline companies, as well as the administration in Hong Kong and London, had espoused high expectations of industry growth in the colony, geopolitical transformations dashed such hopes as quickly as they had engendered them. The Communist takeover of mainland China severed aviation ties between Hong Kong and the various Chinese

---

115.   TNA, BT245/1005.

cities. Demands for infrastructure development subsided, but apprehension about the defensibility of a Deep Bay airport became readily apparent. Drastic reductions in air traffic through Hong Kong eliminated any urgency in constructing a new airport. British officials took their time to deliberate on the choice of its location and to adjust the scale of the investment. Originally condemned as impractical for modern aviation, Kai Tak resurfaced as a feasible option for its distance from mainland aerial forces.

British authorities had to recalibrate their commitment not merely in response to the changing geopolitics to the north of Hong Kong but also in the context of the age of decolonization. In the 1950s, the British government strove to reassert its interventionist role around the world.[116] Hong Kong featured prominently in the British vision of its regional power.[117] Located on the other side of the globe, the colony allowed Britain to assert its power in a region where local actors successively claimed autonomy from their previous colonial masters. For the British metropole, Hong Kong served both as a crucial geopolitical stronghold as well as a site for its articulation of colonial hierarchy. In the world of aviation, Hong Kong operators needed to be British owned (or at least owned by a Commonwealth conglomerate).

The metropole strove to retain control through its control of aviation politics and BOAC. As British conglomerates in Hong Kong competed for aviation rights, the colonial machinery mediated. Yet the metropole knew the inevitability of sharing its power with local British interests in the colony. In its folder titled "Hong Kong: Local Air Service" for the year 1948, the Colonial Office included an extract from *The Aeroplane*. The article called attention to how British colonial governments in East and Southeast Asia prejudiced local British interests, especially among the longstanding shipping hongs. "I only hope that emphasis on the shipping side does not submerge the air transport expansion too much," noted the article. "The adaptations and compromises needed for local colonial air transport pioneering," continued the article, had better come from BOAC

---

116. Lynn, *British Empire*, 7. Howe argued that "Britain in the 1950s was post-many things . . . [but] was far still from being post-imperial" ("When (If Ever)," 234).

117. See Husain, *Mapping the End*, for an in-depth discussion of the British approach to the Middle East and South Asia.

and "some of the British independent aviation companies," which were better equipped to provide "technical advice" and "tend[ed] to be more versatile in their ideas." This article resonated with the Colonial Office. Alongside the section on the power of local British concerns, a handwritten note in the margin reads, "How true!"[118]

Longstanding British shipping interests endeavored to shape the budding industry of commercial aviation. Yet, Hong Kong airlines had to be subjugated to BOAC, a carrier that controlled the passage of passengers, mail, and cargo along the trunk routes of the old imperial network. Even at the level of regional traffic, BOAC yielded to local operators only because of its own resource constraints and only after chronic losses. Cathay Pacific emerged as the Hong Kong airline, to the chagrin of the British titan. BOAC had hoped for the formation of a regional conglomerate that would put air traffic of the former colonies in the region under its auspices. In view of the events of the 1950s, Cathay Pacific was justified in accusing BOAC of putting British national interests ahead of colonial requirements.

As the rivalry among British interests raged on, the development of commercial aviation demanded infrastructure investments. If prewar development reflects the impact of a new technology on global traffic flows, postwar transformations of commercial aviation in Hong Kong underscore the unsettled geopolitical landscape, the continued demands for infrastructural upgrades to suit sustained technological enhancements, as well as the unrelenting negotiations of commercial interests in both the colony and the metropole. Alongside a revived and reduced scheme, investments came to be underwritten by another wave of geopolitical development that swept the region.

Hong Kong, along with the rest of the region, became engulfed in Cold War struggles for territorial and aerial control. Geopolitical factors redirected the development of air routes, producing a corridor that ran on the periphery of mainland China, linking the region to connections with long-haul traffic from North America and Europe. From this configuration emerged a region that encompassed South Korea and Japan to the north of Hong Kong, Taiwan and the Philippines to the east, and other urban hubs in Southeast Asia. Combined with the long-haul con-

118.   TNA, CO 937/69/4.

nections, this traffic pattern formed the basis for the expanding airways around Hong Kong. The region thus formulated also served as a platform for economic and cultural exchange, generating for Hong Kong both a catchment area of its commercial aviation business as well as a pool of competing airlines, which Cathay Pacific would seek to differentiate itself from. In the subsequent decades, Cathay Pacific would contend not only with these budding carriers but also with British interests, primarily BOAC, its shareholder and one-time rival. This historical backdrop would condition the airline's strategy to position itself in the evolving industry as commercial aviation continued to extend its reach.

CHAPTER 3

Branding Hong Kong

Fashioning Cathay's Pacific

Discover the many faces of the Orient on Cathay Pacific Airways.

As the British airline of Hong Kong for 25 years, we're very conscious of "face." Like the prettiest faces of 9 exotic lands smiling at you in 23 languages. Including English, of course. Like our British million-mile jet pilots and skilled maintenance crews. Our Swiss chefs and their tempting international cuisine. Complimentary cocktails. Our helpful, multi-lingual agents at airports and downtown ticket offices all over the Orient. Our international passengers who fly with us to and from 14 major cities. More often, more pleasurably, than on any other airline. Discover our book of Orient vacation ideas. Free—because we have to entice them to the Orient before we can entice them to fly Cathay Pacific.

—Cathay Pacific advertisement in the United States, 1971[1]

Since absorbing its only local rival in 1959, Cathay Pacific Airways flew as the local carrier of British Hong Kong. During that period of phenomenal economic growth, Hong Kong developed into an international air traffic hub. Intensified engagement with the markets in the West broadened the international reach of Hong Kong and cultivated the city's centrality in the network of regional and global traffic. In the competitive marketplace of commercial aviation, how did Cathay Pacific position itself vis-à-vis its regional rivals and the heavyweights from Europe and North America? What did it mean to represent British Hong Kong in footprint and brand?

1. Swire HK Archive, Swire HKAS 391044158720.

Fashioning Cathay's Pacific

Cathay Pacific's strategy to stake out its turf mirrored Hong Kong's geopolitical position during the period of postwar reconstruction and political realignment. Retreating from its early ambition that stretched nebulously from "Cathay" in Marco Polo's travel accounts to the vast oceanic space of the "Pacific," Hong Kong's burgeoning carrier pragmatically carved out a regional space that reflected Hong Kong's position on one side of the Cold War divide. As the industry of commercial aviation expanded in the 1960s, the budding carrier articulated its business ambitions and fashioned its services in a bid to differentiate itself not only from the titans in the field but also from the other fledgling regional carriers. As Cathay Pacific continued to articulate its corporate identity, the airline represented Hong Kong in association with the region and the rest of the world and fostered a special brand of cosmopolitanism for the city.

This chapter explores first Cathay Pacific's schematic representation of its turf in relation to other operators, and proceeds to a discussion of the airline's staffing of its frontline personnel—the cabin crew. Through its staffing policy, Cathay Pacific transformed the looks of its cabin crew, in particular, during the successive rounds of uniform changes for its female members. By the early 1970s, Cathay Pacific had fashioned a cosmopolitan brand and crafted a service proposition of an evolving "Orient" from its base in Hong Kong. This process underscores both the competitive positioning of Cathay Pacific as well as the local, regional, and global dynamics of the era.

Cathay Pacific projected its presence beyond the home base of Hong Kong during the transformative years of the airline as well as during the city's economic takeoff. To attract its geographically expanding clientele, Cathay Pacific brought to bear cultural elements that echoed the business exigencies of the airline's network and customer base. The blending of these elements with which the airline fashioned its flight attendants also conditioned the manifestation of cosmopolitan aspirations in Hong Kong. This brand of cosmopolitanism extended the city's worldliness beyond its Chinese heritage, embraced regional ingredients from other ports with which the city interfaced, and cultivated a service mentality towards a Western market into which Hong Kong sought to integrate.[2]

---

2. Cosmopolitanism has attracted scholarly attention as individual orientations that transcend the local and the national to express a feeling of belonging to a world

In deploying cosmopolitanism for commercial interest, Cathay Pacific sidestepped the issue of the nation in Hong Kong.[3] The notion of the nation had posed a problem for Cathay Pacific as the airline introduced ethnically inspired uniform design. In that first round of uniform design, the airline's emphasis on Chinese heritage subsumed Hong Kong's uniqueness under the culture of the greater China area. The Chinese cultural heritage that Hong Kong could boast afforded Cathay Pacific a business resource and profit opportunity; however, casting a cosmopolitan image against this larger cultural backdrop for the Hong Kong–based carrier required careful maneuvering. A cosmopolitan imagination is usually predicated on an openness to differences; the business strategy of Cathay Pacific took it one step further. Taking its cosmopolitan presentation as critical to the airline's success in the world of commercial aviation, Cathay Pacific persistently pursued such positioning to court international travelers.

Cathay Pacific did not create the various iterations of cosmopolitan images in a vacuum. During the period of its rapid growth from the 1960s onward, the uniform of female flight attendants provided the airline with one of the most productive avenues to express its cosmopolitan brand identity.[4] Taking a longitudinal approach, this chapter examines the different iterations of Cathay Pacific uniforms for its female cabin crew to reveal cultural considerations and practical issues behind the airline's choice in the context of the rapid development of late-colonial Hong Kong. Its uniform designs, just like the local conditions of Hong Kong

community (see, for example, Cheah and Robbins, *Cosmopolitics*, for a philosophical discussion). Yet, corporate branding and positioning in a similar cosmopolitan vein has received little coverage.

3. As a framework in social and political theory and research, cosmopolitanism extends the analysis beyond the national (see Beck and Grande, "Varieties of Second Modernity").

4. The uniform has also been the focus of many scholarly analyses. Haise and Rucker surveyed flight attendants for their satisfaction with their uniform and their assessment of the uniform's role in image creation (Haise and Rucker, "Flight Attendant Uniform"). Zhang, Ngo, and Wang have applied scientific principles with the aim of modifying the Vietnamese national costume for comfort and beauty as airline stewardess uniforms (Zhang et al., "Optimizing Sleeves Pattern"). Exploring the issue from a cultural perspective, Black has explained the function of the flight attendant uniform and the message it conveyed in the case of Qantas (Black, "Lines of Flight").

Fashioning Cathay's Pacific

and the commercial interests of Cathay Pacific, grew out of a web of connections that facilitated the growth of the city and the airline during this period. Cathay Pacific's uniform designs for its air hostesses provides a unique angle to explore the airline's (and Hong Kong's) search for a cosmopolitan look that reflected local sensitivities against the backdrop of regional and global developments.

Mapping Cathay's Pacific[5]

As a new technology that increasingly penetrated the masses through commercialization, aviation intensified the flows of goods and people on an ever-expanding scale. In the process, industry titans such as Pan Am and BOAC strove to broaden their global reach while regional startups struggled to establish their footholds. The crisscrossing route maps of the world's airlines coalesced into a dynamic pattern that represented not only a schematic rendition of intensifying international air traffic but also the aspirational expressions of corporate identity and strategy. While BOAC adopted from its predecessor the Speedbird emblem as a stylized image of flight service, Pan Am presented a series of logo designs that incorporated cartographical depictions in line with its extending flight network.

Though already a carrier with a global network, Pan Am redesigned its logo to reflect its expanding network and growing ambitions. The airline's first globe logo in 1928 depicted only the southern tip of the United States and Latin America. By the 1940s, the cartographical depiction had extended across the Atlantic. Not until 1950 did the company change its name officially from Pan American Airways to Pan American World Airways. Subsequently, the company adopted a full globe trademark to symbolize its worldwide operations (fig. 3.1).[6]

In Asia, Pan Am attracted corporate admirers, which imitated its logo design as they too sought to broaden their reach. Philippine Air Lines, touting its network as the "Route of the Orient Star," adopted in the 1950s as its logo a winged globe that depicted an expanded map of the Philippines

---

5. I thank Jane Ferguson for suggesting the term "Cathay's Pacific."
6. Pan Am, Series 12, Box 1, Folder 1.

FIGURE 3.1. Different generations of Pan Am logos. Source: Pan Am, Series 12, Box 1, Folder 1.

centered on an oval-shaped world that stretched from Europe and Africa to the left and the Americas to the right.[7]

As a growing airline with an inchoate route map in the early 1950s, Cathay Pacific also mimicked Pan Am's cartographically inspired logo. Cathay Pacific's first aircraft, *Betsy* (a DC-3 registered as VR-HDB in 1946), now on display in the Hong Kong Science Museum, features a map that encompasses the southeastern part of China with Taiwan across the strait, the Philippines, Indo-China, Malaya, Borneo, and the northern tip of Australia. The replica of Cathay Pacific's other original DC-3, *Niki*, which guards Cathay City, the airline's headquarters in Hong Kong, presents a more nebulous landscape reminiscent of Pangea (figs. 3.2 a and b).

These early logos of Cathay Pacific reflect the evolving business plans of the airline in the 1940s and 1950s. In 1948, the airline carried the tagline of "Serving the Orient" with a globe-shaped logo that included the Americas.[8] In those early days, not only had the airline not unified its business insignia, but the rendition of its corporate name in Chinese also remained fluid. A coaster from the airline's early history features a Chinese company name that includes a reference to "Pacific" (later dropped) and reflects a grander ambition toward a world that stretches from Asia to Australia and the Americas (fig. 3.3). When the airline's route network achieved

7. TNA, FO371/93180.
8. Swire HK Archive, CPA/7/8/51.

FIGURES 3.2 a and b. *Betsy* and *Niki*, Cathay Pacific's first planes. Source: Author.

FIGURE 3.3. A Cathay Pacific coaster from the 1950s. Source: Swire HK Archive, JSS. CX2011.0003.

more definition by 1960, its corporate logo reflected a more circumscribed geographical reach in keeping with its service offerings (fig. 3.4).

Prescribed by British sanction of its scheduled routes, Cathay Pacific crafted its definition of the "Pacific" it served. The authorities also regulated many features of such sanctioned services: frequency, fares, and reciprocal rights. Therefore, the Hong Kong–based carrier operated within tight guidelines with competition not only from the local carriers at the other end of the aerial connection but also at times from BOAC.

The marketing tagline in the 1950s, "C[athay] P[acific] A[irways] Serves the Far East,"[9] had given way by 1965 to the claim of Cathay Pacific as "the largest regional carrier operating scheduled services in North and South East Asia," a region known as "the Orient." To assert its dominance

9. Swire HK Archive, JSS.CX.2010.00002; Swire HK Archive, JSS.CX.2011.00005; Swire HK Archive, JSS.CX.2011.00008; Swire HK Archive, JSS.CX.2011.00028.

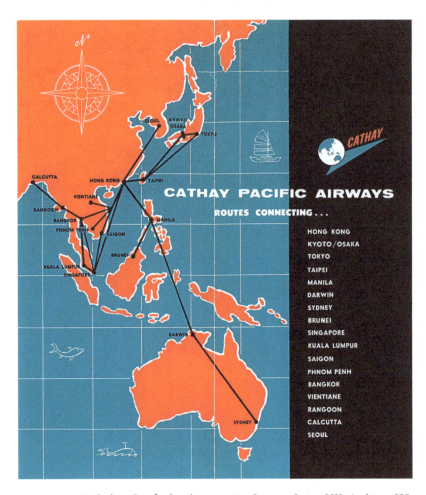

FIGURE 3.4. A Cathay Pacific brochure, 1960. Source: Swire HK Archive, JSS. CX2011.0003.

in this region, the airline advertised in 1964, "Cathay Pacific is the Orient's most experienced airline. From Singapore to Seoul . . . from Calcutta to Tokyo, Cathay Pacific has the fastest, most frequent flights of any airline in the Orient." The advertisement concluded thus: "Cathay Pacific . . . the airline that knows the Orient best."[10] In the process, Cathay Pacific refor-

10. Swire HK Archive, JSS.CX.2011.00012.

Fashioning Cathay's Pacific 109

mulated its claim of turf from the ill-defined "Far East" in the 1950s to the equally nebulous but more magical "Orient" in the 1960s. The "Far East" was "farther" than the "Near East" or "Middle East" from the perspective of "the West." Even though the "Orient" was defined as "North and South East Asia," perhaps to exclude what might be construed as the western part of Asia (as far as Turkey) and the far eastern edge of the Asian continent, the airline's reach had extended to Calcutta in the West and Tokyo in the East. Defying any defined logic of distance to take on the tone of otherness, the airline invited visits to the wonders of the foreign "Orient." The "Orient" not only governed for this period the scope of Cathay Pacific's sky routes but also became the resource pool from which the airline drew its alluring service crew and inspirations for its product offerings. This set the stage for the airline in the articulation of its cosmopolitanism, as well as Hong Kong's, as the city deepened and broadened its ties regionally and globally during the ensuing period of its economic takeoff.

Staffing Cosmopolitanism for Cathay's Pacific

Cathay Pacific had paid close attention to its staff profile since the early days. In its first-ever newsletter issue, published in 1958 for the airline's "friends and . . . passengers," the airline that touted itself as "a British Airline with British pilots" elaborated not only on its British private ownership but also on its establishment of "a new hostess school in order to select and train some more CPA Hostesses" for its expanding services. Instead of providing further details on the selection and training processes, the newsletter included the picture of a certain Miss Katherine Cheuk, "one of CPA's young and attractive 'older hands'."[11] Before the end of the year, the fourth issue of the newsletter would provide readers with a glimpse of the private lives of its hostesses.

The airline purchased its first jet plane, the *Electra*, in late 1958. This new aircraft heralded an era of passenger comfort. The first-class sleeper seats provided extra space for a leg extension of over six feet, allowing "the

---

11. Swire HK Archive, CPA/7/4/1/1/1 Newsletter [1958].

Branding Hong Kong

tallest of handsome men . . . [to] sleep soundly on board."[12] Juxtaposed with the article on Lockheed's macho turboprop *Electra*, "the fastest jet-and-propeller airliner in the world," a picture of two "charming CPA ground hostesses, Miss Alice Cheng . . . and Miss Mary Lewis," graced the back cover, posing "amongst the colourful flower stalls in Hong Kong" where they "spent some leisure time shopping."[13]

Although cabin service attendants had become predominantly female in the period after World War II, in the early decades of commercial aviation, male workers were considered more appropriate for the job. Aboard the US juggernaut, female employees entered the cabin crew only toward the end of 1943, when the military draft forced Pan Am to transform the profile of its all-male clipper crews. "Seven pretty girls" were the "first of their sex to 'win their wings'" when, "on the basis of ability, aptitude, and appearance," they joined the flight crews of the Pan Am clippers operating out of Miami, breaking "the long-standing unwritten tradition of Pan American's 17-year history of all-male flight crews on international routes." Each of the female recruits undertook the assignment of flight duty, which "in effect 'relieve a man for war duty' since they will supplement the present staff of male stewards." Even as wartime manpower shortages brought them into the Pan Am cabin, these women remained limited to certain flights because "it was thought unsafe for women to fly to Europe during the war." As the turmoil of World War II subsided, equal rights emerged as an issue. "The male pursers didn't like having us," recalled a Pan Am stewardess pioneer. "They did half the work for twice the pay."[14]

12. Swire HK Archive, CPA/7/4/1/1/4 Newsletter, November–December 1958; Swire HK Archive, CPA/7/4/1/1/7 Newsletter [Feb 1960].

13. Swire HK Archive, CPA/7/4/1/1/4 Newsletter November–December 1958. In contrast to most other female workers whose entrance greatly expanded the Asian workforce in this period, flight attendants came from a vastly different socioeconomic stratum and recorded a drastically divergent work experience (see, for example, Tsurumi, *Factory Girls*).

14. Pan Am, Series 1: Corporate and General, 1920–1994, Sub-Series 5: Financial and Statistical, 1922–1992, Sub-Series 1: Traffic and Sales, 1930–1984, Box 1, Folder 1 255 5 1930–1954 Traffic/Sales; Pan Am, Series 10: Personnel, 1912–2005, Sub-Series 2: Flight Attendants, 1930–2005, Box 1, Folder 1 257 22: Women ground Job, 1941–1960, *New Horizons,* October—December 1944; Pan Am, Series 10: Personnel, 1912–2005, Sub-Series 2: Flight Attendants, 1930–2005, Box 1, Folder 1 291 1: *Clipper* magazine, Stewardesses, 1967–1976, Pan Am *Clipper* vol. 24 No. 9, October 1, 1962, and vol. 6, No. 10,

Fashioning Cathay's Pacific

In its early days in postwar Hong Kong, Cathay Pacific was quick to adopt the employment of cabin crew of mixed sexes. For Cathay Pacific, news coverage often provided the names of these female flight attendants, or air or flight hostesses (sometimes stewardesses) as they were called. Their names provide some clues on their cultural or ethnic backgrounds: Western oriented because of their English given names; of Chinese or Western descent based on their surnames. New recruits to the crew in mid-1959 added another dimension to the crew's profile. "Three new Japanese stewardesses, employed by CPA in Tokyo" reported to duty in Hong Kong for a three-week training course. The report listed the three additions as "Miss Sachiko Sato[,] 23; Suma Takeuchi, 25; and Taruko Azuma, 25."[15] These new additions to the Cathay Pacific team were promptly featured in the Staff Jottings section of the November 1959 edition of the airline's newsletter as the "three attractive Japanese girls." Chinese, Western, and Japanese, these "attractive" female cabin staff members provided the perfect complement to Cathay Pacific's new *Electra* aircraft, the "Speedy Monsters" as the same issue of the newsletter called them.[16] The beastly "monsters" of new technology were to complement the "beauties" in the cabin crew and provide the perfect combination for the fledgling airline.

Despite its tremendous growth in the previous decade and the consolidation with its Hong Kong rival, Cathay Pacific's cabin crew remained a relatively intimate crowd in the closing days of the 1950s. Supervising its some fifty cabin attendants was Miss Josephine Cheng, whose duties included managing the attendants' uniforms, equipment, and roster schedules. With an impressive background (including a BA in Business Administration from Lingnan University in Canton), "Jo" joined Cathay Pacific because "the world of business did not really hold sufficient attraction,

---

October 1980; Pan Am, Series 10: Personnel, 1912–2005, Sub-Series 2: Flight Attendants, 1930–2005, Box 1, Folder 1 291 5: Stewards and Stewardesses, 1957–1970. Ironically, the entrance of female workers to cabin service broke the occupational barrier of sex segregation, but the glass ceiling endured (Hesse-Biber and Carter, *Working Women*, 54–60).

15. Swire HK Archive, CPA/7/4/2/1/1 News In Brief, July 15, 1959.

16. Swire HK Archive, CPA/7/4/1/1/5 Newsletter, November 1959. Chinese newspapers also noted the arrival of Cathay Pacific's flight attendants of Japanese descent (*WKYP*, July 9, 1959, 15; *KSMN*, February 14, 1960, 4; *WKYP*, February 14, 1960, 15; *WKYP*, March 25, 1960, 15).

whilst the possibility of becoming an air hostess—meeting people, going places—was irresistible!"[17] She started as the airline's "youngest hostess in 1948."[18] From the "small local-area airline" at the time of her joining, Cathay Pacific had become in ten years an airline that could boast "an international reputation," taking Jo "as far afield as Tokyo and Sydney in the course of her duties."[19]

In the late 1950s and early 1960s, the airline's clientele came predominantly from the United States and Britain. Although Cathay Pacific served many businesspeople besides tourists, the Hong Kong Tourist Association's report gives a good estimate of the profile of the airline's passengers. Tourism was a modest but fast-growing industry in the British colony, which attracted 163,661 visitors in 1960, a growth of nearly 60 percent in two years (fig. 3.5). Although the representation of British and Americans had decreased, the two countries still commanded over two-thirds of the market in 1960. Japan and the Philippines were a distant third and fourth at 8 percent and 7 percent, respectively.[20]

To its heavily Western clientele, how did the Hong Kong–based airline project its brand? Along with the entire airline industry, Cathay Pacific was eager to portray the glamorous life of its female cabin crew. Featured in the same issue as Jo was a picture of two "CPA hostesses" at Manila Airport, Miss Sachiko Sato, now kimono clad, and *cheongsam*-attired Miss Eleanor So, greeted by a travel manager of the Philippino Tours & Travel Association.[21] Their glamorous life extended beyond work too. The January 1961 issue of the newsletter reported that Rose Tam, a "pretty CPA Air Hostess," came second in the Miss Hong Kong contest (part of the Miss Universe pageant) and was chosen as "a most attractive cover girl" for the new Hong Kong publication, the *Sing Tao Pictorial*, where she appeared on a full-color cover and received two pages of coverage on her life at work and at home.[22]

---

17. Swire HK Archive, CPA/7/4/1/1/6 Newsletter, December 1959.
18. Swire HK Archive, CPA/7/4/1/1/19 News, November 6, 1961.
19. Swire HK Archive, CPA/7/4/1/1/6 Newsletter, December 1959.
20. *Hong Kong Tourist Association Annual Reports*, 1957/58, 15; *Hong Kong Tourist Association Annual Reports*, 1960/61, 16.
21. Swire HK Archive, CPA/7/4/1/1/6 Newsletter, December 1959.
22. Swire HK Archive, CPA/7/4/1/1/9 Newsletter, January 1961.

Fashioning Cathay's Pacific

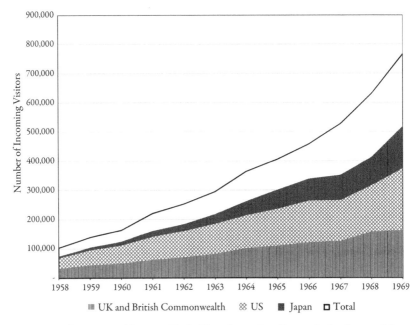

FIGURE 3.5. Incoming visitors to Hong Kong by nationality, 1958–69. Source: Hong Kong Tourist Association.

Despite all that presentation of glamor, the airline did suffer "severe depletion" of its cabin staff, "mainly because of matrimony." A condition of service was that the female cabin crew member be single, so as to be "free of domestic worries so that she can be blithe about her job." The reason for depletion? Their "excellent" circumstances if they were to be looking for a marriage partner. "One Cathay hostess flew for three months and then turned in her resignation," the airline noted in a 1961 report. "She married a passenger. So have five others." That was a drain of over ten percent of its entire cabin crew.[23] Although it might be tempting to characterize this marriage-based employment condition as backward for the Hong Kong–based airline, it is important to remember the prevalence even in the US labor market of "marriage bars"—policies against the hiring or retention of

---

23. Swire HK Archive, CPA/7/4/1/1/19 News, November 6, 1961.

114 *Branding Hong Kong*

married women. Marriage bars continued in the United States until 1950, and even after that, such practices persisted for flight attendants.[24]

Staff departure through matrimony meant that the airline constantly had to replenish its workforce. The same report mentioned that fourteen pupils ("nine young women and five young men; average age 21") were undergoing the twenty-three-day training to become "flight hostesses and flight pursers." The new recruits included "three Japanese girls, an Indian girl, three girls from Singapore, a Filipino, [and] two Chinese girls." Compared to the vibrant diversity of the female crew, the male members were all "young Chinese men."[25]

The airline had chosen these trainees "on the basis of self-reliance, enthusiasm, and an engaging personality." The only Hong Kong–based airline and the only airline to conduct cabin crew training in Hong Kong, Cathay Pacific had handpicked these fourteen trainees from a pool of eight-five applicants in 1961. Competition was even more intense in the previous year when eager job candidates totaled three hundred and forty. The airline interviewed all applicants and questioned them about "schooling and language claims." The lucky ones who proceeded to appear before the selection board needed to write an essay. Aspiring pursers had to write one in Chinese as well (hence, "young *Chinese* men" accounted for all the male recruits destined for the purser job).[26]

The airline's emphasis on languages did not end with these difficult interview tasks. During training, recruits listened to tape recordings of their own voices and received scrutiny of their presentation of public announcements and welcoming speeches. "You said that as though you were not quite sure whether it *was* a pleasure," many were told. "We must not neglect speech defects," supervisor Jo Cheng reminded them. "After all, English is not our native tongue."[27] The airline aimed not only to address the inherent language deficiency that had stemmed from the re-

---

24.   Goldin, *Understanding the Gender Gap*, 174–75, cites United Airlines' continued legal battle to uphold its "no-marriage" rule well into the 1980s.

25.   Swire HK Archive, CPA/7/4/1/1/19 News, November 6, 1961; *WKYP*, October 15, 1961, 15.

26.   Swire HK Archive, CPA/7/4/1/1/19 News, November 6, 1961 (emphasis added).

27.   Swire HK Archive, CPA/7/4/1/1/19 News, November 6, 1961.

cruits' non-English mother tongue but also to inculcate a service attitude that might not come naturally.

The training program indoctrinated the recruits in the Cathay Pacific "creed" of individual attention toward each passenger, which meant that the recruits had to "learn to accomplish demanding cabin tasks efficiently and master the art of serving passengers with seemingly tireless attentiveness." In addition to understanding pressurization, technical issues, and time calculation, they learned to handle documentary work, route information, and bar service in the air. They needed to know how to serve a cup of tea ("there are six ways of doing this, and one of them is the wrong way, according to Miss Cheng"). They needed to be more than what the average passenger expected of them—"a charming being who fetches and carries." Indeed, they should help parents struggling with bags and babies. There was more, as Jo Cheng reminded them: "Some people may think that what you do could be done by an amah . . . but we know better. You are employed because you are very much more." They were not to act like "the more regrettable variety of hospital nurse, shaking the limp traveler back to consciousness to serve him tea." The "speciality" of Cathay Pacific was that "its *girls* are never guilty of passenger bullying" (emphasis added; this gender-specific reference is particular to a class of mixed members). Instead, they were "to be there with whatever is needed before it can be requested." They were to be guided by "three essentials": a reliable stock of information, courtesy, and anticipation ("whatever was needed should arrive before [a passenger rings] the bell").[28]

The 1961 news report closed with passengers' accolades for "the smiling air of serenity" inflight. A female "travel agent from Detroit" was said to have written, "Your staff seem to be trained to make each person feel as though he or she is a personal guest." An Australian noted, "This flight is really the friendliest and best serviced that I have experienced on a journey around the world." A Texan couple remarked, "We have traveled the world for the past 16 years. We have been on every aircraft of the major airlines in Europe and the USA. Never have we enjoyed such service, such cordiality and such attentiveness as we have on your aircraft from Singapore to Hong Kong."[29] Two years after becoming Hong Kong's only

28.　Swire HK Archive, CPA/7/4/1/1/19 News, November 6, 1961.
29.　Swire HK Archive, CPA/7/4/1/1/19 News, November 6, 1961.

Branding Hong Kong

airline, Cathay Pacific took pains to craft its internationally competitive inflight service to attract the business of experienced (Western) travelers. Its selection and training of a distinguished cabin crew formed a crucial aspect of this corporate strategy.

By 1964, Cathay Pacific was ready to articulate openly its hiring and retention policy for its female cabin crew. In the fifth issue of *Orient Travel*, a monthly news magazine the airline launched in October 1963 "to interest not only members of the travel and tourist trade but the travel-conscious public alike," Cathay Pacific unveiled the alluring "Life in the Clouds" for its female cabin staff. "'How can I become an Air Hostess?' So many girls all over the world looking for an exciting career ask this question," began the article, alluding to the perception of a glamorous life that held "universal appeal for girls of all ages." Cathay Pacific, which offered training courses for "flight hostesses" (and pursers), maintained a policy of hiring "hostesses of many nationalities" to handle "the cosmopolitan nature of the passengers flying its routes in Southeast Asia." Boasting "a more multi-lingual assortment of flight hostesses than the majority of airlines," Cathay Pacific provided a cabin experience in which its passengers were rarely deprived of a "cabin attendant on his flight who is able to converse with him in the language he understands best." Bilingual in English and in at least one Asian language spoken in Cathay Pacific's service zone, its flight hostesses were also required to have "an attractive appearance[,] although even more important than looks is a pleasant personality, a good education and good health." The airline selected hostesses "from girls between the ages of 20 and 25[,] and they may continue to fly until the age of 35." These policies resulted in an average "flying life" of a Cathay Pacific hostess of four years, already exceeding the worldwide average of eighteen months. Cathay Pacific offered a family atmosphere, and "the girls" seemed to enjoy their work and would stay longer "but most of them leave to get married." During their service at Cathay Pacific, these hostesses were trained and expected to deliver cabin service "comparable to that of a first class hotel."[30]

---

30. Swire HK Archive, CPA/7/4/6/5 Cathay Pacific Airways—*Orient Travel* Monthly News Magazine, February 1964. The mandatory marital status and forced retirement age of flight attendants were an international issue, garnering the attention of the print media in Hong Kong. In 1968, local Chinese newspapers reported that on the

Fashioning Cathay's Pacific

While striving to maintain its service standard, Cathay Pacific endeavored to present an appealing array of hostesses for its flyers. If the diversity of the surnames did not make it obvious, Cathay Pacific explained its hiring policy in an article titled "Attractive Girls from Far Corners of The Far East" in the November 1964 newsletter. The article opened with a news writer supposedly "dazzled" by Cathay Pacific not limiting its choice of cabin attendants by nationalities, "unlike most airlines." In addition to being chosen for physical beauty, Cathay Pacific's cabin attendants were "chosen for ability to speak at least one of the languages of the regions served by Cathay Pacific (besides managing fluent English)" and were "selected for their tact, grace and beauty," the article claimed. To prove its case that Cathay Pacific scoured for beautiful talent far and wide, the article noted that the latest recruits "winnowed from a host of eager applicants" included "three Thai girls from Bangkok, five girls from Taiwan, four from Singapore[,] thirteen from Osaka and an Indian girl from Hongkong."[31]

In an attempt to convey the reach of the airline and the diversity of its female cabin crew as represented by the languages they spoke, the airline published a cartoon in its travel magazine with "Kathy the air hostess" (depicted as a voluptuous figure tightly clad in Cathay Pacific uniform) saying, "With all these fast jet flights, I've no sooner finished takeoff instructions in all Cathay's languages, than it's time to start the arrival information!" On the facing page, the Chinese version repeated the cartoon with Air Hostess Ah Tai conveying a similar message. The only difference between the two versions was that in Chinese Ah Tai referred to the various "Asian dialects" instead of "Cathay's languages" (figs. 3.6 a and b).[32] As bold as the claim was in the English version that the various languages were "Cathay's" despite their disparate linguistic origins, the Chinese version characterized the service zone of the airline as an

---

basis of gender equality, new regulations in the United States prohibited legislation requiring female crew members to quit their jobs upon marriage, or upon turning the age of thirty-three (*KSEN*, August 11, 1968, 2; *KSMN*, August 12, 1968, 2).

31. Swire HK Archive, CPA/7/4/1/1/71 Newsletter, November 4, 1964. Local media also paid attention to the arrival of trainees from such places as Japan, Taiwan, and Thailand to join local hires for training in Hong Kong (*KSEN*, November 24, 1964, 3).

32. Swire HK Archive, CPA/7/4/6/28 Cathay Pacific Airways—*Orient Travel Monthly News Magazine*, December 1966.

FIGURES 3.6 a and b. Cartoons boasting that the cosmopolitan female cabin crew could barely deliver the multilingual announcements before the fast jets finished the flights. Source: Swire HK Archive, CPA/7/4/6/28 Cathay Pacific Airways—*Orient Travel* Monthly News Magazine December 1966.

Asian region flattened to only dialectic variation because of the talents of Cathay's hostesses.

The attractive polyglots whom Cathay Pacific had recruited from its service area in the Orient showcased a Western orientation toward modern definitions of beautification. Not only were they fluent English speakers, many with English given names, but they were also vanguards in the adoption of Western cosmetics. In 1963, British cosmetic company Cyclax dispatched a makeup professional to Hong Kong to show Cathay

Pacific flight hostesses how to apply makeup according to the climatic conditions of the locale the airline served.[33] The first issue of *Orient Travel* readily reported on that event, which was hosted by Lane Crawford, "Hong Kong foremost department store." The article was eager to note that one of the beauty consultants was "responsible for Queen Elizabeth's make-up for her coronation."[34]

In 1965, a beauty course at Kai Tak airport enthralled more than fifty Cathay Pacific female staff members. Instructing them on the art of makeup was Diane Cheng, a former flight hostess with the airline, who had joined Revlon as a beauty consultant. She demonstrated to the eager attendees the use of makeup and nail polish that Revlon had donated. Explaining the choice of colors, she taught her students, "A passenger may not be aware of exactly what it is, if a girl wears an unsuitable lipstick or nail colour, but he certainly knows something is not right." Complementing her lessons, a representative from Shaw Brothers Studio, "a friendly Californian wise in the ways of Hollywood make-up methods and grooming," provided instructions focused on "subtley [sic] aided" natural looks. Wincing at the mention of "mascara-stiffened lashes, or brazenly pencilled brows," he cited a survey that claimed that only 5 percent of girls in the Far East applied makeup correctly. He photographed each girl and discussed the structure of her face individually. "A completely natural looking make-up, bringing out the best features, is essential for those who are in constant contact with the public," the beauty instructor said. The beauty products might be readily available in the commercial marketplace, but correct applications required connoisseurship that would tailor the look to one's natural features, a skill he found wanting among "girls in the Far East."[35]

The airline's female staff sustained their immense interest in this subject. When an Australian beauty consultant visited Hong Kong and gave makeup demonstrations and answered questions on cosmetics, six

33. *WKYP*, October 15, 1963, 11.

34. Swire HK Archive, CPA/7/4/6/1 Cathay Pacific Airways—*Orient Travel* Monthly News Magazine, October 1963.

35. Swire HK Archive, CPA/7/4/6/15 Cathay Pacific Airways—*Orient Travel* Monthly News Magazine, July 1965; Swire HK Archive, CPA/7/4/1/1/77 Newsletter, March 30, 1965.

Cathay Pacific flight hostesses eagerly took advantage of the makeup advice she offered.[36] Offering a viewpoint contrary to that offered by the Shaw Brothers representative, a flight hostess who visited the American Airlines training school at Fort Worth, Texas, and United's training center at O'Hare Airport in Chicago, Illinois, reported that false eyelashes were "as routine as lipstick." Expressing her admiration of her American counterpart, she claimed, "US hostesses set a very high standard when it comes to grooming."[37] A 1968 edition of the airline's newsletter featured a picture of ground hostesses attending a makeup class at Kai Tak, learning "how to keep false eyelashes in place."[38] The year after, Check-Flight Hostess Jenny Tung was featured for her instructions on eyebrow makeup: "Always shape them upwards. . . . Otherwise you might look melancholy."[39]

The June 1967 issue of *Cathay News* offered the most expressive analysis of the linkage between Cathay Pacific's hiring policy and the fascination with makeup. "Many believe that airlines only hire pretty girls as air hostesses and ground staff. That's a misunderstanding," the Chinese article claimed. "Actually, most airlines have the same hiring criteria as other companies: conduct, personality, and work aptitude. Appearance does not play an important role." However, the article did allow that in general, airline staff almost uniformly had pleasant facial features. The reason, the article maintained, was simple: they knew how to apply makeup and paid attention to their attire and posture. Therefore, Cathay Pacific held regular training on makeup and comportment for its female staff.[40]

The careful selection and training of its female cabin staff had allowed Cathay Pacific to grow its crew to keep pace with business expansion. By 1967, the airline boasted a total of seventy-five female cabin members from various Asian countries, all fluent in English in addition to their native tongue. Adjusting to the operating environment after Japan's liberaliza-

---

36. Swire HK Archive, CPA/7/4/6/17 Cathay Pacific Airways—*Orient Travel Monthly News Magazine* 1965.

37. Swire HK Archive, CPA/7/4/1/1/85 Newsletter, August 24, 1967.

38. Swire HK Archive, CPA/7/4/1/1/97 Newsletter, November 25, 1968.

39. Swire HK Archive, CPA/7/4/1/1/105 Newsletter, September 22, 1969.

40. Swire HK Archive, CPA/7/4/6/33 *Cathay News* [June 1967]. A Chinese news article also reported that cosmetics were an important element of the rigorous training for female flight attendants (*KSEN*, March 6, 1967, 3).

tion of outbound travel in 1964, twenty-nine of Cathay Pacific's female cabin crew were of Japanese nationality, exceeding the twenty-three from Hong Kong. Six were from Thailand, five each were from Malaysia and the Philippines, three were Korean, and two each were from India and Taiwan.[41]

In 1965, a local English newspaper in Hong Kong featured Somruedee Tongtaam, a Thai "air hostess" who had worked at Cathay Pacific for nearly a year. A speaker of "English and French as well as Thai," Somruedee flew between Hong Kong and Singapore, Manila, Calcutta, and Japan. She had followed her aunt into the industry. A lady-in-waiting to a Thai princess "whom she accompanied to England for the coronation of Queen Elizabeth II," Somruedee's aunt was "the first-ever Thai air hostess." Paralleling her linguistic repertoire, Somruedee cooked both Thai and European dishes, a skill that she showcased in entertaining guests. The article reported that Somruedee liked to wear her national Thai dress, which she complemented with the appropriate jewelry depending on the formality of the occasion.[42] In Cathay Pacific's cabin crew, Somruedee's pedigree might have been among the most exquisite. Yet, the language requirement the airline demanded of its female crew members guaranteed a certain level of distinguished background. Even for those who did not come with tastes as refined as Somruedee's from their upbringing, Cathay Pacific offered help in enhancing their grace, charm, and beauty.

The airline industry is not unique in putting female staff at the forefront to shape customer perceptions.[43] Nor was Cathay Pacific alone in mobilizing this notion of Asian femininity for the service of a budding air carrier. Other Asian carriers also crafted visually appealing female cabin crews who served as their frontline interface with passengers.[44] To compete in a global marketplace dominated by Western ideologies, practices, and expectations, Asian carriers constructed ambiguous notions of

---

41. Swire HK Archive, CPA/7/4/6/33 *Cathay News* [June 1967]; *KSMN*, April 4, 1967, 6. In 1963, Cathay Pacific decided to end its agency agreement with BOAC and opened its own office in Tokyo. In March 1964, Cathay Pacific opened its own offices in Kyoto and Nagoya as well (JSS, 13/10/1, minutes for meeting in April 1964).

42. *SCMP*, November 2, 1965, 4.

43. See, for example, Barnes and Newton, "Women, Uniforms."

44. Arnold, "For the Singapore Girl."

122   *Branding Hong Kong*

Asianness, often by conscripting women's bodies.[45] Despite its status as Hong Kong's sole airline operator, Cathay Pacific crafted its frontline female personnel for cosmopolitanism. Its female cabin crew reflected the region in which Cathay Pacific's clients trafficked. Hong Kong was an important, but not dominant, element of this composition, which was to project the cosmopolitanism of Cathay's Pacific. Ethnicities alone would not suffice. This visually appealing female staff had to converse with the airline's English-speaking clients in addition to charming them in their mother tongues. Their natural appearances needed to be augmented by the latest glamor-enhancing techniques that Cathay Pacific arranged for them from such places as the epicenter of show business.

This array of Asian women in the cabin stood in sharp contrast to the profile of the cockpit crew. In one of its earliest newsletters, Cathay Pacific made the appeal to "fly with the Pilots with the most experience in the Far East." The airline's chief pilot in 1958 had earned the Queen's Commendation for his services in the air. Another senior captain the airline featured was a RAAF veteran.[46] "A lot of them were aces from the Second World War," recalled a Cathay hostess from the 1940s.[47] By 1961, Cathay Pacific had shed its earlier tagline of "A British Airline with British Pilots."[48] However, British dominance in the cockpit had not yielded to Asian participation. Instead, Cathay Pacific pilots continued to be dominated by Westerners, mostly of British Commonwealth origin. In 1967, the airline touted the experience of its cockpit staff in an article titled "Some average!" Its pilots "from Britain, Australia and New Zealand" clocked on average "350,000 miles around the Orient each year."[49]

This budding Hong Kong flag carrier needed to succeed in the world of multinational businesses. For this up-and-coming carrier, the global business of air travel did not flatten or homogenize consumer preferences. Instead, such preferences varied according to the business purpose or function involved. Accordingly, Cathay Pacific sourced the best from dif-

---

45.   Obendorf, "Consuls, Consorts or Courtesans?," 35–39.
46.   Swire HK Archive, CPA7/4/1/1/2 July–August 1958.
47.   Swire HK Archive, *CX World* Issue 124 (July 2006).
48.   Swire HK Archive, CPA/7/4/1/2/9 Newsletter, [Jan] 1961.
49.   Swire HK Archive, CPA/7/4/1/2/27 Newsletter, [Feb 1967].

Fashioning Cathay's Pacific

ferent regions of the world. The staff of a cosmopolitan airline thus comprised a Western (British Commonwealth) cockpit crew to man Western machinery and a team of Western-minded Pan-Asian beauties (augmented by Chinese men in leadership roles as flight pursers) to provide the personal touch of "Oriental" customer service as Cathay Pacific carried its passengers around the region. Alongside modern (read: Western) engineering and technological know-how, Cathay Pacific fashioned cosmopolitan "beauties" who featured invented notions of local aesthetics.[50]

Orientalizing Cosmopolitanism through the Uniform of Cathay's Female Cabin Crew

By the 1960s, the luxury of air travel on Cathay Pacific was a far cry from the experience in the years immediately after World War II. Having inherited flying equipment and personnel from the military, Cathay Pacific in the immediate aftermath of World War II, along with many startups in commercial aviation in Asia, did not modify drastically the cabin experience for its early flyers. Not only did the founders purchase the first planes from the military surplus, but the cockpit crew also received their training from flying warplanes.

It was not until the late 1950s that Cathay Pacific invested in commercial aircraft and radically transformed the travel experience with the flyer in mind. The demilitarization of air travel also involved the makeover of the hostess outfit. In the 1940s and 1950s, Cathay hostesses wore conservatively designed uniforms (figs. 3.7 a and b). They ranged from a combination of a white blouse and a blue suit to a one-piece linen dress— "shortsleeves, button pockets at the hips, military stripes on the shoulder, and a skirt that soared close to five inches above the ankles." "It was like a St. John Ambulance nurse's dress," recalled a male cabin crew member from that era.[51] The "big switch" came in 1962 when Cathay hostesses shed the military-inspired gear that they had donned long after

50. See Jones, *Beauty Imagined*, for a discussion of the coexistence of globalization with tribalization in definitions of beauty for the purpose of global business.

51. Swire HK Archive, CPA/7/4/1/1/99 Newsletter, March 24, 1969.

FIGURES 3.7 a and b. The wide-ranging looks of Cathay Pacific female cabin crew, 1940s through 1962. Sources: Swire HK Archive, M-11(a)-03; Swire HK Archive, HK-2019-92.

Fashioning Cathay's Pacific   125

World War II had concluded and put on a uniform in keeping with the airline's "Oriental" flair.

In 1962 Cathay Pacific announced "a distinct change in shade and style" for the airline. Their flight hostesses were to get rid of the "unassertive workday outfits of dark blue," a "serviceable dark blue adopted by successive cabin crew since the airline began 17 years ago." With this makeover, flight hostesses changed into a new ensemble, "a dashing blend of East and West," designed by Rudella Shull, "an American fashion expert." A graduate of the Traphagen School of Fashion in New York, then "working for a textile firm in Kowloon," Shull produced the winning design picked from among five submissions by a panel of twelve "experienced judges."[52] "Definitely a departure from the usual air hostess uniform," the rose-colored uniform was reported to be a "traffic stopper."[53] It was not a mere adoption of a vibrant, eye-catching color. The demilitarization of the cabin crew also needed to look for a style becoming the character of the carrier.

For the airline that crisscrossed the Orient, the uniform was Chinese inspired. The flight hostesses came to be clothed in a form-fitting, crisp white cotton blouse that came with a Chinese-style collar. Together with the "cerise-piped" mandarin collar, three "Chinese-style buttons" from neck to shoulder embellished the blouse in "the traditional Chinese manner." The lower half of the ensemble was a Chinese-style skirt, with "slits on both sides . . . gently rounded." Underlying the visual attraction of this ensemble was "a special concealed tie" that held the blouse securely in place at the waist "no matter how hard the air hostess exerts herself removing meals from aircraft ovens and bending over serving trolleys." The Chinese-style top-and-bottom combo was modified, through a concealed feature, for functionality in a work setting guided by Western technology and a Western service mentality. Over this thoughtfully crafted Chinese inner set, the hostesses could don a collarless jacket, cut with

52. Swire HK Archive, CPA/7/4/1/1/45 Newsletter, September 14, 1962; *SCMP*, December 2, 1963, 42; *WKYP*, September 15, 1962, 15. Cathay Pacific's retention of a fashion designer followed closely the model of Pan Am, which introduced "the jet age look" for its flight stewardess in 1959 through designer Don Loper, the "famous Beverly Hills couturier" who emphasized "femininity as well as utility" (Pan Am, Series 10: Personnel, 1912–2005, Sub-Series 2: Flight Attendants, 1930–2005, Box 1, Folder 1 257 29).

53. Swire HK Archive, CPA/7/4/1/1/45 Newsletter, September 14, 1962.

126 *Branding Hong Kong*

"contoured arc seam above the waist." The "snappy suits of fine wool gabardine," which suggested a distinctly Western texture and modern look, came fully lined in matching silk and were to complement the chinoiserie blouse and skirt.[54]

Shull, who claimed that she had always considered airline uniforms "too stereotyped and dull—lacking imagination," explained that Cathay Pacific had allowed her full latitude in this uniform design. Putting herself in the place of the hostesses, she considered it important to design a "fashionable, dashing and feminine" ensemble. Shull said that she drew inspiration from "the Chinese cheongsam" (rendered as *qipao* in Putonghua), which she adapted for the skirt and blouse. The outerwear was an adaptation of the Eisenhower jacket. She did not believe that the jacket would become dated—paired with the *cheongsam*-inspired, mandarin-collared blouse and skirt combination, it "automatically creates its own style . . . definitely a case where East meets West in an airline uniform." "Certainly," she added, "one could say that this has an international flavor, and its own individual style."[55]

For Cathay Pacific, this East-meets-West ensemble disarmed its air hostesses who had been burdened with military-inspired outfits. The new design infused elements of style and femininity to the airline's cabin service. Its handpicked female cabin service staff would get to don this sartorial creation, which was designed specifically "for the small bone structure of Cathay Pacific's Asian girls." To be handcrafted by the famous Shanghai tailors, the airline customized the uniform "to the individual measurements."[56] More than just mimicking the details of Chinese costume, the blouse and skirt that formed the inner layer of the new uni-

54. Swire HK Archive, CPA/7/4/1/1/45 Newsletter, September 14, 1962. See Silberstein, "Fashioning the Foreign," on the history of the use of British woolens in China. For an extended discussion on the use of European fabric in Asia, see Pyun, "Hybrid Dandyism."

55. Swire HK Archive, CPA/7/4/1/1/45 Newsletter, September 14, 1962; *SCMP*, December 2, 1963, 42.

56. Swire HK Archive, CPA/7/4/1/1/45 Newsletter, September 14, 1962; *SCMP*, December 2, 1963, 42. These fine craftsmen hailed from the Chinese city that once hosted many Western visitors to Asia in the prewar era. On tourism in Hong Kong induced by geopolitics during this period, see Mark, "Vietnam War Tourists." Katon Lee has written about "suit tourism" in Hong Kong ("Suit Up," chap. 5).

Fashioning Cathay's Pacific

127

form were to hug the figure and reveal the contours of Cathay Pacific's petite Asian hostesses. The more generous fit of the Eisenhower jacket, the very name of which readily evokes the power dynamics of a new world order in the Cold War period, provided a Western flair to the Chinese-inspired core.

Cathay Pacific was not alone in driving the airline fashion world. The airline had noticed around the time of its uniform reform that Malayan Airways had replaced the high heels, beige blouse, and skirt with a new hostess uniform comprising wooden-soled sandals and an adaptation of the traditional sarong.[57] This period also registered heightened interest in the nationalities of crew members as represented in the costumes they wore. In a 1963 fashion parade of air hostesses, the highlight was "the girls, all from different countries, paraded first in their national dress."[58]

If in this era the Western gaze from beyond had essentialized what it categorized as distinct ethnicities in Asia,[59] the styling of Cathay Pacific's Asian cabin crew had self-orientalized its staff members. Fashion facilitates bodily representation, and many indigenous subjects in Asia had deployed fashion in their nation-building efforts often to express their resistance to colonial regimes.[60] It is noteworthy that the female body often served as the site of fashioning ethnicity. In the case of China, pronouncements on clothing, along with other practices of everyday life, contributed to the creation of a new sense of Chineseness.[61] The development of a national costume also foregrounded the woman's body as the

57. Swire HK Archive, CPA/7/4/1/1/14 Newsletter, September 11, 1961.

58. Swire HK Archive, CPA/7/4/6/1 Cathay Pacific Airways—*Orient Travel Monthly News Magazine*, October 1963.

59. In a process she calls "Cold War Orientalism," Christina Klein has demonstrated how American cultural production in the Cold War period presented Asia to the US audience in the hope of promoting acceptance of nonwhites and America's internationalist role in Asia (Klein, *Cold War Orientalism*).

60. For Burma, see Ikeya, "Modern Burmese Woman." For India, see Tarlo, "Problem of What to Wear"; Tarlo, *Clothing Matters*; Bhatia, "Fashioning Women." For Sri Lanka, see Wickramasinghe, *Dressing the Colonised Body*. For Java, see Taylor, "Costume and Gender." For the Philippines, see Roces, "Gender, Nation"; Roces, "Dress, Status, and Identity."

61. Harrison, *Making of the Republican Citizen*. For a discussion that covers an extended period, see Finnane, *Changing Clothes in China*.

128   *Branding Hong Kong*

site of social contention in the early twentieth century.[62] As a national costume,[63] the *cheongsam*, which became increasingly figure hugging, accentuated the desirability of the modern Chinese woman.[64] The uniform redesign of Cathay Pacific in 1962 was but an attempt to appropriate this cultural icon for its commercial interests.[65]

For Cathay Pacific, the reform was simpler because the carrier only needed to focus on its female cabin crew. It was not necessary to utilize sartorial devices to differentiate its male cabin staff from the cockpit crew—its Caucasian pilots were readily distinguishable from its Asian cabin servicemen. Unlike many Asian airlines that had hired pilots of local descent, Cathay Pacific maintained a cockpit crew exclusively of Western pilots, as indicated in the 1967 newsletter in the previous section. "When I joined the company in 1956 the male attendant's uniform was identical to that of the pilots," recalled the airline's first male flight attendant. By the time he reached the rank of senior purser, he earned three stripes on his shoulders, the same as a senior first officer.[66] Male cabin attendants kept their military-inspired uniform. By March 1964 the airline issued all "flight pursers on all routes" an additional ensemble of jackets, bow ties, and cummerbunds of different colors for cocktail and meal services to "[keep] up with the girls in their striking rose-red uniforms."[67] The complementary outfit for the male cabin colleagues served to accessorize the hostesses' uniform as much as the matching color of nail polish chosen "to harmonise with the costume colour" of the women.[68]

62. Ng, "Gendered by Design."

63. For a description of the cultural politics and gender dynamics that resulted in the national costume of *qipao/cheongsam*, see Finnane, "What Should Chinese Women Wear?"

64. The sensualizing outfit also came with the attendant anxiety of moral decay, at least earlier in the twentieth century (see Edwards, "Policing the Modern Woman").

65. Shifting the focus to the beginning of the twenty-first century, Chew notes the influence of cultural producers and celebrities in the resurgence of the *qipao* (Chew, "Contemporary Re-Emergence").

66. Swire HK Archive, *CX World* Issue 67 (October 2001).

67. Swire HK Archive, CPA/7/4/6/5 Cathay Pacific Airways—*Orient Travel* Monthly News Magazine, February 1964; Swire HK Archive, CPA/7/4/1/1/58 Newsletter, February 24, 1962.

68. Swire HK Archive, CPA/7/4/6/1 Cathay Pacific Airways—*Orient Travel* Monthly News Magazine, October 1963; Swire HK Archive, CPA/7/4/6/34 *Cathay News* No. 34 [1965].

In the implementation of a new look for Cathay Pacific, the focus remained on its air hostesses' costume-inspired uniform (figs. 3.8 a, b, and c).

Within weeks of its introduction, the new hostess uniform was reported to be "blossoming all over Asia," as hostesses donning the new outfit kept their hats and gloves on until airborne.[69] Newspapers in Hong Kong, both English and Chinese, offered their accolades on the new design.[70] In a July 1963 article featuring the jet-setting lifestyle of air hostesses, the bilingual magazine *Woman Today* featured on its front page a Cathay Pacific hostess in the new uniform, calling it a "clear rose ensemble designed on a compromising East-meets-West style to give maximum freedom for strenuous work and maximum femininity of the cheongsam."[71] Besides blending the best of East and West, the style was to "harmonise with Cathay Pacific's Asian cabin attendants" who in 1962 included contingents from Hong Kong, Japan, and India and were soon to include "girls from Thailand, Malaya and the Philippines."[72]

Although the Chinese accents expressed in the new uniform offered visual appeal, the uniform had to contain the bursting array of ethnicities Cathay Pacific's hostesses sought to represent. Their Japanese crew members presented the uniform's visuality with a difficult challenge. The Japanese market swelled with the relaxation of travel restrictions on its citizens in 1964, and in 1965, Japan was second only to the United States in the count of visitors incoming to Hong Kong (fig. 3.5).[73] Responding to this changing profile of its clientele, Cathay Pacific introduced its Kantai service as "a special way of saying welcome" to its expanding Japanese clientele. For this important market, the airline selected a "special platoon of Japanese-speaking Cathay Pacific attendants." To identify themselves, these crew members wore Japanese-flag shoulder patches for easy identification. Onboard, these hostesses changed out of their Cathay Pacific uniforms and put on their kimonos when they served Japanese

---

69. Swire HK Archive, CPA/7/4/1/1/46 Newsletter, September 24, 1962.

70. *WKYP*, September 15, 1962, 15; *SCMP*, December 2, 1963, 42.

71. Swire HK Archive, CPA/CE/6/13/7/(14) *Woman Today* 现代女性, July 1963.

72. Swire HK Archive, CPA/7/4/1/1/37 Newsletter, May 28, 1962.

73. *New York Times*, February 9, 1964, 1; *Hong Kong Tourist Association Annual Report*, 1965–66, "Statistics of Incoming Visitors (Non-Chinese) to Hong Kong by Nationality in 1965."

FIGURES 3.8 a, b, and c. (a) and (b) The 1962 uniform for the female cabin crew. Sources: Swire HK Archive, CPA/7/4/6/43; Swire HK Archive, CPA/7/9/2/1/A. (c) The 1962 uniforms for the male and female cabin crew. Source: Swire HK Archive, CPA/7/9/2/2/1/2.

passengers.[74] As Cathay Pacific marked the 1965 introduction of its daily service between Hong Kong, Taiwan, and Japan, the apparel's consistency had given way to diversity, with some hostesses "dressed in their national dress, the Kimono, while others wore their Cathay Pacific uniforms."[75] The airline continued this practice of mixed attire in its promotion events. When the company sent its flight hostesses on overseas tours as promoters to add "to the lure of Far East Travel," it dressed its "fascinating flight hostesses in their rose coloured uniforms or in national dress. . . . Cherry Ho, Tomoko Nagahata and Keiko Mayuzumi returned home from Tokyo . . . after taking care of thousands of inquiries about Hong Kong."[76]

The issue of the uniform extended beyond this special apparel arrangement for this fast-growing clientele. As the airline featured its pride of "attractive girls from far corners of the Far East," Cathay Pacific endeavored to showcase visually the diversity of their ethnic background. In so doing, the airline had to forgo the visual uniformity of company apparel. As the airline's passenger services manager surveyed his "photogenic flock" in 1964 with the largest number yet of overseas recruits, the new hostesses arrayed themselves in their national costumes. That, too, proved problematic as some of the recruits fell under the same "national costume" classification. Besides the kimono, the group sartorially featured their ethnicities, with "the five Taipei recruits, all in glamorous cheongsams" hardly distinguishable from the local Hong Kong Chinese (fig. 3.9).[77]

This issue of uniformity is not unique to Cathay Pacific's search for an outfit for its female staff. In her exploration of "Australian fashion," Jennifer Craik asserts that the search for a unique fashion for a place is rife with cultural politics.[78] Although Hong Kong was not as multicultural as

---

74. Swire HK Archive, CPA/7/4/6/4 Cathay Pacific Airways—*Orient Travel Monthly News Magazine*, January 1964; Swire HK Archive, CPA/7/4/1/1/60 Newsletter, February 24, 1964.

75. Swire HK Archive, CPA/7/4/6/11 Cathay Pacific Airways—*Orient Travel Monthly News Magazine*, March 1965.

76. Swire HK Archive, CPA/7/4/1/1/79 Newsletter, May 10, 1965.

77. Swire HK Archive, CPA/7/4/1/1/71 Newsletter, November 4, 1964; Swire HK Archive, *CX World* Issue 53 (August 2000). Cathay Pacific continued to feature in public relations events its flight attendants dressed in kimonos as the Japanese market remained vibrant (*KSMN*, November 26, 1967, 4).

78. Craik, "Is Australian Fashion?"

FIGURE 3.9. Cathay Pacific showcased its multiethnic cabin crew in national costumes. Source: Swire HK Archive, CPA/7/9/2/1/5/15.

Australia, Cathay Pacific's multiethnic cabin crew made it difficult to underscore a single distinct feature that would uniformly represent the cosmopolitan background Cathay Pacific had sought to accentuate in staffing. Furthermore, the presentation of the entire female cabin crew in a single style entailed a process of negotiation of the role of the body the uniform was to clad.[79] Although the practical modifications to the *cheongsam*-inspired outfit might have resolved the incongruence of work discipline and femininity for women in uniform,[80] this ensemble, which evidently drew inspiration from notions of Chinese ethnic designs, did not resonate equally with the female cabin crew of diverse backgrounds.

The ethnic particularities in Hong Kong offered Cathay Pacific an entrée to fashioning a unique uniform for its air hostesses in the 1960s. The uniform that the airline introduced in 1962 accentuated the Chinese heritage of many who worked for the Hong Kong airline, but this Chinese heri-

79. See, for comparison, the development of school uniforms for boys and girls in Japan (Namba, "School Uniform Reforms").
80. Craik, "Cultural Politics."

tage, shared by its competitors in Singapore, Taiwan, and elsewhere, failed to differentiate the airline in a region in which Chinese culture played a pervasive role. The issue became more complex as Cathay Pacific projected in its hiring practices not only its base in Hong Kong but also its service area in "the Orient." The *cheongsam*-inspired uniform muted the array of "Oriental" heritages the airline strove to showcase through its female cabin crew.[81] When the airline needed to represent specific ethnicities, its hostesses had to shed the uniform that evoked Chinese sentiments and change into national costumes according to their individual backgrounds.[82]

Cathay Pacific aimed to project a cosmopolitan image, which shined through its air hostesses' pan-regional representation. However, presenting this brand of cosmopolitanism in a uniform manner required the design of a standard outfit that bound together the cabin staff this Hong Kong airline offered. Deliberate multiculturalism came into conflict with the coerced homogenization of an ethnically heavy package design. By the end of the decade, Cathay Pacific would embark on another uniform redesign that proposed a brand-new expression of its cosmopolitan offerings that articulated a restructured vernacular and local identity for the Hong Kong carrier.

Branding Pragmatism in a Logistics Industry

As in the development of other global brands, Cathay Pacific rejuvenated its corporate image through continual adjustments.[83] The successive rounds of uniform redesign reflected Cathay Pacific's keen efforts to refresh its brand in response to the changing business environment.

---

81.   Similarly, not all "Singapore Girls" serving on Singapore Airlines are "Singaporeans" (Obendorf, "Consuls, consorts or Courtesans?," 45). When Singapore established its own carrier independently from Malaysia in 1972, what came to be known as Singapore Airlines (SIA) kept the familiar *batik sarong kebaya* Pierre Balmain had designed for Malaysia-Singapore Airlines in 1968. In spite of its independent status, SIA continued to fashion the Singapore Girls by cladding them in tight-fitting Malay evening gowns (*Straits Times*, July 30, 1968, 10; *Straits Times*, July 27, 1972, 2).

82.   On board Malaysia-Singapore Airlines in the late 1960s, female cabin crew members of Japanese descent also shed their *sarong kebaya* and donned pastel-colored kimonos (*Straits Times*, August 1, 1968, 5).

83.   Da Silva Lopes and Casson, "Entrepreneurship and the Development."

134 *Branding Hong Kong*

After the 1962 introduction of the *cheongsam*-inspired outfit, the next round of uniform redesign began as a component of a corporate image makeover that the airline initiated in 1968.[84] The number of visitors Hong Kong welcomed in 1968 had grown six-fold in a decade to 618,410, of which the United States accounted for 26 percent and Japan, 16 percent.[85] Except for its "firecracker red" color, the new uniform introduced in March 1969 cast off all connotations of Chinese ethnicity. The ensemble included an "A-line dress and jacket in terylene, with mushroom shaped hat, and matching red topcoat of waterproofed English gabardine." The outfit looked decidedly not "Oriental," and the airline described this ensemble with an emphasis on functionality. "Hostesses, like *astronauts*, need the right type of clothing for flexibility and mobility" (emphasis added). This set, designed by an Australian fashion panel and "tailored by [clothing manufacturer] Lai Wah of Kowloon," was "not flamboyant" but instead fulfilled "the practical need of having to look reassuringly trim and appropriate in all climates and all working conditions." The overcoat was "satin lined with detachable padding for cold weather." For each flight hostess, the airline issued a "dress, jacket, overcoat, hat, serving smock, handbag, high-heeled shoes, low-heeled shoes (for cabin service), gloves, language pin, and cap badge." The handbag was designed with its contents in mind: "money wallet, powder, lipstick, mirror (this came with the bag), tube of handcream, gloves . . . [and] the most important items of all, passport and medical papers, in a zippered compartment."[86] Complementing the set, the airline augmented the ensemble a few months after its introduction with a "raincape of tetron, with nylon lining [that] matches the firecracker red Cathay Pacific uniform and pixie hats." This raincoat was to "entirely cover and protect [the] uniform, pixie hat, and overcoat, from winter weather without looking bulky" (figs. 3.10 a and b).[87] Thoughtfully designed for functionality, the entire ensemble took full advantage of materials that new technologies afforded the sophisticated

84. Swire HK Archive, CPA/7/4/1/1/92 Newsletter, June 18, 1968; Swire HK Archive, CPA/7/4/1/1/94 Newsletter, August 19, 1968.

85. *Hong Kong Tourist Association Annual Report*, 1968–69, 23.

86. Swire HK Archive, CPA/7/4/1/1/99 Newsletter, March 24, 1969; *SCMP*, June 29, 1969, 32.

87. Swire HK Archive, CPA/7/4/1/1/105 Newsletter, September 22, 1969.

FIGURES 3.10 a and b. The 1969 uniform redesign. Source: Swire HK Archive, CPA/7/9/2/1/5/15EN; Swire HK Archive, JHK/7/5/1/7.

136 *Branding Hong Kong*

and fashionably conscious airline. The crew welcomed these changes to the design, "feeling glad to say goodbye to the tuck-in blouse that kept popping out."[88]

Echoing its emphasis on functionality, the airline launched the new uniform by highlighting the logistical practicalities of the process: an "order list" of "446 complete outfits (136 flight hostesses, 123 for outports and 187 for Hongkong)"; "one thousand five hundred square feet of leather from Australia" for the beige shoes and handbags; 4,500 yards of terylene imported from the United Kingdom for the dress and jacket; 1,500 yards of "matching gabardine . . . from Bradford" for the overcoats.[89] Playing an instrumental role in this uniform overhaul was "Lai Wah of Kowloon," one of Hong Kong's own garment manufacturers as the city developed into a powerhouse in the global clothing industry. "The change in plumage was all part of an entirely new look planned for Cathay Pacific," the airline explained. This "new graphic image," as it was known in "advertising dialect," required the careful orchestration of the airline's operational staff, who oversaw the logistics of this overhaul.[90] This emphasis on logistics resonates with the bustling business of travel in Hong Kong—the number of arrivals to Hong Kong had expanded by double-digit percentages in each year of the 1960s to 765,213 in 1969.[91] Growing faster

88. Swire HK Archive, CPA/7/4/1/1/99 Newsletter, March 24, 1969. Interestingly, as Cathay Pacific abandoned its Oriental motif, other airlines came to include ethnic elements in their uniforms in the 1970s, some more subtly than others. Qantas incorporated the brilliant colors of Australian flowers, feathers, and petals in the prints of Emilio Pucci; British Caledonian Airways included the subtle colorways of traditional tartans; Japan Air Lines incorporated ethnic touches in Hanae Mori's design; Singapore Airlines showcased Oriental patterns and colors in the Pierre Balmain design; and Air India featured colorful saris (Pan Am, Series 10: Personnel, 1912–2005, Sub-Series 2: Flight Attendants, 1930–2005, Box 1, Folder 1 291 7).

89. Swire HK Archive, CPA/7/4/1/1/99 Newsletter, March 24, 1969.

90. Swire HK Archive, CPA/7/4/6/37 *Cathay News* 43 April–June 1969. A description of Cathay Pacific's 1969 ensemble echoed that of the Malaysia-Singapore Airlines' uniform introduced in 1968: "designed by Pierre Balmain, the French couturier," but "bears the distinctive label: Made-in-Malaysia/Singapore"; "not only are the uniforms tailored in Singapore but the accessories—shoes and handbag—are also locally manufactured" (*Straits Times*, August 1, 1968, 32).

91. *Hong Kong Tourist Association Annual Report*, 1969–70.

Fashioning Cathay's Pacific

than the overall industry, Cathay Pacific carried 457,964 passengers in 1969, an eight-fold increase over the course of a decade.[92]

Although it was unique in its move to shed the Oriental emphasis in its previous uniform, Cathay Pacific was not alone in its emphasis on practicality in this new design in the late 1960s. The airline industry's stress on functionality was so intense that a travel magazine from that period cautioned against designs that came down excessively on the practical side. "The passenger must still be allowed to distinguish a hostess from a high-flying fashion robot." Cathay Pacific considered itself to have struck an appropriate compromise. The new uniform issued to "all female staff members dealing directly with the public" was to deliver "the major impact of Cathay Pacific's new look." "When you see the 1962 uniform together with the new one," said the airline's passenger service manager, "you appreciate that we have something completely different. But at the same time, it doesn't depart too much, so that we retain our identification." Another Cathay Pacific executive was reported to have said that the new uniform "made the girls look younger and even prettier" and "made him feel younger too!" Compared to the 1962 "cheongsam-style skirt and jacket," this new ensemble was "a bloodysight smarter." So the airline promoted it.[93]

This makeover transformed the Oriental beauties cloaked uniformly in *cheongsam*-inspired garb into a crew of attractively and pragmatically dressed hostesses. The outfit that came to be known as the "mini" reflected not only the fashion trends of the late 1960s but also the increasing logistical requirements of running an airline. The timing of its appearance and the comparison to astronauts in the flexibility and mobility it offered was no coincidence. Air hostesses began to greet passengers in this new uniform in 1969, the year of the first manned moon landing in the space race. As the 1962 uniform proved to be too constricting for its multiethnic cabin crew, the new Cathay Pacific packaging of cosmopolitan beauty had grown to transcend racial categories in the practicalities of the age of technology.

92. Swire HK Archive, Cathay Pacific Airways Limited Report of the Directors and Statement of Accounts for the Year Ended 30th June 1962 and 1969.

93. Swire HK Archive, CPA/7/4/1/1/99 Newsletter, March 24, 1969; Swire HK Archive, CPA/7/4/6/37 *Cathay News* 43 April–June 1969.

138   *Branding Hong Kong*

Although freed from its ethnic connotations and enhanced for functional requirements in cabin service, this 1969 design proved no more enduring than its predecessor. In its endeavor to create a distinct mark of sophistication, Cathay Pacific launched yet another uniform reform in 1974. By then, visitors to Hong Kong were coming from a mix of geographies. Japan led the tally, followed by roughly equal shares of North America and Southeast Asia, with the United Kingdom and the rest of Europe, Australia, and New Zealand trailing behind.[94] In this round of uniform redesign, Cathay Pacific introduced what it called its "traditional Tung Hoi—Eastern Seas—pattern with magnetic and eye catching swirls of red, yellow and blues." Printed on a trim, fitted cotton blouse, these vibrant colors intensified the visual impact of the two-piece suit of "delicious burnt orange—the latest fashion colour to emerge." The new uniform was considered "graceful and sophisticated and, in keeping with the times, uncluttered and free." Continuing the practical focus of the previous design, the cut of this new set "allow[ed] for ease of movement," as did the four pleats and the front and the back of the skirt, "stitched slightly below hip level" for "a graceful swirling movement." Similarly accessorized and complemented with numerous practical items like the 1969 outfit, this generation of the Cathay Pacific uniform had to look distinctive nonetheless. It came with a "bowler styled hat with its provocative brim . . . turned up in the back and down in the front."[95] To accentuate the iconic design, the scarf repeated the Eastern Seas motif. The airline was intent on building its brand as the made-in-Hong-Kong ensemble "feature[d] one more essential detail—the Cathay Pacific Airways brevet worn on the suit jacket, the coat, smock and hat" (figs. 3.11 a and b).[96] Since the last makeover, the management of uniform logistics had only become more complex. Concurrent with the introduction of this new look, Cathay Pacific recruited a uniform controller, with "thorough knowl-

94. *Hong Kong Tourist Association Annual Report*, 1974–75, 4.

95. Following the airline fashion of the times once again, Cathay Pacific's new headwear for its female cabin crew chased Pan Am's lead to replace the old high-crowned pillbox with bowler hats when Frank Smith, chief designer for the women's wear firm Evan Picone, redesigned the ensemble for its stewardesses as the airline launched the Boeing 747 on the world's air routes (Pan Am, Series 10: Personnel, 1912–2005, Sub-Series 2: Flight Attendants, 1930–2005, Box 1, Folder 1 291 10).

96. Swire HK Archive, CPA/7/4/1/1/141 Newsletter, October 1974.

Fashioning Cathay's Pacific   139

FIGURES 3.11 a and b. The 1974 uniform. Sources: Swire HK Archive, CPA/7/9/2/2/2/17; Swire HK Archive, HK/2017/22.

edge of textile composition, textures, colour fastness and strength tests" to lead a functionally dedicated "section of four people."[97]

All these deliberate efforts to create distinctive visuals for the Cathay Pacific brand came with a designer price. The mastermind behind this Eastern Seas ensemble was none other than Pierre Balmain, the French designer responsible for among other designs, the iconic Singapore Girl dressed in the *sarong kebaya* for Singapore Airlines.[98] Balmain also became the first of many brand-name fashion designers to create a distinctive look for the Cathay Pacific cabin crew. Hermès would replace the Balmain

---

97. *SCMP*, July 22, 1974, 24.
98. The *South China Morning Post* readily featured an article that juxtaposed the two Balmain-designed uniforms (*SCMP*, September 7, 1974, 8).

140   *Branding Hong Kong*

design in 1983,[99] only to be supplanted by Nina Ricci in 1990.[100] For the rest of Hong Kong's colonial days, French haute couture dictated the rendition of Cathay Pacific's cosmopolitanism in its cabin crew uniform.

The airline industry altered people's perception of time and space, at least in an aspirational sense. Equally significantly, the transformative industry refashioned beauty time and again by repackaging its handpicked female cabin crew. The iconic images of this frontline representation of Cathay Pacific aimed at conveying a Hong Kong–based cosmopolitan image that was to embody its multiethnic service region. Yet, little of this array of this beauty was "natural"; in fact, its presentation was highly engineered. Cathay Pacific's successive rounds of uniform overhauls underscore the historically contingent nature of visuality as a construct and point to the power of businesses to shape people's perception of the world as they generated profit opportunities.[101]

Just as individuals express their identities through socially and culturally mediated fashion,[102] Cathay Pacific continuously transformed its corporate brand by adorning its female cabin crew in an evolving style that reflected its business exigencies at particular moments in time. The fashion icon that Cathay Pacific presented in each era also signaled the positioning of the airline, as well as the city it represented, in the region and the world of commercial aviation.

A Cosmopolitan Service Based in Hong Kong

In its humble beginning in the aftermath of World War II, Cathay Pacific Airways began its business with chartered flights around Hong Kong to and from destinations drawn primarily from Southeast Asian cities.

---

99. Swire HK Archive, CPA/7/4/1/1/171 Newsletter, November 1982; *SCMP*, January 13, 1983, 1; *WKYP*, January 13, 1983, 8; *TKP*, January 29, 1983, 7; *WKYP*, May 2, 1983, 6. A Chinese news report called the 1983 design "elegant, luxurious, and ever more charming" (*WKYP*, May 26, 1983, 10).

100. Swire HK Archive, CPA/7/4/1/1/186 *Cathay News* No. 48, March 1990.

101. For an analysis of corporate inventions of visuality in the food industry, see Hisano, *Visualizing Taste*.

102. Crane, *Fashion and Its Social Agendas*.

Fashioning Cathay's Pacific 141

By 1971, Cathay Pacific had matured into the flag carrier of colonial Hong Kong. That year, the airline posted the advertisement in the United States with which this chapter began. "Discover the many faces of the Orient on Cathay Pacific Airways," read the title of the advertisement (fig. 3.12). The advertisement featured a group photograph of sixteen Cathay Pacific employees (cockpit and cabin crew, ground staff, engineers, and chefs) in uniform. Also included was an artistic rendition of the airline's network that linked the cities it served into a petal pattern.[103]

As the advertisement demonstrates, Cathay Pacific claimed its base of Hong Kong but projected an image of a larger catchment area in the nebulous region of "the Orient." Its services, the advertisement maintained, combined the best of all worlds with the local sensitivities of Hong Kong and the region. With the backing of the political regime ("the British airline"), it offered European know-how, experience, and technological reliability ("our British million-mile jet pilots and skilled maintenance crews"). Not privileging the European power that governed Hong Kong, the airline embraced other Western elements renowned for their cultural distinction ("Swiss chefs and their tempting international cuisine").[104] More than merely presenting the world's best to its customers, the airline exhibited a careful understanding of its target customers (Americans, in this case) as it served "complimentary cocktails" that the Western palate would appreciate.

Beyond the warm embrace of this Western comfort and world-class luxury, Cathay Pacific offered its customers "Oriental" charm formulated to suit its customers' needs. Underscoring its claim to understanding Asian cultural peculiarities of "face," Cathay Pacific presented the appeal of "the prettiest faces of 9 exotic lands smiling at you." From the clearly enumerated but loosely defined territories ("lands") that exuded the enigmatic allure of foreignness, the airline had handpicked the most visually pleasing and trained them to serve its customers with a friendly smile. The "23 languages" these "prettiest faces" spoke accentuated the airline's assertion of its local awareness, but such command of native knowledge would not have been useful to the Western traveler if not for the pretty

103. Swire HK Archive, Swire HKAS 391044158720.

104. At the grande dame, the Peninsula Hotel, all the chefs were Swiss in the 1950s (Hong Kong Heritage Project, Interview of Felix Bieger, August 3, 2007).

FIGURE 3.12. "Discover the many faces of the Orient on Cathay Pacific Airways." Source: Swire HK Archive, HK/2015/34.

Fashioning Cathay's Pacific   143

faces' fluency in the traveler's own tongue—English.[105] These exotic but understanding service staff paralleled the "14 major cities" to and from which Cathay Pacific carried its passengers. Besides logistical benefits ("more often"), Cathay Pacific promised to fly its passengers "more pleasurably" than other airlines. The name of the airline, Cathay, invoked magic as much as the region it served, the Orient. To entice travelers to fly Cathay Pacific, the airline enticed them to explore not just the airline's home base of Hong Kong but its broader home turf of the Orient. As travelers embarked on their journey to "discover the many faces of the Orient," they began their encounter with this mystical region with the comfort of Western amenities and in the company of the locals who came equipped with an understanding of their needs.

Aviation facilitates a formidable escape from the local as it transports travelers and connects them to an ever-broadening horizon. Yet all flights are grounded at their points of departure and arrival. As Cathay Pacific and other airlines extended their aerial reach for the expanding privileged class, they profited from, and operated with, aspiring worldviews rooted in the transnational particularities of Hong Kong.[106]

105. Cathay Pacific modeled its advertising language after Pan Am, which in November 1965 claimed that the airline's stewardesses came "from 39 lands and speak 43 languages—at last count" (Pan Am, Series 13: Technical Operations, 1919–1991, Sub-Series 9: Training and Education, 1937–1989, Box 2, Folder 1 392 21). Like Pan Am at the time, which US government regulations had limited to international routes and excluded from domestic traffic, Cathay Pacific, whose tiny home turf of Hong Kong precluded any domestic operations, had to capitalize on the appeal of its international connections. However, unlike Pan Am, which could sell the appeal of its foreign connections to US domestic customers, Cathay Pacific advertised to the international market outside of Hong Kong its ability to provide connections throughout "the Orient" once the visitor from the West reached this mystical region. In contrast to Cathay Pacific, Singapore Airlines doubled down on its local image. The airline focused on hiring Singaporeans as cabin crew, also stating in 1977 its intention to not recruit multinationally: "'Singapore Girl' who has helped make Singapore Airlines a great way to fly, will remain Singaporean, come what may" (*Straits Times*, December 13, 1977, 11).

106. This aspect of cosmopolitanism echoes what Appiah called "rooted cosmopolitan," which entails an attachment to one's own home with its cultural particularities while deriving pleasure from the presence of places and people different from one's own (Appiah, "Cosmopolitan Patriots"). In contrast to Appiah, however, my interpretation of "rootedness" is not based on political allegiance.

Cosmopolitanism encompassed geographical reach, assertion of socioeconomic prowess, and claims to cultural sophistication.[107] These elements of cosmopolitanism proved fluid and elastic throughout this period. In the business of aviation, cosmopolitanism proved instrumental and impacted the industry and its participants. The expansion of Cathay Pacific through the early 1970s reveals the process by which colonial powers, together with the colonized, produced a distinct sense of cosmopolitanism emanating from Hong Kong. Together, forces within and beyond the city of Hong Kong established Kai Tak as a cosmopolitan airport that belonged not just to Hong Kong but to the region and the world. As the analysis shows, although emanating from a colony, this special brand of cosmopolitanism did not simply center on the periphery. Instead, its evolving definition drew energy from various regional and global nexuses as Hong Kong became a solid anchor in international airways.[108]

Commercial aviation did begin as a Western invention. Rapid technological developments allowed Cathay Pacific and other participants in Asia to follow closely in the footsteps of the juggernauts in Europe and North America, especially as the Asian economies gained traction. As different airlines populated the air routes, territorial patterns emerged. Reciprocal route arrangements necessitated differentiation among competing airlines, and the female flight crew provided a visible point of reference for airlines to heighten their differences and gain distinction.

Just as the European and North American airlines leveraged the technological legacy of military equipment from World War II, Cathay Pacific owed its beginning to wartime equipment and the opportunities pre-

---

107. Bruce Robbins included this sense of a privileged person that the word "cosmopolitan" evokes in his comparative framework of cosmopolitanisms (Robbins, "Comparative Cosmopolitanisms"). My interest also parallels what Ulf Hannerz has referred to as the elite cosmopolitans who, in a Bourdieuan manner, derive symbolic capital by asserting cosmopolitan knowledge and taste in the game of social and cultural distinction (Hannerz, "Cosmopolitanism").

108. Breckenridge et al. have argued for the appreciation of cosmopolitanisms in plural and warned against the privileging of the Eurocentric forms of the concept (Breckenridge et al., *Cosmopolitanism*). Echoing Arjun Appadurai's exhortation to explore different modes of cultural exchange and expression, this analysis seeks to understand the creativity and the agency of the local even as its development is conditioned by regional and global forces (Appadurai, "Disjuncture and Difference").

sented by the logistical needs of postwar redevelopment. It was no surprise that the airline's flight attendants in its early years donned uniforms modeled after military outfits. As a first step in developing a brand that set it apart from competition, the introduction of the 1962 uniform was a signature move for Cathay Pacific. The design of the outfit was supposed to accentuate local characteristics while echoing international trends in its representation of Cathay Pacific's cosmopolitanism from its home base in Hong Kong. Nevertheless, the ethnically inflected, local Hong Kong–inspired appeal conflicted with the hiring practices of Cathay Pacific, which sought to array its selection of beauties from all the markets it served. This incongruence resulted in the dual system of standard airline uniforms *and* national costumes for the female cabin crew. The emphasis on racial differences also entailed complications from the predominant Chinese ethnicity that Hong Kong shared with other cities in the region.

By 1969, Cathay Pacific had redirected its sartorial emphasis on cosmopolitanism from ethnically motivated uniform design to the representation of broader themes in the industry. Not only was the "mini" fashionably appropriate for the era, but the uniform, through its choice of material and its stress on overcoming logistical challenges, also fittingly accentuated the jet-setting lifestyle in the technologically advanced space age. Distinctive local identity was expressed in the choice of firecracker red as much as in the manufacturing process. This generation of uniform was made in Hong Kong (or more specifically, by Lai Wah of Kowloon), as compared to its 1962 predecessor, which showcased "the hand-detailed workmanship of the famous Shanghai tailors."[109] This description highlighted not only the process of industrial development in Hong Kong but also the crystallization of a local identity that blended together the various groups of Chinese migrants who called Hong Kong home. Aptly representing Cathay Pacific's home base as the 1970s began, the uniform combined foreign design, imported (and technologically enabled) materials, and local manufacturing.

The cosmopolitan appeal of the Cathay Pacific uniform was no longer the strained presentation of discrete ethnicities but the articulation of a special amalgamation of global elements available in Hong Kong.

109. Swire HK Archive, CPA/7/4/1/1/45 Newsletter, September 14, 1962.

146   *Branding Hong Kong*

The end product had yet to become a creation of an entirely localized process from conceptualization, materialization, and representation. From its introduction in 1969, the uniform remained a celebration of the many local, regional, and international ingredients that conspired to formulate cosmopolitan Hong Kong. The reliance on French designers in this and the next two generations of Cathay Pacific uniforms highlighted not just Hong Kong's aspiration toward high culture as dictated by the West but equally importantly the financial wherewithal the city was developing to consume such symbols of sophistication.

From the *cheongsam*-inspired ensemble to the designer-brand suits of the later generations, the Cathay Pacific uniform also signaled Hong Kong's transformation from a Chinese migrant city in search of an economic agenda to a cosmopolitan nexus for international business. Not unlike the attire of the population at large, the Cathay Pacific uniform assumed a form that would not look out of place in any city in the West. Instead of the calculated appeal of "Oriental" beauty that the 1962 uniform exuded, the suits that constituted Cathay Pacific's most ostensible representation to the world conveyed a clear message that the airline, and Hong Kong, meant business—not just business but business with designer names.

Subsequent rounds of brand-name-designed uniforms in the 1970s and beyond pointed to a convergence on modernity as defined by European standards. By contrast, the uniforms of 1962 and 1969 showcased a search for a cosmopolitan look for the cabin crew of Cathay Pacific that manifested the local particularities of Hong Kong and business requirements of the airline. As the advertisement shows, as late as 1971, the various cosmopolitan elements the airline wished to present remained discrete, sitting somewhat awkwardly alongside one another. The juxtaposition of these elements, before their subordination into a coherent presentation that fit the narrative of global convergence, underscores the endurance of local factors.

Reflecting regional and global trends, Cathay Pacific's uniforms for its female cabin crew underwent makeovers that illustrated the demilitarization of postwar reconstruction, the search for ethnic identities in Asian decolonization, the industrialization of select Asian cities at the height of the Cold War, and the rise of a middle class that aspired to European culture. Accentuating Hong Kong particularities, the flag carrier's uniforms revealed its ambitions for its positioning in greater China, its transformations into a manufacturing powerhouse by leveraging technological breakthroughs, and its emergence as an international business hub. The process did not come to a conclusion there. As the city encountered geopolitical challenges towards the end of the British rule, Cathay Pacific continued to refashion its cosmopolitan looks just as Hong Kong refined its positioning in the region and the world.

CHAPTER 4

Upgrading Hong Kong

The Colony Takes Flight

I can see no prospect of being able to earmark sufficient funds from our own reserves (depleted by sterling's devaluation) without some assistance from HMG.... Our own airline [Cathay Pacific] can manage for years ahead with present facilities; and a case can be made out for being content with regional services rather than intercontinental ones, the more particularly as our interests are especially subordinated to BOAC's in the field.

—Hong Kong Financial Secretary John Cowperthwaite to the Commonwealth Office on funding the upgrade of Kai Tak, July 15, 1968[1]

As the 1960s drew to a close, economic expansion and demands for technological improvement strained the aviation infrastructure of Hong Kong. Dwellers in the colony tolerated the intensifying noise from aircraft movement while at the same time celebrating the city's economic takeoff that these aerial disruptions represented. At the level of policy-making, colonial officials found themselves in a new position of representing an overperforming colonial outpost. Although the colony had relied on British government funding for its aviation development in the aftermath of World War II, commercial aviation in Hong Kong had since experienced a meteoric rise as the city developed into a manufacturing powerhouse and built a burgeoning economy. Keenly aware of the rise of the colony, the British government was eager to extract benefits from one of its last remaining imperial strongholds. At the same time, technological improvements in commercial aviation generated for the budding air

---

1. TNA, BT 245/1703.

hub both excitement in the potential to catch up with the world's leaders and anxiety about lagging behind should the colony fail to finance the upgrade.

Despite consistent double-digit growth in passenger miles in the postwar period, the commercial aviation industry continued to be dominated by North American airlines, which claimed market shares of around 60 percent well into the late 1950s. The market in South and East Asia delivered promising if erratic growth, especially after the reemergence of Japanese participants in 1952.[2] From this humble beginning, Hong Kong, alongside the budding economies in East and Southeast Asia, expanded at a phenomenal rate, necessitating investment in aviation infrastructure for sustained development.

Between 1960 and 1975, the civil aviation industry experienced transformative growth on a global scale. In Hong Kong, although geopolitics rerouted air traffic from time to time, the air hub at Kai Tak handled passenger traffic that doubled every four to six years during this period (fig. 4.1). In spite of the construction that was completed in the late 1950s, the capacity and facilities of Kai Tak were stretched to their limits even as the airport made continuous improvements in response to the increased traffic and escalating technological demands.

In addition to passenger traffic, the colony experienced phenomenal, albeit at times erratic, economic growth. Redirecting its activities as an entrepôt, Hong Kong ushered in an era of export-led economic growth.[3] Its GDP (gross domestic product) registered healthy growth from the late 1960s, soaring from HK$13.4 billion in 1968 to HK$35.3 billion in 1974. In line with the overall GDP increase, export value registered breathtaking growth in most years during this period. Reflecting Cold War dynamics, the United States was the biggest buyer of Hong Kong's exports, followed by the United Kingdom and the ascending trade partners West Germany and Japan.[4] Air cargo assumed the critical role of transporting high-value export items and demonstrated consistent expansion even in the face of global crisis (fig. 4.2). Languishing at depressed levels throughout the 1950s after a precipitous fall in 1950, air cargo did not play

2. Davies, *History of the World's Airlines*, table 52.
3. Koo, "Role of Export Expansion."
4. *Hong Kong Annual Digest of Statistics, 1978 Edition*, 76, 91.

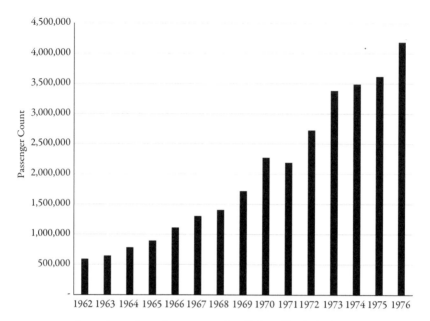

FIGURE 4.1. Annual number of passengers handled at Kai Tak Airport, 1962–76. Sources: *Hong Kong Statistics, 1947–1967*, 122; *Hong Kong Annual Digest of Statistics, 1978 Edition*, 113.

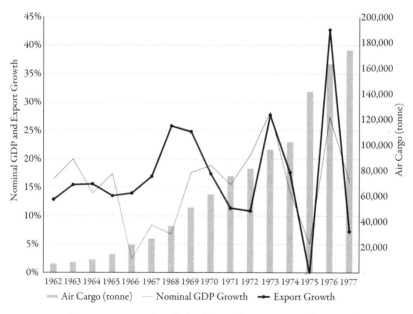

FIGURE 4.2. Air cargo tonnage handled in Hong Kong, 1962–77. Sources: Hong Kong Census and Statistics Department, Table 030: Gross Domestic Product (GDP); *Hong Kong Statistics, 1947–1967*, 88, 122; *Hong Kong Annual Digest of Statistics, 1978 Edition*, 73, 91, 114.

The Colony Takes Flight

a significant role in civil aviation in Hong Kong. Returning to its previous high by 1962 and logging sustained growth, cargo finally became a notable contributor to the airline business in Hong Kong.[5] For Cathay Pacific, revenues from freight and cargo came to exceed 10 percent of the total in 1973.[6]

Sustained sponsorship by a persistent colonial government underwrote the continued growth of Kai Tak as a hub for commercial aviation. This understanding stands in sharp contrast with the conventional notion of Hong Kong's economy being laissez-faire. The contradictory stances of the British and colonial governments also provide a more nuanced picture of state intervention in the context of a decolonizing Asia, parts of which were undergoing phenomenal growth. Geopolitical shifts engendered economic opportunities for Hong Kong. As the metropole and the colony were not marching to the same economic drumbeat, Hong Kong's continued development as an aviation hub, along with its economic takeoff, was miraculous not merely for its magnitude of growth but also for the want of support in the metropole at critical junctures.

This next phase of Kai Tak's expansion endured years of power struggle between Hong Kong and London, witnessed the turmoil of the global economy, and had to respond to technological breakthroughs that transformed the calculations of commercial aviation.

Accelerated Traffic Growth

The expansion of commercial aviation intensified throughout the 1960s. Although commercial aviation continued to be cost prohibitive for the majority of Hong Kong residents, intensifying traffic flows became increasingly integrated into their everyday lives. Not only did the more frequent landings and takeoffs impinge on the senses of those living close to Kai Tak and along flight routes over the city's sky, but aspiring flyers also came to satisfy their curiosity of this novel mode of transportation

5. *HKDCA*, 1963–1964, 33.
6. Swire HK Archive, Cathay Pacific Airways Limited Report of the Directors and Statement of Accounts for the year ended 30th June 1973.

Upgrading Hong Kong

152

with a visit to the observation platform at Kai Tak. Between 1964 and 1965, the year for which the colonial government presented the first report on traffic to the platform, 893,835 passed through the turnstiles and watched, in close range on the observation platform, airplanes arriving at and departing from Kai Tak. That was an average of almost 2,500 visitors a day, an impressive attendance figure for a population that did not reach 3.7 million until 1966.[7] The daily average grew 13 percent and 16 percent in the following two years.[8] For industrializing and modernizing Hong Kong, the race to the skies was not as remote a dream as it was immediately after the war.

The intensifying flow in the sky was spectacular indeed. Not only was Hong Kong registering phenomenal growth in air passenger traffic, but air cargo also demonstrated tremendous potential by the early 1960s. Cargo traffic, however, was asymmetrically distributed with outbound freight from Hong Kong dwarfing inbound flows. For the year ending March 1965, air cargo registered a 26 percent increase compared with the previous year; it was the fourth straight year of double-digit growth. While inbound and outbound cargo growth showed comparable year-over-year increase, the inbound volume of 3.1 metric tons paled in comparison with the 7.7 metric tons of outbound volume.[9] In both directions, the British dominated cargo traffic through BOAC and Cathay Pacific, but North American representation quickly emerged as a formidable rival.[10] As cargo traffic surged over 42 percent from 1965 to 1966, the phenomenal growth caught the attention of the industry, and the *Aeroplane and Commercial Aviation News* reported in March 1966 that Kai Tak "experienced a huge boom in freight."[11]

Explosive growth in cargo volume, along with the smaller yet respectable increase in passenger traffic (15 percent), posed considerable problems for the capacity of the airport and the related facilities. The colonial authorities attempted to cope with the unanticipated but welcome traffic

---

7. *HKDCA*, 1964–1965, 11; Barnett, *Population Projections*, 1.
8. *HKDCA*, 1965–1966, 11; *HKDCA*, 1966–67, 11.
9. *HKDCA*, 1964–1965, 23, 31.
10. *HKDCA*, 1964–1965, 63–66.
11. TNA, CO937/581; HKPRO, HKMS189-1-225 (duplicated from TNA, FCO 40/360).

increase with modifications to the airport and reported that it had become "apparent that major development of the building will shortly be required if it is to keep pace with the rapidly increasing demand." Traffic volume was not the only concern either. Aviation technology was witnessing groundbreaking enhancements, and Hong Kong had to upgrade its airport facilities "in the light of the latest information becoming available on the characteristics of the 'Jumbo' and supersonic aircraft." The upgrades required included not only the extension of the runway but also the development of an air traffic control system "in anticipation of operations by these aircraft within the next five years," reported the colonial authorities in 1966.[12] "Hongkong needs master plan for jumbo jets," read a newspaper headline as it reported the warning by an international aviation expert for the need to update Hong Kong's infrastructure for new technologies lest it "be bypassed and lose considerable tourism revenue."[13] "H.K. 'must have longer runway in 1970'," said another news article."[14]

Hong Kong was indeed behind in runway construction. In 1961, at a cost of US$10 million, Manila had extended its runway from 7,500 to 11,000 feet for the latest jet models.[15] The local media in Hong Kong was abuzz with talks of plans for a runway extension and general upgrade at Kai Tak to cope with traffic growth and to keep pace with technological advancements.[16] The competition was on—one newspaper was quick to compare Kai Tak's not-so-old 8,350-foot runway, which paled in comparison with the 11,400-foot one in Kuala Lumpur, "Far East's best."[17]

At the air hub of Hong Kong, the race was on for the United States and Japan, both in terms of capacity and technology improvement. In the long-haul segment, North American carriers extended their penetration into the Hong Kong market. From 1966 to 1967, Trans World Airlines

12.   *HKDCA*, 1965–1966, 1. In the same year, the Legislative Council discussed the benefits to tourism "with the advent of jumbo and supersonic jets" (LegCo, December 21, 1966, 454).

13.   *SCMP*, November 10, 1966, 32.

14.   *SCMP*, November 17, 1966, 7.

15.   *WKYP*, June 20, 1961, 15.

16.   *SCMP*, August 5, 1965, 7; *SCMP*, September 3, 1965, 6; *SCMP*, November 6, 1965, 1; *KSEN*, September 24, 1965, 3; *KSMN*, September 25, 1965, 10; *TKP*, October 7, 1965, 4; *TKP*, November 7, 1965, 4; *KSMN*, November 7, 1965, 5.

17.   *KSEN*, December 12, 1965, 3.

154 *Upgrading Hong Kong*

and Northwest Orient Airlines joined Pan Am in connecting Hong Kong to the United States, with the former routing through the Middle East and Europe and the latter via Japan and the Pacific.[18] Keeping pace with its North American rivals, Japan expanded its network. With its commencement of round-the-world service on March 7, 1967, Japan Air Lines was firmly installed as one of the three "world's major airlines" servicing the port of call of Hong Kong. Remarkably, Japan Air Lines' penetration into the air cargo market in Hong Kong also developed so rapidly that by 1967, the newcomer had outpaced BOAC, Cathay Pacific, and Pan Am in total volume.[19]

Traffic growth at Kai Tak was therefore multifaceted. Complementing consistent double-digit growth in passenger count since 1952, the meteoric rise in freight volume in the 1960s indicative of Hong Kong's surging manufacturing power fueled traffic at Kai Tak.[20] Air routes connecting Hong Kong to the region and beyond shifted to reflect the changing needs of the network to which Hong Kong connected politically and economically.

The colonial government noted the "dramatic increase in the growth of air travel throughout the world" but was quick to note also that Hong Kong had benefited from a higher-than-average share of the growth in the Far East in all aspects—aircraft movement (24 percent growth from 1966 to 1967), passenger traffic (26 percent), and air cargo volume (51.5 percent). For "the unprecedented growth in the use of air transport for the carriage of freight," the local government credited demand to "locally manufactured products such as textiles, plastics and electronic equipment."[21] George Ross, trading company taipan and the representative for the Hong Kong General Chamber of Commerce in the Legislative Council, spoke of the Hong Kong airport being "over-crowded at peak periods" and called for expansion and upgrade.[22] The 1962 Hong

18. *HKDCA*, 1966–1967, 16.

19. *HKDCA*, 1966–1967, 42–45. Japan was also a critical market for Cathay Pacific as it represented more than half of the carrier's earnings in certain periods in 1966 (TNA, BT 245/1060).

20. *HKDCA*, 1966–1967, 27.

21. *HKDCA*, 1966–1967, 1; HKPRO, HKMS189-1-225 (duplicated from TNA, FCO 40/360).

22. LegCo, March 15, 1967, 176.

FIGURE 4.3. Kai Tak airport terminal. Source: *Hong Kong Yearbook 1962* (Hong Kong: Government Press, 1963).

Kong Yearbook featured a photograph of crowds around the check-in counters at Kai Tak (fig. 4.3) that could well have been confused with a scene of the hustle and bustle of a wet market in the city, except that the people pictured were more impressively attired. Against this backdrop of aviation development and industry growth on a global scale and the rising demands for Hong Kong products to be shipped by planes, the colonial authorities had to plan for improvements and modifications to the aviation facilities, cautioning that "quite naturally the enormous cost of developments" was not to be "overlooked at the planning stage."[23]

Compared with the previous round of discussions on airport development, the colony's appeal for facility enhancements, which began in 1966, focused squarely on the existing airport location. The colonial government formed a "Terminal Building and Terminal Area Planning Committee" to investigate "Kai Tak Terminal Building Extension." In addition, the colonial authorities secured financial approval for "Kai Tak Airport Runway Extension—Feasibility and Cost Survey" with a particular view

23. *HKDCA*, 1966–1967, 1.

156 *Upgrading Hong Kong*

toward the requirements of supersonic and "jumbo" jets.[24] Probably taking a cue from the experience after the Communist takeover of the mainland, Hong Kong's colonial government operating at the height of the Cold War confined the infrastructure project to the more defensible aerial space over the colony. In such tight confines, the colonial administration had to expand its aviation infrastructure to usher Hong Kong into the age of the jumbo jet.

Who Pays for Growth?

Discussions on the financing of the airport expansion project started in the mid-1960s. In October 1965, the British civil aviation representative (Far East) stationed in Hong Kong wrote to the Ministry of Aviation in London, and reported that the Hong Kong governor David Trench had raised the question of extending the runway "to a little over 10,000." The proposal for Hong Kong, according to this civil aviation representative, should allow the resulting runway to not interfere with shipping, but the issue remained subject to the deliberation of those involved in managing the harbor. Trench was convinced of the necessity of the runway and wished to begin construction as quickly as possible. Although he could not be precise, the governor "mentioned the figure of £1.6 million for this runway extension." Apparently, Trench had conceded that the extension would serve Hong Kong's interest primarily as it would allow the colony to maintain its function "as a major trunk-stop (thus bringing in tourists both for stretched versions of current subsonic aircraft and for later supersonics)." However, the governor also found it "impracticable" for the colony to borrow funds for its "uncertain political future." He recalled that the British government had granted the colony an interest-free loan for the first runway extension project and appeared to be "more than satisfied with a further interest-free loan" for the new extension.[25] The

24. *HKDCA*, 1966–1967, 12–13.

25. TNA, BT 245/1703; HKPRO, HKRS1764-1-5, HKRS1689-1-201. Cambridge-educated, Trench joined the Colonial Service in 1938 as a cadet. After a stint in the military during World War II, Trench served for a decade in Hong Kong in areas of defense, finance, and labor, culminating in an appointment as deputy colonial secretary

public in Hong Kong was quick to suggest that it was appropriate to seek British financial support.[26]

In the next couple of years, the project took more concrete shape. Estimates of the construction costs, which remained vague, had risen, and the colonial government seemed to have changed the form of subsidy it requested from London. To finance the project, Trench wrote to the authorities in London. In a letter dated March 10, 1967, Trench detailed for Secretary of State for Commonwealth Affairs Herbert Bowden the rationale and costs of the project. Trench rehearsed growth statistics of aircraft movements, passengers, and air cargo in the previous ten years (with annual rates of 12, 21.5, and 19 percent, respectively). These trends, Trench noted, were expected to continue into the ensuing decade. Even with the expansion project, which spanned from 1956 to 1962, Kai Tak was becoming overloaded. The colonial government had thus prepared for the airport's further development. In particular, the governor underscored the necessity of a runway extension. An extended runway was necessary not only to accommodate "long-range" jets, the "Jumbo" aircraft, "stretched" versions of existing airliners, and the touted Concorde, but also to provide a more adequate margin of safety for aircraft operations in adverse weather conditions, argued Trench.[27] The governor was not alone in his plans to accommodate the latest in aviation technologies. The Legislative Council noted that new aircraft would "soon out-date the runway": "The stretched DC-8 will be here in 1968, the 'Jumbo' Jets in 1969 or 1970, the Concord [sic] in 1971 and probably supersonic aircraft by 1973."[28] The local media also projected an outlook into 1970s and looked forward to welcoming the new aircraft.[29]

---

in 1959. Following his service as high commissioner for the western Pacific, Trench returned to Hong Kong as governor in 1964. He had hoped to work on social issues in Hong Kong but found himself generally thwarted by the bureaucracy (including Financial Secretary Cowperthwaite [as discussed in this chapter]) and the business elite. Yet, under his governorship (1964–71), the economy of Hong Kong grew more than 50 percent in real terms (Goodstadt, "Trench").

26. *SCMP*, November 16, 1966, 1; *WKYP*, November 17, 1966, 5; *KSMN*, November 17, 1966, 6.

27. TNA, FCO 14/75.

28. LegCo, March 15, 1967, 176.

29. *WKYP*, December 28, 1966, 5; *WKYP*, March 17, 1967, 13; *WKYP*, May 4, 1967, 4.

The governor could not provide precise cost estimates for all the construction work at that early stage but indicated that a 2,500-foot extension of the runway would likely require £3 million (HK$48 million) and the rest of the facilities upgrade, another £7 million.[30] Trench concurred with the British government that Kai Tak, like other airports in the United Kingdom, should pay for itself and not receive any subsidy from the Hong Kong taxpayer. The colonial government expected the airport to break even on operations in 1968 and 1969, yet a substantial accumulated deficit of HK$26 million remained. To justify the investment on the basis of its business prospects, Hong Kong officials prepared a "profitability forecast."[31]

The colonial government claimed that it could not afford to meet the capital expenditure of this massive project as it was expecting to narrow its deficit in 1966 and 1967 with only the additional taxation that had recently been introduced. Recalling that the British government had granted a £3 million interest-free loan to Hong Kong for the last Kai Tak expansion,[32] of which £1.2 million had been repaid, Trench asserted, "It would appear to be equitable that HMG should make a substantial contribution by way of an outright grant to the cost of this development."[33] George Ross argued in the Legislative Council in 1967, "That Hong Kong be up-to-date in this field is of equal importance both to us and to Britain who controls our traffic rights; and there may well be a case for seeking financial aid from the British Government." Acknowledging that it was not clear whether the Hong Kong authorities should push to extend the runway "some 1,800 feet to provide for the immediate future, or 2,500 feet to meet all known contingencies," Ross suggested bringing pressure to bear on London nonetheless by engaging with the issue of landing rights.[34] At this stage of the discussion, not only were the cost estimates far from certain, but the form of British contribution to be requested also varied from situation to situation.

---

30. Before the devaluation on November 23, 1967, the official rate for conversion to pound sterling remained at HK$16 to £1. The rate fell to HK$14.5 to £1 after the devaluation (Schenk, "Empire Strikes Back," 570).

31. TNA, FCO 14/75.

32. HKPRO, HKRS1764-1-5; HKPRO, HKRS931-6-189.

33. TNA, FCO 14/75.

34. LegCo, March 15, 1967, 176.

Table 4.1 Peer comparison of existing and projected runway length

Airport	Existing Length of Runway (ft)	Length Firmly Planned or Under Construction (ft)
Hong Kong (Kai Tak)	8,350	
Tokyo (Haneda)	10,335	11,000
Tokyo (Narita)	-	13,000
Seoul	8,100	
Taipei	8,530	10,000
Manila	11,000	
Djakarta	8,120	11,480
Singapore	9,500	11,000
Bangkok	9,840	
Saigon	10,000	
Colombo	6,013	11,000
Calcutta	8,700	
New Delhi	10,600	
Bombay	10,925	
Kuala Lumpur	11,400	
Brunei	6,299	12,000
Rangoon	8,100	
Phnom Penh	9,842	
Vientiane	6,560	13,120

Source: "Extension of Kai Tak Airport Runway" prepared by the Economic Branch, Colonial Secretariat (HKPRO, HKRS1689-1-202).

In conjunction with the various departments involved, the Economic Branch of the Colonial Secretariat prepared a report on the proposed extension of the runway. The report warned of the "consequences of not extending the runway": Kai Tak would "diminish in importance as a major trunk stop for international airlines"; Hong Kong would be reduced to dependence on "local or regional services for link-ups with major international services"; tourism would be impacted; and the development of the air cargo business hampered. Among the budding Asian airports, the rivalry of runway length was palpable—the Economic Branch prepared a comparison of runway length, "existing" and "firmly planned or under construction" (see table 4.1).[35]

35. HKPRO, HKRS1689-1-202.

Kai Tak was not yet handicapped by its runway length, but the expansion of its peers' was quickly leaving Hong Kong's airport in the dust. The report stated that it would only be reasonable to undertake the project so long as "problems of finance" could be resolved.

The timing was horrible. In November 1967, the UK government devalued the British pound by 14 percent against the US dollar. To add insult to injury, the unfortunate development in London followed large-scale riots in Hong Kong, a spillover from the turmoil in the mainland.[36] The riots had shaken British confidence in Hong Kong's prospect of sustaining phenomenal growth.[37] Precisely at that juncture, Hong Kong was "entering a period of heavy calls for capital financing, at a time when the devaluation of sterling has reduced reserves by about £30 million."[38] In the devaluation, not only did the Hong Kong government suffer a loss of £30 million, but banks in the colony also took an additional hit of £20 to 25 million.[39]

The devaluation of the sterling had come to underscore the betrayal of colonies. This debacle was especially damaging to Hong Kong. Required by the colonial monetary system to hold 100 percent reserves in sterling, Hong Kong, after its successful industrialization, had amassed sterling balances second only to Kuwait. On the one hand, this large balance inflicted tremendous pain on Hong Kong. On the other hand, the colonial government's threat of diversifying Hong Kong's reserve after the debacle exposed British vulnerability and forced the metropole to renegotiate Hong Kong's link to the pound sterling and the British system.[40] In the request for a British loan for the runway project, the damaging impact of the sterling devaluation on the finances of the colony was to become a common refrain.[41]

The project was critical if Kai Tak was to remain "a major aviation centre in S.E. Asia and the Far East," Trench argued. BOAC, the British flag carrier, had long benefited from its traffic rights through the British colony. Just as important were rights to and through Hong Kong, which

36. Cheung, *Hong Kong's Watershed*.
37. Schenk, "Empire Strikes Back," 561.
38. HKPRO, HKRS1689-1-202.
39. Schenk, "Empire Strikes Back," 569.
40. Schenk, "Empire Strikes Back," 552–54.
41. HKPRO, HKRS1689-1-202.

in the negotiations of international air service agreements, had provided "an essential bargaining counter . . . second in importance to none, except possibly London," noted the governor. Astutely he cited as an example concessions made by the US government to the United Kingdom to allow BOAC flights in and through the United States "in return for substantial traffic rights to and through Hong Kong" that the UK government granted to US carriers. "Incidentally," the regional interests of Cathay Pacific, "Hong Kong's own flag-carrier," suffered because of British control of Hong Kong traffic rights. Trench also pointed out the conflicting interests of London and Hong Kong: restrictions of traffic rights to and through Hong Kong by foreign countries to enhance BOAC's interests went against the colony's own interests in traffic maximization. "Numerous examples can be quoted," Trench stressed, "for example, NWA was for many years prior to 1966 excluded from Hong Kong whilst SAS is still excluded to this day."[42]

On a more positive note, the extended runway, along with the associated upgraded facilities, would serve BOAC, a principal user of Kai Tak. Cathay Pacific, whose focus was not in long-haul traffic, might not require the extended facilities quite as readily. It would also be in the best interest of the British government not to allow Hong Kong to be relegated to a second-class airport, wrote Trench. "The value of Hong Kong to airlines as an aviation centre and therefore to BOAC as a bargaining counter would be seriously diminished," Trench argued. "It would remain a useful bargaining counter only in purely regional negotiations." British investment in further upgrading Kai Tak would serve the political purpose of enhancing British prestige in the eyes of the Hong Kong community, Trench contended.[43]

Trench thought it "appropriate and equitable if HMG could now agree in principle to contribute to the runway extension improvements and associated apron and taxiway improvements by way of an outright grant on a dollar-for-dollar basis." He suggested that the colonial government and the British government share equally the costs of these items,

---

42. TNA, FCO 14/75. Public interest in the SAS case endured and prompted an inquiry in 1968 into the division between authorities in the metropole and the colony over traffic rights in Hong Kong (*SCMP*, December 21, 1968, 20; HKPRO, HKRS276-1-1-1).

43. TNA, FCO 14/75.

which translated into a request of £2.25 million from the British government. The colony would still shoulder the majority of the costs. Trench's team in Hong Kong would still need to raise some £7.75 million for their share of the runway apron, taxiway extensions, and for other works.[44]

This discussion over the cost sharing of the Kai Tak project was a major component of an ongoing dispute between the colony and the metropole. In the field of civil aviation, London's control of Hong Kong's landing rights was a significant issue. In 1967, the bureaucracy in the metropole presented a defense for its policy. The British government was responsible for matters of international air traffic rights concerning Hong Kong and under the terms of the Convention on International Civil Aviation in Chicago, was answerable internationally in its discharge of these responsibilities. The British government claimed that it gave full weight to Hong Kong interests, which were "in no way treated as generically inferior to the other constituents of the ensemble of British interests." In their internal exchange, the London officials conceded that the British government handled civil aviation in Hong Kong "as with other British places, as a responsibility as well as an asset." Cathay Pacific received its designated routes in accordance with the bilateral air service agreements made by the British government. The officials in the metropole disagreed with the argument that the British government profited from such agreements at the expense of Hong Kong, dismissing any possibility of Hong Kong acting on its own: "Hong Kong is not a nation and does not itself function internationally." Its "existing territorial status" was only tolerated by China for its economic function, this 1967 document argued. "If Hong Kong were to strike out for independent nationhood it would presumably promptly cease to exist." In the absence (and impossibility) of a national government, Hong Kong could not have negotiated for itself its own international air service agreements. In a detailed commentary on Hong Kong, an officer in the division for aviation overseas policy in the Board of Trade asserted that "to talk of a national airline and support as with any other Government is to talk in terms of an independent national state. A priori this language is irrelevant."[45] From London's perspective, Hong Kong was reliant on the British government

44. TNA, FCO 14/75.
45. TNA, BT245/1402.

and should not attempt to exact from London a larger share of gains in the aviation business.

The British government was anxious to assert its control over civil aviation in Hong Kong. In January 1967, the head of the Aviation and Telecommunications Department in the Commonwealth Office expressed concerns on the part of the Board of Trade "about the way the Hong Kong Air Transport Licensing Authority were liable to interpret their powers under the Hong Kong Air Transport Licensing Regulations." The Board of Trade wanted to see that scheduled services by foreign airlines in Hong Kong receive their permission "under the Air Navigation Order in the United Kingdom" and "if it would make it more acceptable to Hong Kong, the permits might be issued by the Governor on the basis of decisions taken in London."[46]

In the following year, negotiations over the airport expansion project continued. In February 1968, in a dispatch to the Civil Aviation Department at the Board of Trade, the Commonwealth Office reported that the Hong Kong governor requested a grant towards the costs of developing Kai Tak, which the Commonwealth Office considered "essential" if Hong Kong was to remain prominent in aviation. Hong Kong had proposed the project to further the colony's "own economic interest," but the governor also noted that the maintenance of a major international airport in Hong Kong was "also substantially in the economic interest of British civil aviation." The Commonwealth Office did not believe that the colony could afford the entire cost of the project and considered it "entirely reasonable that the United Kingdom should pay something towards the cost of maintaining and increasing the value of this considerable asset to United Kingdom aviation interests." The Board of Trade was therefore requested to consider extending a £4 million grant for the project, which was then projected to cost £12.5 million lest the facilities at Hong Kong should fall below the necessary standards for foreign operators.[47]

In a letter dated July 15, 1968, Hong Kong's Financial Secretary John Cowperthwaite wrote to the Commonwealth Office stating plainly, "Let me say straight away that what we are now seeking is a loan, on terms to

46. TNA, FCO 14/35.
47. TNA, BT 245/1703.

be negotiated, for at least half the cost of the extension." He appealed to the authorities in London to help equip Kai Tak to satisfy the needs of passenger and cargo traffic in the 1970s. The colonial coffers would not suffice, partly because of the pain London inflicted: "I can see no prospect of being able to earmark sufficient funds from our own reserves (depleted by sterling's devaluation) without some assistance from HMG." Equally importantly, the extended runway would benefit BOAC disproportionately: "Our own airline [Cathay Pacific] can manage for years ahead with present facilities; and a case can be made out for being content with regional services rather than inter-continental ones, the more particularly as our interests are especially subordinated to BOAC's in the field." To substantiate his case, he presented a "profitability forecast" that extended into 1984.[48]

The focus then shifted away from Cathay Pacific as Hong Kong officials argued over the basis of British responsibilities: the British government was responsible for Hong Kong civil aviation matters and should honor the obligation of Kai Tak maintenance "at a standard capable of meeting reasonable requirements in accordance with [International Civil Aviation Organization] Regional Plans." Also highlighted was Hong Kong's primacy in the entire British framework. The British government understood that the only weapon at its disposal "to hold and gain British airline rights abroad" was "the value of the rights it controls and can bargain with." By the late 1960s, "next to London, Hong Kong is at present the most valuable international traffic point remaining within United Kingdom responsibility." The colony was important not only for the revenue BOAC derived for its services to and from Hong Kong but more so "through the important concessions in traffic rights secured for British airlines in exchange for traffic rights at Hong Kong granted to airlines of other countries in air service negotiations."[49]

Such discussions were not confined to official negotiations, and the media reported on the delay in London's response. In August 1968, the *Far Eastern Economic Review* carried an article titled "Hong Kong, A Valuable Pawn." The reporter's stance differed from the colonial authorities that pledged financial hardship. "Money is not really the question for this

48. TNA, BT 245/1703.
49. TNA, FCO 14/426; TNA, BT 245/1372.

wealthy Colony," he wrote. The real issue lay in the mounting conflict between the colony and the metropole: "It is becoming increasingly apparent here, what is good for Britain is not necessarily good for Hongkong." This conflict became evident in the dispute over "highly desirable landing rights in Hongkong," on the basis of which the colonial authorities petitioned for London's contribution to work at Kai Tak. On this issue of landing rights, the author noted, "In a vastly diminished British Empire, Hongkong is one of the few pawns left to London with which to obtain reciprocal privileges."[50]

Yet, there was no commitment from London to help with the financing of the project, which remained on hold despite repeated media coverage.[51] On August 23, 1968, the financial secretary reported to the Legislative Council, "The possible extension of the runway is not in the Public Works programme at the moment; certain further information has been requested and furnished to Her Majesty's Government in London and we are hopeful, if not yet confident, that some financial assistance will be forthcoming if we decide to lengthen it."[52]

Making the Most of Its Colonial Holdings

Their denial in public and in their exchange with the colonial authorities notwithstanding, officials in the metropole were keenly aware of the value of air traffic rights at Hong Kong. The Dutch had been asking to establish an aviation foothold in Hong Kong since 1955, but the British had not identified something of comparable value in return from the Dutch for that exchange. In 1967, the Netherlands pressed again for traffic rights at Hong Kong for KLM.[53]

Commonwealth Office officials claimed that their aim was to develop air services to Hong Kong, but that did "not necessitate an 'open skies'

---

50. Polsky, "Hong Kong," 411–12.

51. *SCMP*, March 30, 1968, 9; *SCMP*, August 26, 1968, 8; *SCMP*, September 4, 1968, 9; *SCMP*, October 25, 1968, 6; *KSMN*, May 27, 1968, 4; *KSEN*, July 9, 1968, 4; *WKYP*, August 17, 1968, 2; *KSMN*, September 4, 1968, 6; *KSEN*, November 10, 1968, 4.

52. LegCo, August 23, 1968, 386–87.

53. TNA, FCO 14/583.

policy." Such development had to be "selective and orderly." Otherwise "unrestricted rights for all and sundry at Hong Kong" would only "reduce its attractiveness as a traffic-centre to most." The process of negotiating air connections, according to these officials in London, followed the "long established practice in most if not all countries to exchange traffic rights only for traffic rights." Although the British government was "always concerned to assist Hong Kong's growth," it would not grant traffic rights "in order solely to assist the Colony's economy," a Board of Trade official admitted.[54]

London officials' claims that they at least took Hong Kong's interests into consideration do not always stand up to scrutiny. In their negotiations with the Dutch government over KLM's traffic rights in Hong Kong, not only did the London officials put aside the practice of exchanging traffic rights for traffic rights only but they also calculated the desirability of the exchange not for the benefit of the colony but for British interests. In his attempt to structure a deal in October 1968, British Minister of State Fred Mulley suggested trading those rights for a Dutch agreement to a £75 million purchase of Chieftain tanks from Britain. The Ministry of Defense wrote to the Board of Trade to support Mulley's proposal.[55] In their response, the Board of Trade acknowledged that "overseas air services are large earners of foreign exchange and depend entirely on traffic rights which we control or can secure in negotiations with other countries" and that Hong Kong was "a vital point . . . for both these purposes." In the end, the Board of Trade disagreed with the proposed exchange, partly to not open themselves to further attempts to exchange export agreements with traffic rights, and more poignantly for the financial disadvantage of the proposed deal. "The net advantage to the balance of payments from the bargain you propose is, in any case, not nearly so large as might at first appear," the Board of Trade wrote in its reply to Mulley. "One frequency a week between Europe and Hong Kong is worth up to half a million pounds," they calculated, and that "would be either directly or indirectly at the expense of the UK." Over the duration of the proposed deal, that would add up to "a large sum to set off against the thirty seven and a half million pounds which might remain from the

54. TNA, FCO 14/397.
55. TNA, BT 245/1802; TNA, T 317/1229.

Chieftain contract."[56] In a separate exchange, BOAC calculated for the Board of Trade that "each frequency operated by a foreign trunk-line operator through Hong Kong would cause us a revenue loss of approximately £200,000 p.a.," and acknowledged the importance of Hong Kong in the British negotiation with the United States over rights in the South Pacific.[57] The value of each traffic right to Hong Kong (and perhaps elsewhere) was carefully calibrated and applied in the calculation of the British grand plan. The colony's interests did not figure prominently in British consideration.

Regardless of British calculations, the Dutch decided against the British Chieftain tanks in favor of Leopard tanks from the Germans. Nonetheless, the British government remained on the lookout for "the necessary non-aviation element" to structure a "potentially suitable *quid pro quo*."[58] In October 1967, London officials deliberated on making "a political gesture" to the Netherlands "at a time of [British] approach to the Common Market." In February 1968, they also thought traffic rights at Hong Kong could "put pressure on KLM to commit themselves to the air bus." The Board of Trade had wished to adhere to the principle of exchanging traffic rights for traffic rights. At the same time, the board "agreed to its breach only under heavy pressure and only as a means of promoting important aircraft sales."[59]

Of the various issues considered, British entry into the European Common Market (European Economic Community [EEC]) proved grave enough a concern to warrant an exception. In May 1969, the British embassy in the Hague wrote to the Foreign and Commonwealth Office, flagging for London's consideration the importance of Dutch support for British entry into the European Common Market.[60] The Embassy sent another dispatch in October 1969 to the Aviation, Marine and Telecommunications Department of the Foreign and Commonwealth Office, indicating that the issue of KLM landing rights at Hong Kong had "got under the skin of the Dutch. He suggested caution in

56. TNA, BT 245/1818; TNA, FCO 14/407.
57. TNA, BT 245/1802.
58. TNA, FCO 14/583.
59. TNA, FCO 14/244.
60. TNA, FCO 14/585.

negotiations in light of "the importance of close Anglo/Dutch relations in the Common Market and other European contexts" and advised against testing the political will of the Dutch to retaliate.[61] In the end, the British desire for Dutch support in its entry into the European Common Market provided the key to KLM's eventual entry into Hong Kong.

Before the final deal was struck, the various interest groups in the British bureaucracy continued to evaluate possible alternatives to maximize their gains for this concession to the Dutch at Hong Kong. In 1970, British negotiators had tried but failed to strike a deal with the Dutch involving a purchase of the British Swingfire anti-tank guided missile.[62] Ultimately, the final offer to the Dutch did not entail "a civil aviation negotiation in the normal sense" because the British did not ask for reciprocal air traffic rights. Instead, diplomats in London noted in 1971 that "an opportunity has now arisen, in connection with hoped-for Dutch support over certain aspects of the negotiations for HMG's entry into the EEC, which has at last enabled ministers to agree to the grant of the desired traffic rights to KLM." The Dutch would also help with modifying the EEC's fisheries regulations in the favor of the British. British officials in the Hague and London took pains to minimize the chances of other EEC members "link[ing] the KLM decision with enlargement negotiations." The Department of Trade and Industry was reminded that "the Ministers concerned, including the Prime Minister, would be extremely displeased if the effect of our gesture to the Dutch was marred by anything" that would suggest British delay of the new Dutch service to Hong Kong.[63] On November 5, 1971, KLM commenced its weekly service between Hong Kong and the Netherlands.[64]

At certain points in their deliberations over the Dutch application for rights at Hong Kong, British officials alluded to the Netherlands' importance in Hong Kong's trade and development.[65] In one instance, the Foreign and Commonwealth Office discussed in May 1969 the possibil-

---

61. TNA, BT 245/1303; TNA, FCO 14/587.

62. TNA, FCO 14/711; TNA, FCO 14/712.

63. TNA, FCO 14/813, T 317/1642; TNA, PREM 15/692. For an extended discussion of the impact of Britain's negotiations to join the EEC on the development in Hong Kong, see Fellows, "Britain, European Economic Community."

64. *HKDCA*, 1971–1972, 10, 42.

65. TNA, FCO 14/585; TNA, FCO, 14/711.

ity of trading traffic rights for KLM in Hong Kong in exchange for "a Dutch financial contribution to the Kai Tak development project." This discussion ended quickly for fear of the Dutch developing "a more or less permanent stake in Kai Tak," and the danger of this arrangement becoming a precedent for British negotiations with other countries. Such overt protection of British interests at the expense of the colony notwithstanding, the foreign and commonwealth secretary considered the granting of KLM's rights at Hong Kong a win-win: that this arrangement both enhanced the Dutch attitude in British negotiations with the EEC and improved British relations with Hong Kong.[66] Contrary to British perceptions, it is evident that throughout the process, British interests prevailed over colonial concerns. Despite the long history of this Dutch application, KLM finally secured the rights to link Hong Kong up with its flight network only because the British government found a non-aviation opportunity that it deemed an appropriate *quid pro quo* for the concession the metropole granted on behalf of its colony.

KLM's service to Hong Kong was but one example of British calculations of the commercial value of Hong Kong as an aviation hub. To assess comprehensively the value of British civil aviation interests in Hong Kong through BOAC, the Board of Trade prepared a detailed analysis in February 1969. This report assigned a specific value to each component of BOAC's activities, including its indirect participation in the colony's aviation from which it derived significant gains.

Although competition in Hong Kong had intensified as the British government had exchanged rights to Hong Kong for other airlines for rights elsewhere, BOAC's passenger revenues in Hong Kong were second only to New York, Toronto, and Johannesburg. From 1967 to 1968, BOAC earned £7 million in net passenger revenues in and out of Hong Kong. This figure does not include BOAC's shares in the gains of partnership agreements. Hong Kong remained pivotal to its agreements with Air India and Qantas, which enabled BOAC's operations between the United Kingdom and the Far East and Australia. Partners in such agreements received a share of the gains based on the amount of "Third/Fourth Freedom" traffic carried on the routes in question, and Hong Kong was a vital point in the UK network for the calculation of such entitlements. To the

66. TNA, T 317/1642.

value of passenger traffic operated directly or indirectly by BOAC through pooling arrangements, the Board added the value of mail carried by air to and from Hong Kong (valued at £1.5 million) and air cargo (valued at £1.75 million). From 1967 to 1968, the total value of BOAC's business to and from Hong Kong was estimated to be in "the order of £14 million."[67]

Beyond these partnership agreements, Hong Kong provided bargaining rights to the British operator. US carriers' rights at Hong Kong were exchanged for British rights in the United States and from there across the Pacific to Japan, Hong Kong, Australia, and New Zealand. Rights for Qantas at Hong Kong secured Qantas's assistance to BOAC in arranging pooled service beyond the United States to Australia. Hong Kong rights were also procured for BOAC services to New Zealand. Rights at Hong Kong granted to Italy, Germany, and Switzerland allowed BOAC to fly to the Far East and Australia via Rome, Frankfurt, and Zurich. Rights granted to Japan Air Lines to fly to London via Hong Kong were also partially responsible for BOAC rights at Tokyo. Besides, BOAC and its sister airline BUA owed operations beyond East Africa to South and Central Africa to the granting of rights at Hong Kong. Foreign rights in Hong Kong were also instrumental to British passage through India and through Thailand; rights to fly over and beyond Indonesia and Burma; rights beyond Iran; rights from the United Arab Republic to India and beyond; and the favored treatment European airlines offered to BOAC for regional rights, and rights beyond by Malaysia and Singapore. Such a vast array of rights secured for the British was only possible because of the high value of flight connections with Hong Kong. The board estimated the earnings of foreign airlines' scheduled services on their trunk routes to and from Hong Kong to be in excess of £20 million per year before the sterling devaluation.[68]

The Board of Trade assessed the value of British rights abroad secured through an exchange for foreign rights in Hong Kong to be £15 million for the year ending March 1968, in addition to the £14 million derived from its business in the British colony.[69] By its own calculation, the British machinery was earning tens of millions of pounds sterling every year

67. TNA, BT 245/1802.
68. TNA, BT 245/1802.
69. TNA, BT 245/1802.

directly and indirectly from commercial aviation in Hong Kong, excluding British interests in the colony itself. The colony was requesting a loan that was a fraction of those annual gains, partly to respond to capacity demands and partly to remain competitive in the midst of a technological upgrade cycle.

The Colony Knew

This discussion of the value of commercial aviation in Hong Kong to the British government needs to be understood in the context of the growing financial autonomy of colonial Hong Kong. The damage Hong Kong sustained in the sterling devaluation of 1967 and the subsequent compromise between the metropole and the colony laid bare to British officials the mounting financial wherewithal of the colony. In the aftermath of the sterling devaluation, the British government signed an agreement with Hong Kong in September 1968, whereby Hong Kong undertook to maintain a minimum portion of the colony's reserves in sterling for five years in exchange for Britain's provision of a US dollar value guarantee for the bulk of Hong Kong's sterling balance.[70] Hong Kong was no longer the poor colony in need of postwar reconstruction in the 1950s. The colony had grown into a pivotal contender in the financial markets denominated in the currency of the metropole.

On March 14, 1968, Legislator George Ross alluded first to "the $400 million we lost from our reserves as a result of Sterling devaluation" before underscoring the paramount importance of the runway extension to Hong Kong's continued prosperity in the modern air age. To Ross, it was only fair for Britain to share the cost of this infrastructure upgrade:

> Britain should certainly help to foot the bill in exchange for the traffic rights which she continues to control. Hong Kong is her second highest card and she certainly has no case for playing the poor relation over this. We have every right to expect, and in fact we should demand a contribution from Britain of at least 25% of the cost.

70. TNA, FCO 14/711.

Ross advocated that Hong Kong demand payment from London toward the cost and also "retain some control over the traffic rights which affect Hong Kong's own airline Cathay Pacific," as well as over its important trading partners to fly in and out of Hong Kong.[71]

The resistance of the British government to commit funds to this round of Kai Tak expansion was but a reflection of a broader, rising British reluctance to shoulder the economic burden of empire.[72] Discussions of the colony's massive sterling reserves circulated in the media.[73] As officials at the treasury discussed the Hong Kong airport project in March 1969, they referred to recent reports of the per capita income in Hong Kong and noted that "her sterling reserves must be colossal." Officials questioned how the colony would qualify for assistance from London, calling it "nonsense to be thinking of lending to her on aid terms." If London were to lend to Hong Kong at all for the project, "she should be charged our hardest terms." The officials proceeded to suggest that Hong Kong should borrow the funds "in the Colony or elsewhere (perhaps on the Eurodollar market?) on normal terms" on the basis of "her wealth" and the desirability of the airport extension project.[74]

In their discussions of Hong Kong's application for a development loan, London officials situated the debate in the context of Britain's trade deficit with its colony. One official considered it "a useful goodwill gesture" at a time when they had just asked the colony "to exercise restraint on certain of her exports to the UK." In addition, the Board of Trade wished to make sure that "any export opportunities generated with the help of the loan should go to the UK."[75]

Resentment grew among senior officials and unofficial members (members of the colony's Executive Council and Legislative Council who were not employees of the Hong Kong government) in Hong Kong.[76] This group in the colony considered it necessary that the British govern-

---

71. LegCo, March 14, 1968, 115, 118; *WKYP*, March 15, 1968, 11.

72. Schenk, "Empire Strikes Back."

73. TNA, BT 245/1802.

74. TNA, T 317/1229.

75. TNA, FCO 14/601.

76. "Unofficials" were drawn from business and professional elites. They included British expatriates, local Chinese, and members of other ethnicities domiciled in Hong Kong (Ure, *Governors, Politics*, chap. 2).

ment acknowledge its interest in Kai Tak and participate in some measure in the financing of the expansion project. Although those in the colony admitted that Hong Kong could fund the project with its own resources, they agreed that the colony should not divert spending from its expanding social services. Complaining about the two-year delay in a response from London after the colony submitted its application, Cowperthwaite contended that if Hong Kong could assign its own traffic rights, the colony "would have no difficulty for Panam [Pan American] to finance the whole of the extension."[77] A trade commissioner reported that the financial secretary was making "blood curdling noises." Cowperthwaite was said to have threatened that "when HMG's refusal to grant the loan become evident, the Hong Kong Government would adopt an entirely different attitude towards granting traffic rights for airlines in Hong Kong."[78]

Priding himself on being a "Hong Kong chauvinist," Cowperthwaite harbored deep suspicions of the British government's intentions toward Hong Kong. A graduate of the University of Edinburgh, the University of St. Andrews, and Cambridge University, he was first appointed to Hong Kong in 1941. The outbreak of World War II diverted him to Sierra Leone. Upon arrival in Hong Kong, he mastered the colony's problem with its growing manufacturing industry. Before his decade-long appointment to the post of financial secretary in 1961, Cowperthwaite served in the Supplies, Trade and Industry Department and occupied the posts of deputy financial secretary, deputy economic secretary, and acting director of commerce and industry. Through "brinkmanship and outright insubordination," he defied London and in the aftermath of the devaluation of the pound in 1967, secured for the colony "the first guarantees against future devaluation ever conceded to a Sterling Area member."[79] His advocacy of more autonomy for Hong Kong from the metropole in the runway project was but a continuation of his disdain of London authority over the burgeoning power of Hong Kong.

Cowperthwaite's fiery rhetoric notwithstanding, civil aviation officials in London were not moved. Complaining about "the rather motley

---

77. TNA, FCO 14/601.
78. TNA, BT/245/1703.
79. Goodstadt, "Cowperthwaite," 108–10.

174   *Upgrading Hong Kong*

collection of facts and guestimates put forward by Hong Kong," London officials retorted that the British government would consider contributing to the project on the basis of cooperation with Hong Kong rather than its campaign of threat. In a letter dated February 3, 1969, a civil aviation department employee noted, "It is not my job to educate [Cowperthwaite] in realpolitik but the final power on traffic rights rests with HMG and he must surely realise that Hong Kong has no chance to go it alone for more than 48 hrs."[80]

Heated rhetoric notwithstanding, officials in London discussed rationally how to handle Hong Kong's request. In a dispatch from the Board of Trade to the Treasury dated April 15, 1969, an aviation official invoked the "colonial formula" that had evolved from discussion of colonial civil aviation in 1947 in the aftermath of the Convention on International Civil Aviation Organization. The "colonial formula" was used to help determine "in what circumstances and to what extent the United Kingdom should assist colonial territories in the provision of civil aviation facilities—with special reference to trunk route interest." Explaining the sharing mechanism specified in the "colonial formula," the document noted a gloss added at the 1947 conference on colonial civil aviation:

> In assessing these liabilities, account would have to be taken not only of the requirements for local air services and the benefits likely to accrue to the Colonies concerned from the trunk service, including benefits arising from disbursement in the Colony in connection with the trunk service and the creation of local employment. It would also be necessary to take into account the benefits to the United Kingdom and its airline operators of the provision of staging facilities in the Colonies.

The document included a calculation of British (BOAC and other UK airlines, excluding Cathay Pacific) shares at Kai Tak: 7 percent of total civil aircraft movement; 6.3 percent of passenger count; 11 percent of freight in 1967 and 1968. Using "an average of about 8% of all civil aircraft business at Kai-Tak," it arrived at a British share of about £1.096 million of the total cost of the project, estimated to be £13.7 million.[81]

80. TNA, BT 245/1703.
81. TNA, FCO 40/244; the "Colonial Formula" received a brief mention in the discussion of this matter in 1967 (TNA, FCO 14/75).

The decision makers in London remained unimpressed. They understood the "buoyant revenue" of the colony even though they also knew of the expenditures earmarked for the alleviation of social and political pressures in the aftermath of the riots in the colony in 1966 and 1967. London officials rehearsed the benefits that Hong Kong would accrue from the project, and the colonial government repeated its appeal on the basis of British civil aviation interests in Hong Kong. Cowperthwaite wrote to the Foreign and Commonwealth Office in May 1969, making his case not merely from the cost to Hong Kong arising from British restriction of traffic to the colony but also from the British responsibility to protect the residents of Hong Kong, who would face "greater risk to life" if the new aircraft were to call at an inadequately upgraded Kai Tak. London officials seemed not inclined to be persuaded by Cowperthwaite's moral argument. Instead, the Foreign and Commonwealth Office reached out to the Export Credits Guarantee Department and calculated that the British export content of the hardware requirements proposed (runway, terminal building, and associated amenities) would be on the order of £6 million. The cool minds in the metropole insisted on a calculation of cold, hard cash. Despite the frequent exchange between the colony and London, the discussions reached an impasse.[82]

Readying for the Jumbo Jet and Future Growth

Doubts had surfaced in the colony in 1968 that Kai Tak would be ready to receive Jumbo jets by "the advent of the Jumbos in the Pacific area" in early 1970. H. J. C. (John) Browne, an executive at Swire and Cathay Pacific, who served as an unofficial member of the Legislative Council of Hong Kong, asked for an update on the British response at the August 23, 1968 meeting of the Legislative Council. He reminded the council of the safety issue of Kai Tak, which had "one of the shortest runways of any major airport in Asia." He further reminded his fellow councilors of the need of a runway extension if Hong Kong was to receive supersonic aircraft in the 1970s.[83] Along with the discontent in the Legislative Council,

82. TNA, FCO 40/244.
83. LegCo, August 23, 1968, 384.

176   *Upgrading Hong Kong*

the local media also joined in the anxious cries for more immediate action.[84]

The committees charged with the development of Kai Tak had recommended projects that would respond to the anticipated demands on passenger facilities up to 1976, as well as a new air cargo complex.[85] Only with continued infrastructure improvement "fitting into a rational programme of development of the air transport system as a whole" could Hong Kong "maintain its proper position among the airports of South East Asia."[86] The colonial government saw an opportunity for Hong Kong in the context of regional geopolitical developments. The authorities in London had yet to be convinced of the soundness of the investments.

Such discussions extended beyond official debates into the mass media. In an article titled "Missing the Jumbos at Kaitak airport," the *Hongkong Standard* reported on October 7, 1968, that despite the modifications on the terminal building, the Kai Tak runway would not be extended in time "even if work on this important aspect of airport extension was to start now." Nor was it clear that the British government would offer the colony the necessary financial assistance for the project, the article reported. The runway, then 8,350 feet long, was "2,000 feet short of the landing distance of the Boeing 747s," noted the reporter. "Hongkong lags behind," the article concluded. Although Kai Tak was "one of the world's busiest airports," it was falling behind other airports in the region "where plans for the Jumbos jet era are already well in advance." Singapore's Paya Lebar airport was said to be extending its runway to 11,000 feet by the end of 1969. It was possible that the jumbos would just bypass Kai Tak. Hong Kong could not afford "to be left behind in the race for the Jumbo jet traffic," warned the article.[87]

Uncertainty and confusion ensued. The *South China Morning Post* quoted H. E. Atkins, a conservative member of Parliament (MP) in a January 21, 1969 article with the headline "Loan for Airport Runway 'Certain'"; the *Hongkong Standard* reported two days later that the "Kaitak loan [was] unsure," citing another conservative MP, Anthony Royle.[88] In

---

84.   *KSEN*, January 6, 1969, 4; *KSEN*, February 23, 1969, 4.
85.   *HKDCA*, 1967–1968, 12–13.
86.   *HKDCA*, 1968–1969, 2.
87.   *Hongkong Standard*, October 7, 1968, 8.
88.   TNA, BT 245/1703.

The Colony Takes Flight

February, the *South China Morning Post* reported that airlines using Kai Tak expressed discontent over the possibility of being "called upon to carry the entire cost of improvements and extensions to the airport" as recent press reports had suggested.[89] The *Daily Telegraph* reported on "Hong Kong's battle with Whitehall over airport" on March 5, 1969, citing Hong Kong's demand for a British loan on the basis of London's restrictive landing rights at Kai Tak for the benefit of BOAC and at the expense of the colony. "If Kaitak loan fails, squeeze the BOAC?," suggested the *Hongkong Standard* on March 11, 1969.[90] In June 1969, the pro-Beijing *Ta Kung Pao* must have been all too delighted to report on the visit to Hong Kong by Malcolm Shepherd of the Ministry of State for Foreign and Commonwealth Affairs, at the conclusion of which he admitted "the inability of the UK" to extend a loan for the runway-extension project.[91]

The colony registered a mounting level of anxiety as the first landing of the jumbo jet approached. Criticizing the British refusal to grant a loan, the *Hongkong Standard* posted a cartoon in June 1969 mocking a "vertical take off" for the "Jumbo Jet" without the financing of an extended runway. Satire aside, the technical requirements of runway length appeared unsettled. As late as April 1969, BOAC was still reporting to the Ministry of Aviation that the British airline could not justify extension to 11,700 feet, but "the argument for extension to 10,800 seems strong based on the 747 operations through the area."[92]

Reminiscent of its role in the British-American rivalry in air travel in the 1930s, Pan Am intervened in the debate on the runway project. William Cowden, regional manager of Pan Am, announced that the first Hong Kong-bound jumbos would fly daily from San Francisco via Tokyo starting in 1969. "Jumbojets [are] to fly in next April," reported the *Hongkong Standard* on June 26, 1969. The first Boeing 747 was to touch down at Kai Tak and usher in "the era of the super transport." How could the jumbo land at Kai Tak without an extended runway? Without the extension, Cowden explained, the jumbo would not be able to carry a full load of passengers. There would be "a penalty of 30 per cent pay load

89. *SCMP*, February 28, 1969, 7.
90. TNA, BT 245/1703.
91. *TKP*, June 8, 1969, 5.
92. TNA, BT 245/1703.

178   *Upgrading Hong Kong*

factor." In return, the US carrier would like to see the landing fees reduced. "It would mean something like using an elephant to do the job of a donkey," Cowden said.[93]

The arrival of the jumbo jet was to force a technological breakthrough and to alleviate the capacity issue at Kai Tak. Kai Tak was bursting at the seams with continued growth in traffic volume in 1969 (by 11 percent in aircraft movements, 25 percent in passenger count, and 48 percent in cargo volume). Such dramatic and sustained increases caused "serious congestion" in the passenger terminal even with the opening of a new extension. In keeping with the age of technologically inspired logistics management,[94] the authorities designed a new system of "passenger flow" and an improved system of "mechanized baggage handling and distribution" scheduled for implementation by April 1970 "to coincide with the arrival of the first Jumbo jets." The major improvement remained, however, the extension of the runway. Planning proceeded even in "the absence of a decision . . . pending the outcome of negotiations with the British Government on assistance on its financing."[95]

British reluctance to provide a loan to the runway extension continued to spark resentment in the colony and beyond. An article in the August 7, 1969 issue of the *Far Eastern Economic Review* spoke of Hong Kong being "caught in the spiral of high-cost jumbo planning" as the colonial and British governments vied for the control of landing rights. During a visit to Hong Kong in June, British Minister of State for the Board of Trade William Rodgers rejected the colony's request for financial assistance "after months of stalling by Whitehall," the article reported. His reason, that Britain had to assist "territories which were in far greater need than wealthy Hongkong," did not seem relevant to those in the colony.[96]

"HK landing rights [are] our own business," read a *South China Morning Post* headline on September 10, 1969. International airlines, tourism professionals, and political observers were said to disagree with Rodgers and asserted the colony's command of landing rights in Hong Kong. On the same day, the newspaper's editorial board noted the differences

93.   TNA, BT 245/1703.
94.   Southern, "Historical Perspective."
95.   *HKDCA*, 1968–1969, 1; LegCo, July 30, 1969, 473, 477, 479.
96.   TNA, BT 245/1703.

between the colony and the metropole: "Hongkong has never been backward about acclaiming its own success in the fields of trade and industrial development." The posture of the colony stood in sharp contrast with the metropole's lack of interest in the Kai Tak project as "the British Government struggl[ed] to keep Sterling afloat and exports rising." If the colony was to foot the bill of the expansion, Hong Kong should assume increasing control over its landing rights. "It is from landing fees that we derive much of our airport income and to permit this to remain under London's control would unjustly restrict our ability to finance this costly project."[97] The Chinese media concurred and underscored the incongruity in the British not extending a loan to Hong Kong while demanding control over traffic rights.[98]

In the meantime, Hong Kong continued to record growth in air traffic in spite of the political turmoil and social unrest of the late 1960s. In particular, cargo traffic recorded growth in excess of 18 percent in the midst of the riots in 1967.[99] To calibrate more precisely the airport's capacity, the colonial authorities measured throughput, reporting an increase in Kai Tak's capacity from "a designed 720 passenger per hour to 1,100 passenger per hour," barely matching the "Standard Busy Rate" in 1967 and 1968. On a number of occasions, the airport recorded peak rates as high as 1,654 passengers per hour.[100] These figures made it evident that the airport was stretched to the limit even with the tremendous improvements beyond its initial design.

It was decision time for Hong Kong. The colony would not wait any longer for the generosity of London. On October 1, 1969, the governor told the Legislative Council that on the matter of the Kai Tak runway, "we must regretfully forget about any loan from Her Majesty's Government in the United Kingdom." The colony had to "decide the issue on its own merits and in the light of the probability of either having to finance the whole project directly ourselves, or perhaps with the help of contractor or other finance."[101]

97. *SCMP*, September 10, 1969, 6, 8.
98. *KSEN*, September 10, 1969, 4.
99. *HKDCA*, 1967–1968, 1.
100. *HKDCA*, 1967–1968, 2.
101. LegCo, October 1, 1969, 11; *KSMN*, October 2, 1969, 4.

In February 1970, although the colony had "not yet had a formal reply to our request of three years ago to Her Majesty's Government in London for a financial contribution to this project," the colonial government arrived at the decision to proceed with a 2,780-foot extension of the runway (to a total of 11,130 feet) just as the colony readied itself for the introduction of the daily service of the Boeing 747 on April 11, 1970. The estimated cost of the runway extension had increased to HK$115 million (close to £10 million). In his budget speech to the Legislative Council, Cowperthwaite indicated that the colonial government would continue to press their "moral case in London for a financial contribution." He further underscored the conflict of British and colonial interests: that the British government desired to restrict the use of Hong Kong's aviation facilities against the colony's interest of maximum utilization.[102] Although financing remained unresolved and the conflict of interests persisted, construction had to begin. A pro-Beijing newspaper was quick to report that construction began without British financial support—"one can only blame [their] home country for not helping."[103]

E. O. Laird of the Hong Kong Department in the Foreign and Commonwealth Office admitted to his colleagues in London that the request for British assistance was doomed. A primary reason was the reversal of fortune between the metropole and the colony. While "the British economy was in a particularly unhealthy state, business and trade in Hong Kong were booming and the Colony already had substantial reserves in London." The colony was registering "increasing surplus—above even the expectations of the Financial Secretary." Laird acknowledged "the entangled question of landing rights at Hong Kong." The Board of Trade, which controlled such rights "at great profit to UK interests," refused to come up with the financing. Laird understood the political arguments for London's aid. In the postwar period, the colony had grown into its own through bootstrapping. More recently, the metropole had dealt Hong Kong some major blows including the sterling devaluation.[104]

---

102. LegCo, February 25, 1970, 359; HKPRO, HKMS189-1-176 (duplicated from TNA, FCO 40/311).

103. *TKP*, February 26, 1970, 4.

104. HKPRO, HKMS189-1-176 (duplicated from TNA, FCO 40/311).

The colony had to count on itself. The construction of the Boeing 747 pier had commenced in May 1969 and was completed by March 1970. Additional assembly and installation work continued. Construction was carried out during the night between October 1969 and March 1970 to widen the main taxiway from 60 to 75 feet in order to accommodate jumbo-type aircraft. The extension of the runway, to be completed in two and a half years, was to be complemented by an "additional taxiway, high-speed turn-offs, aircraft sequencing or by-pass areas and a fire sub-station." To respond to the rapid changes in aviation technology and sustained traffic growth, the Hong Kong government needed to upgrade comprehensively the city's airport infrastructure. The long-awaited decision on runway extension was not to foreclose further discussions, cautioned the colonial authorities. "To ensure that Hong Kong will keep pace with the needs of this fast growing industry of civil aviation," the government needed to schedule the necessary developments and phase the construction carefully, they advised.[105]

On the subject of "Hong Kong—Extension of Kai Tak Airport," the Treasury prepared a document dated May 4, 1972. Its conclusion read, "Hong Kong subsequently made their own arrangements to find funds and the runway has been extended. The FCO cannot have any grouse against the Treasury on grounds of delay, still less that we took an unreasonable line on any of their proposals."[106]

The Arrival of the Jumbo and the Growing Autonomy of the Colonial Hub

The concerted efforts of the colonial authorities paid off. In 1970, Hong Kong welcomed scheduled Boeing 747 services to Kai Tak. On April 11, 1970, the first Boeing 747, operated by Pan Am, landed at Kai Tak from the United States via Japan. "The Big Bird arrives in HK," reported a local English newspaper in Hong Kong. Although some construction work was still ongoing, progress was sufficiently advanced for "a smooth

---

105. LegCo, February 25, 1970, 359; *HKDCA*, 1969–1970, 1, 2, 5.
106. TNA, T 317/1642.

Upgrading Hong Kong

and efficient operation of this very large aircraft at Kai Tak."[107] In the passenger terminal building, the pier custom built for the jumbo jet was ready for use, and all passengers on the inaugural flight were cleared in thirty-five minutes. Kai Tak was "ushering in the era of jumbo jet," reported a local Chinese newspaper.[108] By March 1971, the airport handled some twenty-two weekly services of these large aircraft without any difficulties. The technology race in the jumbo jet age proved unrelenting. In January 1971, Northwest Orient Airlines introduced scheduled Boeing 747 services to and from Japan and the United States. In February 1971, Japan Air Lines sent a Boeing 747 to Hong Kong for crew training.[109]

Once the excitement over the approval for runway extension and the arrival of the 747 abated, civil aviation in Hong Kong experienced an unanticipated hiatus in its rapid growth. A slowdown of the industry and the global economy allowed Hong Kong some breathing room in its rush for infrastructure upgrades. The year 1972 saw a noticeable deceleration in the growth of scheduled airline business worldwide. Accordingly, Hong Kong witnessed traffic growth at rates slower than what had become customary in previous years. There was a singular exception: Japan's traffic through Hong Kong continued to grow during this period of industry weakness.[110]

In fact, not only did Japan generate air traffic for Hong Kong, but Japan Air Lines also asserted its influence in the development of aviation infrastructure in Hong Kong. In 1969, the Japanese carrier introduced to Kai Tak a "containerization" system to handle aircraft baggage. This system reduced the loading and unloading time while minimizing damage to baggage.[111] By August 1970, cargo business had matured sufficiently for Japan Air Lines to commence its scheduled service between Japan and Hong Kong.[112]

---

107. *HKDCA*, 1969–1970, 1; Pan Am, Series 5, Sub-Series 1, Sub-Series 6, Box 3, Folder 1 187 14; *SCMP*, April 12, 1970, 2. A pro-Beijing newspaper eagerly noted that the much-touted flight was "almost an hour behind schedule" (*TKP*, April 12, 1970, 4).

108. *WKYP*, May 11, 1970, 4.

109. *HKDCA*, 1970–1971, 1, 10, 11, 21.

110. *HKDCA*, 1971–1972, 1, 70–75.

111. *HKDCA*, 1969–1970, 7.

112. *HKDCA*, 1970–1971, 11.

The impact of the large-capacity aircraft became apparent. By 1973, the number of airlines operating 747 services through Hong Kong had expanded to six with eighty movements of these large-capacity aircraft every week. The number of aircraft movements in Hong Kong remained static while passenger count increased over 22 percent to over three million from 1972 to 1973. Even with a reported growth of 8.1 percent, air cargo results were considered disappointing compared with the worldwide increase of 17 percent. The colonial authorities attributed the results to a decrease in chartered cargo aircraft flying from Southeast Asia to the United States via Hong Kong.[113] Before the United States' retreat from Southeast Asia, Hong Kong had begun to feel the economic influence of shifting Cold War dynamics.

Even as the worldwide oil crisis depressed the air transport industry, both passenger traffic and air cargo flow through Hong Kong continued to grow from 1973 to 1974, each increasing by nearly 20 percent. In the following year, as the crisis continued, passenger count was flat (the worst in 23 years) and air cargo increased a meager 3 percent from 1974 to 1975. Cargo traffic remained asymmetrical as air-carried export was 1.54 times the import in 1973 to 1974, increasing to over 2 times the import in 1974 to 1975. In 1973 to 1974, air cargo accounted for 16 percent by value of total cargo exported from Hong Kong and 27 percent of re-exported cargo.[114]

In addition to incoming and outgoing passenger traffic, the colonial authorities began to report the number of transit passengers, which climbed 18 percent to 490,273 for the year ending March 1974. The year also registered a passenger count of 1,961 for helicopters. The reach of air traffic continued to penetrate the city as 5,482 passengers satisfied their curiosity on "Round Hong Kong" helicopter tours, in addition to some 2,000 daily visitors to the airport terrace. In general, passenger traffic had grown so much that the colonial authorities projected optimistically that the number of passengers Kai Tak handled was to exceed the colony's population in 1975. From 1974 to 1975, the terminal building was operating at a standard busy rate of 30 percent over its design capacity. During peak times, Kai Tak handled more than 3,300 passengers and 28 aircraft

113.   *HKDCA*, 1972–1973, 1, 17.
114.   *HKDCA*, 1973–1974, 1, 8; *HKDCA*, 1974–1975, 6, Appendix I.

184   *Upgrading Hong Kong*

movements per hour. While construction-related nighttime closures ceased in August 1973, the city had to impose tight restrictions on aircraft movements between midnight and 6:30 in the morning to reduce noise disturbance.[115] By March 1975, Hong Kong was connected by direct flights to 67 "major world cities," 36 of which were nonstop services. Kai Tak received 950 scheduled services a week operated by 31 airlines. An additional 30 nonscheduled carriers provided some 48 weekly services. That total translates to an average of over 140 takeoffs and landings daily. More widebody aircraft comprised 20 percent of total aircraft movements in 1974 and 1975, a 60 percent increase from the previous year.[116] In light of the resilient vibrancy of both passenger and cargo traffic, it was particularly disappointing that construction projects experienced repeated delays.[117] As the colony waited for the completion of the extension project, the Legislative Council continued to explore cost-benefit trade-offs for investments and Kai Tak, and plans were afoot for continued improvements.[118]

At long last, the entire extended runway became available for use in December 1975. The new air cargo complex also commenced operation in the following month. The complex was to handle two hundred and fifty thousand tons of cargo every year and was expandable to double that capacity to match the demand expected by 1985. The introduction of the complex provided a timely intervention for the tremendous growth in air cargo of 46 percent in 1975 to 1976 as the air transport industry in Hong Kong made a strong recovery. Air freight had grown to some 25 percent of all exports, and 30 percent of re-exports, by value.[119] By weight, air cargo only accounted for less than 1 percent of total tonnage of commer-

115.   *HKDCA*, 1973–1974, 4, 6, 7, 21; *HKDCA*, 1974–1975, 13. The noise issue at Kai Tak was discussed in the Legislative Council on March 29, 1973 (LegCo, March 29, 1973, 673; LegCo, May 23, 1973, 810–11).

116.   *HKDCA*, 1974–1975, 1, 6, 12.

117.   Weather-related delays pushed the completion target date for the runway extension to 1973–74. As a contractor ran into financial difficulties in the following year, the project suffered from further construction delays. However, the authorities made progress on the technological front, installing a new approach-guidance system for landing over the Kowloon peninsula. This new system contributed significantly to the safety and efficiency of aircraft movements in Hong Kong (*HKDCA*, 1973–1974, 1, 4; *HKDCA*, 1974–1975, 1).

118.   LegCo, October 31, 1974, 98–99; LegCo, November 14, 1974, 198–99.

119.   *HKDCA*, 1975–1976, 1, 17.

The Colony Takes Flight   185

cial cargo carried, indicating the reliance on air transportation for the highest-value export items.[120]

Hong Kong was justifiably proud. When the project finally began after years of delay, legislator Wilfred Wong expressed his elation that the colonial government decided "to extend the airport runway, now called Hong Kong International Airport, with our money rather than borrowed money." Wong called that "a resolute step toward fulfilling a function leading to future prosperity."[121] Not only did the runway extension project reshape Kai Tak physically, but the undertaking also transformed the financial model of aviation infrastructure investment in Hong Kong. In their annual departmental report, the director of civil aviation had previously included in the finance section only "departmental expenditure[s]." As the colonial government began to make considerable investments in Kai Tak, the annual report for 1967–68 specified that the reported figure did "not reflect loan repayment, or the capital expenditure involved in [the] Airport works project."[122] The annual report for 1968–69 further indicated that the government had started a new phase of extensive airport development, which was to involve "very high capital expenditure." This move was to inaugurate a new mentality toward infrastructure investment in the new age:

> It seems in fact that this has become the typical pattern of modern airport financing, with relatively short periods of consolidation, during which the airport operating account approaches balance, followed immediately by further heavy capital investment in extensive development projects necessitated not only by increasing traffic, but also by the pace of technological advancement in the aviation industry.[123]

Calling it "modern airport financing," the colonial government recognized the immensity and cyclicality of infrastructure investment for commercial aviation, regulated by the pace of technological progress.

---

120. *Hong Kong Annual Digest of Statistics, 1978 Edition*, 114.
121. LegCo, March 11, 1970, 419–20.
122. *HKDCA*, 1966–1967, 17; *HKDCA*, 1967–1968, 17.
123. *HKDCA*, 1968–1969, 21.

186 *Upgrading Hong Kong*

Putting this recognition into practice, the director of civil aviation began to report capital expenditures after the colonial government announced its commitment to the runway extension and associated project. For 1970 and 1971, the director reported a capital expenditure of HK$36 million with another HK$65 million approved for the following year.[124] Of that approved amount, the project actually spent HK$53 million, and the government approved another HK$63 million for the 1972–73 financial year "with approximately $200 million of capital expenditure forcast [sic] during the next four years."[125]

This practice of "modern airport financing" in Hong Kong entailed not merely the anticipation of capital outlay but also the cultivation of financial discipline in infrastructure investment. The director of civil aviation began to disclose returns on the capital invested in the airport, which in 1971–72 amounted to 9 percent. In the following year, an increase in revenue provided "a satisfactory return on capital" of 9.5 percent but did not appear to meet the benchmarks set. The Hong Kong figures fell short of "the achieved target of 14% set by the British Airports Authority for the combined return on the five major United Kingdom airports under their administration." Besides lagging behind the benchmark calibrated according to British standards in commercial aviation, the results were also "considerably below the return permitted to public utilities and similar enterprises in Hong Kong." As a government investment in public infrastructure, Kai Tak needed to measure up to other public projects. However, recognizing that the colony was in the investment phase of the cycle, the director expected steady improvements in the near term. The report also recognized "the recent and budgeted heavy capital expenditure for the present development programme" and disclosed specifically fixed assets of HK$778 million on March 31, 1973, of which "UK Govt. Loan and Grants" constituted only HK$8 million.[126]

As the colonial government continued to make heavy capital investments in Kai Tak (HK$61 million in 1972–73 and HK$83 million in 1973–74), the director of civil aviation imposed increasingly stringent assessments. The return of 12 percent in 1973–74 "hardly provides an ade-

---

124. *HKDCA*, 1970–1971, 27.
125. *HKDCA*, 1971–1972, 28.
126. *HKDCA*, 1971–1972, 28; *HKDCA*, 1972–1973, 3, 26, 71.

quate return on the capital employed, bearing in mind prevailing commercial standards and the heavy commitments for capital expenditure."[127] This appraisal held his operations to standards beyond returns on British aviation infrastructure and local public utilities. The government's investment in Kai Tak was to be held accountable to "prevailing commercial standards": in other words, market rates of returns.

Kai Tak achieved its goal in 1976. In the same year the runway extension was completed, the director of civil aviation reported a "return on capital" of 14.17 percent, basically meeting "the target of 15% operating return on average net fixed assets."[128] Although meeting the targeted returns was sufficiently laudable, the report found its more significant bragging rights on its last page: the balance sheet listed that financing by "United Kingdom Government Loan and Grant" had fallen to zero.[129]

The rapid expansion of air cargo volume and its share of total Hong Kong exports underscores its criticality to the colony's economy. The colonial government noted its concern about the asymmetry between the inbound and outbound air cargo traffic: "Exports by air exceed imports by a factor of between two to three, thereby denying air carriers the revenue potential for the inbound journey." When the global environment experienced heavy worldwide demand for air cargo capacity, Hong Kong ran the risk of capacity shortage to meet its export demands.[130] Interpreted differently, this asymmetry in air cargo traffic also signifies the success of Hong Kong not only in export growth but also in producing higher-value exports to be transported by air freight. At the same time, while imports into Hong Kong also recorded healthy increases, residents in the colony did not generally demand high-value items that needed to be transported by air.

From 1975 to 1976, passenger count also recovered, surpassing the four million mark for the first time. Although the figure was 9 percent shy of the city's population and did not reach the estimate that the government had set optimistically, the goal was within reach.[131] Arrivals and

127.   *HKDCA*, 1973–1974, 20.
128.   *HKDCA*, 1975–1976, 1, 17.
129.   *HKDCA*, 1975–1976, 1, 17, Appendix XI.
130.   *HKDCA*, 1975–1976, 18.
131.   *HKDCA*, 1975–1976, 1; *Hong Kong By-Census 1976*, 1.

188   *Upgrading Hong Kong*

departures by air were only 10 percent less than the corresponding figures by sea.[132] Transit passengers through Kai Tak also increased 20 percent compared with the previous year. Shifting geopolitics resulted in four airlines (Air Cambodge, Air Vietnam, Air Lao, and Union of Burma Airways) ceasing operations to Hong Kong in May 1975. At the same time, wide-body aircraft movements grew to over 30 percent of international movements in 1975 and 1976.[133]

"Kai Tak: How an Airport Was Built," read the headline of a *South China Morning Post* article featuring a picture of the Hong Kong airport runway extension that came into operation on June 1, 1974. The article proudly described the transformation of a "'level patch of grass' in 1928" into "one of the world's busiest air terminals."[134] The long history of Kai Tak was certainly impressive. Yet more impressive had been the colony's struggle since the late 1960s to update the airport for the phenomenal traffic growth and technological requirements for new equipment. At a final cost of HK$170 million, the project which commenced in 1970 extended the runway to 11,130 feet, a length the Civil Aviation Department deemed "the most suitable and economically feasible for Hong Kong."[135]

To capitalize on economic opportunities and to facilitate the city's growth, colonial officials were eager to upgrade civil aviation infrastructure and invest in the sustained expansion projected for the coming years. In particular, the advent of the jumbo jet age made it imperative that the runway at Kai Tak be extended. Not without painstaking negotiations with the metropole, the colony eventually prevailed by investing its own capital. These investments in difficult times paid off. The enhanced technological capability and enlarged capacity at Kai Tak responded in a timely manner to corresponding advancements in aviation science, as well as to the mounting trading volume of Hong Kong in the process of economic takeoff. Diverging from previous practice, the colony aimed to be self-reliant financially in its infrastructure development and conse-

132. *Hong Kong Annual Digest of Statistics, 1978 Edition*, 113.
133. *HKDCA*, 1975–1976, 1, 7.
134. *SCMP*, February 17, 1975, 15.
135. HKPRO, HKMS189-2-32 (duplicated from TNA, FCO 40/578).

The Colony Takes Flight

quently asserted enhanced autonomy in local operations and international negotiations.

In an era of accelerating decolonization and diminished British power worldwide, British control over Hong Kong was precarious.[136] In the world of commercial aviation, the United Kingdom negotiated a bilateral agreement with India in 1947 upon the independence of the latter,[137] just as the industry rose from the ashes of World War II. In the case of Malaya, in addition to resolving the issue of commercial aviation diplomatically,[138] the United Kingdom and the former colonial polity structured a commercial relationship to maintain their ties—BOAC continued to hold more than 33 percent of Malayan Airways shares in 1964,[139] seven years after the formation of the independent Federation of Malaya.

In what has been called a process of "informal devolution," the metropole permitted Hong Kong "a degree of freedom from London's control without precedent in British imperial history."[140] Since the 1950s, London had allowed the colonial administration in Hong Kong ample latitude in its fiscal and financial policies.[141] Yet, it is too tempting to overstate the change in this period. Colonial administration in Hong Kong had long been characterized by the lack of detailed supervision, with colonial officials pursuing "their pet projects and peeves, their individual competencies and zeal."[142] Notable in this particular period was John Cowperthwaite, who exercised "virtually complete control of the colony's finances" during his tenure as financial secretary from 1961 to 1971. Cowperthwaite was "brilliant, well-trained in economics, suffered no fools, and was highly principled."[143] In discussions about upgrading Hong Kong's aviation infrastructure, Cowperthwaite had no qualms about criticizing London's policies over the expansion of Kai Tak.

136. Mark, "Lack of Means?"
137. British Airways Archives, "O Series," Geographical, 2962, 3382.
138. British Airways Archives, "Old Series," Geographical, 10000–10004.
139. TNA, BT245/1060.
140. Goodstadt, *Uneasy Partners*, 49.
141. Goodstadt, *Uneasy Partners*, chap. 3.
142. Bickers, "Loose Ties That Bound," 49.
143. Welsh, *History of Hong Kong*, 461.

Hong Kong is often compared with other "Little Dragons" in their trajectories of industrialization. There is not a monocausal account of their paths to industrialization,[144] and historical contingencies played an important role in their connections to regional and global networks. The Cold War context and a certain level of historical legacy similarly wired these cities on the flight map of East and Southeast Asia. However, their enduring, if not anchoring, presence along the major routes required strategic maneuvers by their political and commercial leaders, as well as the continued support of their patron states. States make a crucial difference in economic development,[145] particularly through their function in infrastructure development for and regulatory sponsorship of commercial aviation. Nonetheless, different political configurations shaped the influence of the state. In contrast to the other Little Dragons, Hong Kong's colonial status throughout most of its economic takeoff added complexities to the definition of the state.

Ronald Robinson and John Gallagher have emphasized "informal empire" as a lower-cost form of imperialism.[146] Although Hong Kong's status until 1997 as a Crown colony made it a part of the "formal empire," the preponderance of the colony was under a ninety-nine-year lease, and its colonial masters did treat the imperial holdings in many respects as a treaty port.[147] For its wider and crucial function as the British presence in China, Hong Kong straddled formal and informal empire.[148] Understanding decolonization as "the breakdown of an international colonial order embracing formal and informal empire, and possessing diplomatic, international-legal, economic, demographic and cultural attributes,"[149] one can comprehend more readily the protracted process by which Hong Kong's interests came to be dissociated from, and even entered into competition with, metropolitan interests.[150] Indeed, there was not a single

144. Vogel, *Four Little Dragons*.
145. Johnson, *Japan, Who Governs?*; Wu, "Taiwan's Developmental State"; Yoon, "Transformations of the Developmental State."
146. Gallagher and Robinson, "Imperialism of Free Trade."
147. Darwin, "Hong Kong," 29–30.
148. Bickers, "Colony's Shifting Position."
149. Darwin, "Hong Kong," 29.
150. On the emergence of such autonomy in an earlier period, see Ure, *Governors, Politics*.

official mindset in the British political machinery but competing interests.[151]

In this complex system of conflicting interests, colonial tension between Hong Kong and London was evident over the financing of the runway project in the late 1960s and early 1970s. To underscore the local autonomy of and interplay among various interests in pre–World War II Shanghai, Isabella Jackson has characterized the configuration of the Shanghai Municipal Council as "transnational colonialism," a system of "locally directed autonomous government by foreigners."[152] Unlike the governance of Shanghai's International Settlement by nonstate actors from various nations, London appointed colonial administrators and bureaucrats for service in Hong Kong. Yet, local factors were equally active in the colony in the age of decolonization as diverging interests of British officials drove a wedge between the metropole and the colony. Vying for enhanced local autonomy, colonial officials co-opted Chinese elites in Hong Kong as they worked against the imperial machinery in London.

For the British colony, laissez-faire might have been a "constructed belief" to justify the lack of British assistance in developing Hong Kong's manufacturing sector, especially in the prewar period.[153] Yet, the colonial regime did follow a noninterventionist model of low taxes and small government, along with restricted social services. This model was predicated on Hong Kong's fiscal autonomy from London's control.[154] The mentality of laissez-faire governed the metropole's supervision of the colony as much as the colonial government's relationship with the business community. Functioning in conjunction with this arrangement, business groups in Hong Kong lobbied internationally for the colony's developing economy at this turning point of its economic takeoff.[155] Such dynamics were on full display as the anxious Hong Kong public, particularly its business community, monitored how the colonial government marshalled resources to upgrade Kai Tak without British support.

151. White, "Business and the Politics," 557.
152. Jackson, *Shaping Modern Shanghai*, 16.
153. Ngo, "Industrial History."
154. Goodstadt, "Fiscal Freedom."
155. Clayton, "Hong Kong."

The demands on continuous improvements were unrelenting. The colony had barely completed its runway extension when officials expressed concerns that Kai Tak could not cope with future demands. Highlighting again the projection that Hong Kong would soon have to handle numbers of arriving and departing passengers that approximated its total population, the director of civil aviation cited in his 1975–76 report a quote from the government's publication *Hong Kong 1976*:

> Hong Kong International Airport is constantly being developed and improved to meet the demands of the civil aviation industry. But because there is little or no space available for further expansion, and it appears inevitable that industry demand will outstrip capability, the Government is considering the feasibility of building a new airport at some future date.[156]

As in previous rounds of airport construction, deliberations over this project would take decades. In the meantime, the colony would continue to wrest control over its air hub and routes in the sky from the metropole. Seismic transformation in the geopolitical landscape would also alter the structure of the industry in Hong Kong before Kai Tak would finally retire from service.

156. *HKDCA*, 1975–1976, 18.

CHAPTER 5

Catapulting Hong Kong

Economic Liberalization and Geopolitical Transformations

Thank you for waiting. Today is a great day. We fly to London.

承蒙久侯　國泰今日首航倫敦

—Cathay Pacific advertising its inaugural flight to London, July 16, 1980[1]

As Hong Kong earned increasing autonomy from the British government, the airline industry underwent structural transformation. Emanating from the British and North American hubs from which the behemoths had long dominated the civil aviation industry, a transnational wave of economic liberalization loosened the regulations of air routes and reduced state control and airline ownership. Calls for deregulation began in the United States in 1978.[2] The trend swept the United Kingdom shortly thereafter with a primary focus on fostering competition and lowering airfares.[3] Many studies of airline deregulation focus on the impact on domestic markets and privilege cost and efficiency in their assessment of the trend.[4] Redirecting the focus to the impact of deregulation on regional carriers eager to penetrate the long-haul market, this chapter

---

1. *SCMP*, July 16, 1980, 4; *WKYP*, July 16, 1980, 13.

2. Dobson, *Flying in the Face*; Kahn, *Economics of Regulation*; Kahn, *Lessons from Deregulation*; Derthick and Quirk, *Politics of Deregulation*.

3. Graham, "Regulation of Deregulation"; Barrett, "Implications of the Ireland-UK Airline Deregulation."

4. See, for example, Graham, "Regulation of Deregulation"; GAO, *Airline Deregulation*; Gaudry and Mayes, *Taking Stock*; Button, *Airline Deregulation*.

194 *Catapulting Hong Kong*

examines the growing reach of Cathay Pacific's network in the midst of this industry upheaval.[5]

As a new technology, commercial aviation wove connections across the skies as a "space of flows" over which budding infrastructure facilitated novel forms of exchange and interaction.[6] Yet, the benefits of a network do not necessarily accrue equally to all connected sites. Power dynamics govern which parties control the conduits of traffic flows. Although Hong Kong had long been linked up with faraway destinations through commercial aviation, its flagship carrier, Cathay Pacific, was confined to being a regional player in Southeast Asia until the city developed the financial and diplomatic wherewithal to send its own planes farther afield.

While the general liberalizing trend in the airline industry provided a necessary condition for the expansion of the route system of Hong Kong's flagship carrier, technological enhancements that eliminated stopovers and allowed for long-haul connections also proved critical to Cathay Pacific's breaking free from its regional confines. A confluence of factors precipitated Cathay Pacific's offering of the long-haul routes: the colonial government's triumph in priming Kai Tak Airport for jumbo jets, the commercial availability of long-haul aircraft, and the mounting leverage the colony was able to muster at a time when deregulation was transforming the airline industry.

The resolution of the dispute between the colony and the metropole over their long-haul connection ended just as another issue arose—this time over connections with the promising market of mainland China. For Hong Kong, an air hub which owed its origin not only to the confluence of traffic from the West but also to its gateway position to China (as discussed in chapter 1), the reopening of mainland China in the Reform Era rekindled aspirations to tap into the vast Chinese market. The assertion of Hong Kong's role in connecting with mainland China over the skies dovetailed with the delicate diplomatic exchange between the United

---

5. Kwong Kai-sun published a prescient study of the possibilities for aviation in Hong Kong in a liberalizing environment (Kwong, *Towards Open Skies*). Sinha provided rare coverage of airline deregulation in Asia (Sinha, *Deregulation and Liberalisation*, chap. 2), but he did not make a single reference to Hong Kong.

6. Castells, *Rise of the Network Society*; Larkin, "Politics and Poetics."

Kingdom and the PRC over Hong Kong. The development of the airways over Hong Kong from the late 1970s to the 1980s underscores the changing global economic times and the shifting geopolitics of the period in which Hong Kong took center stage.

Enhanced Autonomy and British Support in Air Traffic Control

When the metropole declined to fund its prospering colonial outpost, Hong Kong assumed responsibility for its own infrastructure development. In the face of this daunting responsibility, the colony responded remarkably effectively as its economic takeoff yielded phenomenal growth and produced the financial wherewithal to fund infrastructure development. Between 1969 and 1984, Hong Kong registered double-digit annual GDP growth in all but one year. The compound annual growth over the period was 19 percent.[7] The colony celebrated its new mentality of independence and discipline not only for the fiscal autonomy it brought but also for the enhanced representation it allowed in negotiating air services agreements on the international stage.

In negotiating such agreements in 1973 and 1974, either the director or deputy director of civil aviation would represent Hong Kong in international talks involving Singapore, Malaysia, Scandinavia, Australia, and Switzerland. The colony's representation expanded in the following year to senior levels at "overseas negotiations, conferences and visits, covering all aspects of aviation."[8] In addition to direct representation in negotiations over commercial aviation, Hong Kong's interests also came to be represented in "summit meetings" between Cathay Pacific and BOAC when the two airlines began in 1972 to organize regularly scheduled conferences at senior levels.[9]

---

7. *Hong Kong Annual Digest of Statistics, 1978 edition*; *Hong Kong Annual Digest of Statistics, 1981 Edition*; *Hong Kong Annual Digest of Statistics, 1990 Edition*.

8. *HKDCA, 1973–1974*, 1; *HKDCA, 1974–1975*, 1.

9. JSS, 13/10/4 Cathay Pacific/British Airways Summit Meetings.

In 1975 and 1976, the colony's Civil Aviation Department expressly "represented the Hong Kong Government" at various meetings to reflect "Hong Kong's interest in all matters pertaining to international civil aviation."[10] As "air services agreements between the United Kingdom and foreign countries involving Hong Kong's interests came under particularly heavy review and re-negotiation" in 1976 and 1977, Hong Kong earned representation on the British negotiating team. In that year, the Civil Aviation Department represented the Hong Kong government at eleven sessions of air services talks overseas, five of them between the UK and US governments, and the others involving the governments of Thailand, Singapore, South Korea, Niugini (Papua New Guinea), India, and Japan. During this period of intense technological change, department representatives also attended eleven other technical conferences "to protect and maintain Hong Kong interests in all aspects of aviation."[11] The colony had come into its own in commercial aviation and had been able to negotiate distance from the metropole, financially, diplomatically, and technically.

Correspondence between Hong Kong and London during the period also indicates the mounting involvement of the colonial authorities in civil aviation matters. In 1977, Hong Kong Governor Murray MacLehose played a prominent role in the negotiation of such issues as air cargo rates between Hong Kong and the United States, air fares, and air traffic connecting Hong Kong with Australia.[12] In the same year, he also participated in discussions about air services between the United Kingdom (including its colonial outpost Hong Kong) and Indonesia, Italy, Japan, Lebanon, Malaysia, and the Netherlands.[13] His involvement only intensified in 1978. In the last days of January 1978, as Chinese New Year approached, MacLehose was dealing with civil aviation matters on an almost daily basis. He intervened in air services negotiations between the United Kingdom and Canada, East Africa, Indonesia, Japan, Malaysia, New Zealand, Papua New Guinea, Singapore, and Thailand, particularly

---

10. *HKDCA*, 1975–1976, 1.
11. *HKDCA*, 1976–1977, 7.
12. TNA, FCO76/1497.
13. TNA, FCO40/791 (duplicated as HKPRO, HKMS189-2-148).

Economic Liberalization and Geopolitical Transformations 197

in negotiations involving Hong Kong.[14] MacLehose's intense involvement in civil aviation matters was understandable in light of local resentment against British control. In 1978, for example, when Cathay Pacific appeared to be being neglected in favor of British interests, some voices in the colony accused the British of disregarding Hong Kong's local interests and questioned "whether Britain should decide matters of vital interest to Hongkong."[15] In fact, the colonial government was asserting so much influence over air services negotiations that at one point in 1979, officials in London expressed concerns that Hong Kong "might do a UDI"—that is, a unilateral declaration of independence—over air traffic rights and insist on handling its own negotiations.[16]

The Hong Kong authorities also identified increasingly with Cathay Pacific, viewing the airline as Hong Kong's own. Articulating Hong Kong's policy toward Cathay Pacific, MacLehose asserted that Hong Kong did not seek to protect Cathay Pacific against competition per se "but only against restrictive policies of foreign countries and malpractices of other airlines to an extent which would preclude provision of viable competing services by C[athay] P[acific] A[irways] despite efficient and economical operation."[17] Paralleling the spirit of liberalization in the capitalist bloc, the Hong Kong government's desire was to represent the colony's flagship carrier not for protectionist purposes but to ensure that it could operate in an environment of fair competition. Addressing his colleagues in London on air services policy on June 22, 1978, MacLehose reiterated that Cathay Pacific was "a commercially motivated airline" and that the Hong Kong government "would not wish to see them operating in any other way." MacLehose and his Hong Kong colleagues were "not interested in CPA 'flying the flag,'" he claimed, and would not like to pressure Cathay Pacific into flying any route that the airline, "in their commercial judgement, consider[ed] to be less rewarding than others which they could mount."[18] In other words, MacLehose asserted that it was his government's responsibility to foster a fair environment in which

---

14. TNA, FCO40/981 (duplicated as HKPRO, HKMS189-2-257).
15. *SCMP*, March 4, 1978, 1.
16. TNA, FCO 40/1073 (duplicated as HKPRO, HKMS189-2-325).
17. TNA, FCO40/981.
18. TNA, FCO40/983 (duplicated as HKPRO, HKMS189-2-259).

198 *Catapulting Hong Kong*

Cathay Pacific could compete. However, it would not be appropriate to pressure the local airline into doing the government's bidding. In this regard, Cathay Pacific's relationship with the Hong Kong government mirrored what the colonial regime desired of its relationship with the British government.

Reaching Down Under

In terms of air routes, this assertion of the Hong Kong government translated into the expansion of Cathay Pacific's reach beyond its initial domain—regional traffic in Southeast Asia centered in Hong Kong. The earliest expansion of the airline's network reinstated Cathay Pacific's services to eastern Australia.[19] At the time of the expansion, the airline had been operating a service to Perth since 1970,[20] providing an early air link between western Australia, Hong Kong, and Japan.[21] It would take a few more years before it resumed its longer-distance service to the more competitive market of eastern Australia. Cathay Pacific had served Sydney between 1959 and 1961, enticing customers with shark fin soup served in turquoise-patterned Chinese bowls. However, as its catering department strove to solve the issue of the exploding Thermos flasks in which the airline transported the soup ("the flasks pop their corks and sometimes break under pressure in the air"), Cathay Pacific yielded to the technical superiority of Australia's Qantas and retreated from the route in 1961.[22]

---

19. For Cathay Pacific's network that had extended to Sydney in the late 1950s and early 1960s, see chap. 2.

20. National Archives of Australia, C3739, 281/7/68.

21. *SCMP*, January 26, 1970, 31, March 29, 1970, 53; *HKDCA*, 1970–1971, 11.

22. TNA, BT245/552; JSS 13/6/1/1; Swire HK Archive, CPA/7/4/1/2/1 Newsletter [July 15, 1959]; Swire HK Archive, CPA/7/4/1/2/11 Newsletter [July 31, 1961]; Swire HK Archive, CPA/7/4/1/1/151 Newsletter [October 1976]. The airline's public relations manager explained in 1970 that Cathay Pacific had "suspended" the service to Sydney after about 18 months because it was "commercially unworthy" (*SCMP*, January 26, 1970, 31). On November 14, 1961, Qantas launched a weekly service linking Sydney with Hong Kong via Darwin and Manila, with continuing service to Tokyo (Qantas Archives, R10 SYD/HKG). See also Bickers, *China Bound*, 349.

Economic Liberalization and Geopolitical Transformations 199

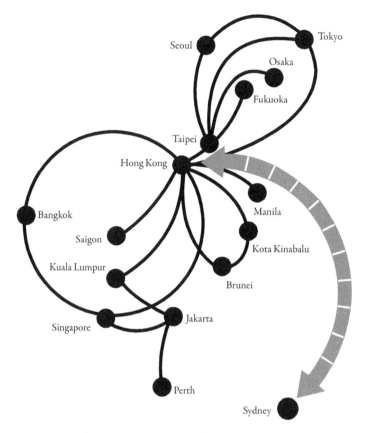

FIGURE 5.1. Schematic representation of Cathay Pacific's expansion beyond its regional network. Source: *SCMP*, June 30, 1974, 5.

Hong Kong's enhanced representation ultimately bore fruit in 1974, when the British team successfully negotiated with the Australian government to allow Cathay Pacific to offer the first nonstop service between Hong Kong and Sydney, Australia's biggest city.[23] Even before securing all of the necessary government approvals, Cathay Pacific presented a new schematic representation of its route map (recall the petal-pattern network of 1971 in fig. 3.12) by adding an oversized arc depicting its new Hong Kong–Sydney connection (fig. 5.1). The Sydney service would indeed

23. *SCMP*, June 21, 1974, 1; *SCMP*, June 24, 1974, 29; *WKYP*, June 21, 1974, 10.

provide "a significant boost" to the airline's route network in terms of size. Calling itself "Asia's most experienced airline," Cathay Pacific proudly announced "the start of the non-stop"—"the only non-stop flight"—on October 21, subject to Government Approval."[24]

Formal approval came in August 1974 as the airline's Managing Director Duncan Bluck announced Cathay Pacific's plans to operate the new route with its Boeing 707s with no intermediate stops in either direction, providing "the fastest flights on the route as well as providing the only non-stop service in both directions." The airline had earned this bragging right after years of equipment upgrades. Since 1971, the airline had at a measured pace been replacing its Convair 880-22M fleet with a Boeing 707-320 B/C fleet capable of covering longer ranges. In 1971, Cathay Pacific had nine aircraft, just one of which was a Boeing 707. By June 1974, eleven of the airline's fleet of eighteen were Boeing 707s. Management had also arranged the sale of its remaining seven Convairs at a moderate profit. With its Boeing 707s, the Cathay Pacific flights were able to leave Hong Kong every Monday, Friday, and Saturday evening, arriving in Sydney the following morning. The return flights to Hong Kong would depart Sydney in the late morning on Tuesday, Saturday, and Sunday, arriving in the evening the same day.[25]

In keeping with its overall flight schedule, this Hong Kong–Sydney pattern, with a duration of nine hours each way, was to reflect what the airline billed as its forte: "We know businessmen like our business-like schedules."[26] In an advertisement for the opening of the new route on October 21, 1974, Cathay Pacific featured a picture of fireworks over "the fantastic new Opera House" (which had opened in October 1973). Unlike Qantas's service, which stopped in Manila, and British Airways', which connected through Darwin, Cathay Pacific was to fly nonstop. The airline placed such an emphasis on its direct flights that the brief advertising copy in the English-language newspaper included four references to

24. *SCMP*, June 30, 1974, 5; *SCMP*, July 15, 1974, 41; *SCMP*, October 20, 1974, 36.

25. *SCMP*, August 2, 1974, 23. Swire HK Archive, Cathay Pacific Airways Limited Report of the Directors and Statement of Accounts for the year ended 30th June 1971, 9; Swire HK Archive, Cathay Pacific Airways Limited Annual Report 1974–1975, 11; Swire HK Archive, Cathay Pacific Airways Limited, Chairman's Statement, November 19, 1975,1.

26. *SCMP*, September 9, 1974, 15.

Economic Liberalization and Geopolitical Transformations 201

FIGURES 5.2 a and b. "Our non-stop flights Hong Kong–Sydney." Source: Swire HK Archive, HK/2017/22. A later edition of the English version appeared in *SCMP*, October 21, 1974, 11; the Chinese version appeared in *WKYP*, October 21, 1974, 21.

"non-stop": "our *non-stop* flights to Sydney"; "*Non-stop* from Hong Kong three times weekly"; "with *non-stop* services all the way"; and "Cathay Pacific's *non-stop* flights to Sydney." For those traveling for leisure purposes, the airline promised to "show you the sights." "And if you're on business, well, what you save in time you can spend on . . . pleasure." "You deserve a break," concluded the advertisement. "But not en route" (fig. 5.2a). The airline also placed advertisements in a local Chinese newspaper with similarly relentless references to its nonstop service (fig. 5.2b).[27]

Reporting on the monumental opening on October 21, 1974, the *South China Morning Post* compared Cathay Pacific's inaugural nonstop

27. *SCMP*, September 27, 1974, 6; *SCMP*, October 14, 1974, 15; *SCMP*, October 21, 1974, 11; *SCMP*, October 28, 1974, 38 (emphasis added); *WKYP*, October 21, 1974, 21.

flight, shorter than nine hours, on the Boeing 707-320C, which offered 154 seats, with its service in 1959 on the *Electra*, then the fastest prop jet in the world. With 55 standard seats and eight sleeper seats, the *Electra* broke the commercial flying record when it covered the 4,300 miles between Hong Kong and Sydney in twelve hours and fifty minutes, some six hours faster than the Super Constellation planes that had previously flown the same route. Impressive as it was, the *Electra* had to make stops in Manila and Darwin en route.[28] A local Chinese newspaper not only carried the story of the inaugural Sydney flight but also presented a two-page special feature to mark the event. The special feature highlighted not only the technological prowess of Cathay Pacific's fleet of predominantly Boeing 707 jetliners but also the airline's plans for continuous equipment upgrades. To complement its latest hardware upgrade, the airline had been presenting its crew in the Pierre Balmain–designed "Tung Hoi" uniform since September, noted the special feature (see chapter 3).[29]

Treated to its vastly improved service, which Cathay Pacific characterized as "business schedules" on the front page of the *South China Morning Post*, were several VIP guests: the chairman of the Hong Kong Urban Council, the deputy director of the Civil Aviation Department, and a prominent filmmaker.[30] It was a celebratory moment for Cathay Pacific to reenter Sydney with its nonstop flights, which, in the airline's words, "heralded the start of major growth in Cathay Pacific's Australian market."[31] In just under six months (from the inaugural flight to the colony's fiscal year end on March 31, 1975), the airline carried 6,369 passengers outbound and 5,616 inbound. In the following year, Cathay Pacific even overtook incumbents British Airways and Qantas in passenger counts in both directions.[32]

28. *SCMP*, October 21, 1974, 37.

29. *WKYP*, October 21, 1974, 18, 22–23.

30. *SCMP*, October 22, 1974, 1; *SCMP*, October 23, 1974, 26.

31. Swire HK Archive, CPA/7/4/1/1/140 Newsletter [July 1974]; Swire HK Archive, CPA/7/4/1/1/151 Newsletter [October 1976]. Cathay Pacific quickly followed with a fare reduction targeting primarily leisurely travelers (*KSMN*, October 23, 1974, 7; *TKP*, December 4, 1974, 5; *SCMP*, December 2, 1974, 11; *SCMP*, December 4, 1974, 1; *SCMP*, December 12, 1974, 17).

32. *HKDCA*, 1974–1975, App. VI; *HKDCA*, 1975–1976, App. VI.

Economic Liberalization and Geopolitical Transformations   203

In 1978, British Airways retreated from the routes linking Hong Kong with Australia and New Zealand, "largely as a result of [the] prompting" of the London authorities. With British Airways' withdrawal from the Hong Kong–eastern Australia route, Cathay Pacific took up the entire British entitlement to capacity on that route. The London authorities bragged that this substantial expansion of Cathay Pacific's activities meant that in three and a half years, the British government had "brought [Cathay Pacific] up from a position in which they were effectively excluded from the Hong Kong-East Australia market to one in which they have the right to provide half the capacity in full parity with Qantas."[33]

Crossing the Pacific

In addition to Cathay Pacific's expansion into the South Pacific toward Australia, the Hong Kong government also engineered the airline's eastward expansion across the Pacific to North America. As early as 1977, the British government had recognized the importance of the trans-Pacific route from Hong Kong. In its air services agreement negotiations with the United States, the British secured rights to Vancouver on the trans-Pacific route for "combination" services (a passenger service combined with a belly-hold freight service). The British negotiators were optimistic about "the longer-term benefits in the form of substantially improved opportunities for an operation by a UK airline from Hong Kong to the US West Coast" even though they recognized that a passenger service on that route would not materialize for a few years.[34] The seeds were sown for Hong Kong's trans-Pacific expansion, both in passenger and cargo opportunities.

Although the trans-Pacific route was deemed an opportunity "some distance in the future," the Hong Kong authorities did not allow the momentum of its discussions to abate. In January 1978, in a telegram to London regarding UK and Canadian air services, MacLehose suggested "as an opening bid, ask for Vancouver, Edmonton and Toronto

---

33.   TNA, FCO40/981; *SCMP*, January 2, 1978, 30.
34.   TNA, FCO76/1498.

204 *Catapulting Hong Kong*

(to mirror the points available to the Canadian designated airline)." The ambitions of MacLehose and Cathay Pacific did not stop with Canada. MacLehose noted that it would be "sensible to secure San Francisco and/or Los Angeles" in conjunction with Vancouver.[35] His colleagues in London responded that "the time had now come to make provision for Cathay Pacific Airways to operate on a route which corresponded to that of C[anadian] P[acific] Air." On the basis of statistics provided by the Hong Kong authorities, the British government was satisfied that this route was growing fast and that "there should easily be room for a second carrier by the time Cathay Pacific were ready to start." The negotiators in London reported that the Canadians were "reasonably receptive" but cautioned against the difficulties of a connection in Japan. Rights beyond Vancouver to the United States could also pose a challenge as the Canadians were not sanguine about the prospects of obtaining corresponding rights beyond Hong Kong.[36]

Negotiations continued for the next couple of years.[37] The chairman of Cathay Pacific reported in his annual statement to shareholders dated April 17, 1979, that the airline had participated in governmental consultations with the United Kingdom and Canada "with the aim of obtaining a reciprocal Hong Kong/Vancouver route to that currently operated by CPAir." He indicated further that another round of negotiations was scheduled for the spring of 1979.[38]

By 1979, Cathay Pacific was the undisputed leader in terms of passenger counts into and out of Hong Kong. With a market share of 27 percent, the Hong Kong airline was way ahead of Japan Air Lines and Singapore Airlines (which each stood at 10 percent). Its traffic remained largely confined to the regional market, however. Although the main North American players Canadian Pacific, Pan Am, and Northwest Orient accounted for only 1 percent, 5 percent, and 2 percent of Hong Kong's

---

35. TNA, FCO40/981.

36. TNA, FCO40/982 (duplicated as HKPRO, HKMS189-2-258).

37. Swire HK Archive, Cathay Pacific Airways Limited, Chairman's Statement, April 17, 1979, 3; Swire HK Archive, Cathay Pacific Airways Limited, Chairman's Statement, April 17, 1980, 3; Swire HK Archive, Cathay Pacific Airways Limited Annual Report 1980, 5; TNA, FCO40/1072 (duplicated as HKPRO, HKMS189-2-324).

38. Swire HK Archive, Cathay Pacific Airways Limited, Chairman's Statement, April 17, 1979, 3.

air passenger traffic, respectively, they held onto their longstanding domination of trans-Pacific traffic to and from Hong Kong, and competed with Cathay Pacific on the Hong Kong–Tokyo route.[39]

The local press reported Cathay Pacific's ambition to further assert its rights at its home base. "Within the framework of reciprocal rights, Cathay Pacific wants to widen its horizon," reported the *South China Morning Post* in January 1979 as it highlighted Canadian Pacific's longstanding monopoly of the route. The Hong Kong carrier was said to have the Boeing 747 jumbo jet in mind for the route.[40] The wide-body 747 was revolutionizing jet air travel in the 1970s as it delivered greater capacity at a lower per-seat cost, thus promising efficiency especially on long-haul international routes.[41] It was believed that Cathay Pacific's jumbos, powered by Rolls-Royce B4 engines, should have no trouble flying nonstop to Vancouver, the newspaper noted, although strong head winds would mean stopping in Anchorage (in the US) or Tokyo on the return journey until the introduction of the more powerful D4 engines.[42]

The industry was abuzz with talk of Cathay Pacific providing a trans-Pacific service.[43] In September 1980, Hong Kong newspapers reported that the British and Canadian governments had reached an agreement to allow a British-designated carrier to fly to the west coast of Canada, ending the monopoly that Canadian Pacific Air Lines had enjoyed over the route since 1949.[44] The director of civil aviation noted progress in his annual report for the year ending March 1981 on a new air services arrangement with Canada that would allow "for the first time a route from Hong Kong to Vancouver to be operated by a British carrier." However, there was an additional hurdle for Cathay Pacific: it would have to compete with other "British" carriers for the route. Laker Airways, a private British operator that emphasized low-cost travel, jumped into the fray. Competition

---

39. *HKDCA*, 1978–1979, App. VI.

40. *SCMP*, January 1, 1979, 18.

41. Bednarek, *Airports, Cities*, 17.

42. *SCMP*, December 29, 1980, 30.

43. *SCMP*, January 22, 1980, 1; *SCMP*, March 1, 1980, 6; *SCMP*, April 26, 1980, 8; *KSEN*, January 22, 1980, 2; *KSMN*, April 11, 1980, 6; *WKYP*, September 9, 1980, 5. Within a year, the media would report Singapore Airlines' plans to fly to the United States via Hong Kong (*SCMP*, September 7, 1981, 38).

44. *KSMN*, September 24, 1980, 6; *WKYP*, September 24, 1980, 9.

206 *Catapulting Hong Kong*

notwithstanding, this inaugural trans-Pacific connection with Hong Kong, previously dominated by North American carriers, bore tremendous significance for the expanding reach of Hong Kong–based air traffic.[45] Objecting to Laker Airways' application to fly from Hong Kong to Vancouver, Los Angeles, San Francisco, Guam, Honolulu, and/or Seattle, as well as to gain rights to Tokyo as an intermediate port of call, Cathay Pacific reminded the licensing authority of its longstanding request—first made in November 1977—for a route from Hong Kong to Vancouver, a request that both the Hong Kong and British governments supported. The Hong Kong carrier also reiterated its intention to serve the US West Coast at the right time. Laker Airways' application posed a threat not only to Cathay Pacific's proposed service to Vancouver but also to its tremendous amount of Tokyo business. The agreement between the British and Japanese governments required that any additional designation of a British operator through Tokyo would have to come at the expense of Cathay Pacific's existing capacity.[46]

In 1981, Hong Kong's Air Transport Licensing Authority (ATLA) deliberated on the two parallel applications.[47] A statutory body established under the Air Transport (Licensing of Air Services) Regulations (Cap.448A), ATLA has been responsible for issuing licenses to Hong Kong airlines for the operation of scheduled air services to and from Hong Kong. With membership that included both British and Chinese residents of Hong Kong, ATLA provided an important channel to assert autonomy. On the application for trans-Pacific routes, ATLA ruled that there was clear evidence of the demand for Hong Kong–Canada services and that such demand was likely to grow. It also stated that it was satisfied that Cathay Pacific was best equipped to offer that service and granted the airline permission to operate the Hong Kong–Tokyo–Vancouver–Seattle–Tokyo–Hong Kong route. However, Cathay Pacific had not provided sufficient evidence to support services to Honolulu, San Francisco, and Los Angeles. Conversely, ATLA found that Laker Airways had not

---

45. *HKDCA*, 1980–1981, 18; *TKP*, April 11, 1980, 5; *TKP*, July 1, 1981, 10; *SCMP*, September 24, 1980, 1; *SCMP*, January 22, 1981, 14.

46. TNA, BT 245/1853; TNA, FCO 40/1081; *SCMP*, June 3, 1981, 23.

47. LegCo, Brief—Civil Aviation Ordinance (Chapter 448); HKPRO, HKRS934-2-49.

Economic Liberalization and Geopolitical Transformations 207

justified its application to Guam, Vancouver, or Seattle but had granted it the Hong Kong–Tokyo–Honolulu–San Francisco–Los Angeles route and the reverse. The long-running saga did not end as expected, however. After the ATLA ruling, the story took a dramatic turn when Laker Airways went into voluntary liquidation on February 17, 1982.[48] The now-defunct airline could not transfer its landing rights in Hong Kong.[49]

The granted stopover provisions notwithstanding, the years of discussions and preparation finally resulted in Cathay Pacific's inauguration on May 1, 1983, of the first-ever nonstop service between Vancouver and Hong Kong. Using its "brand-new, Rolls-Royce powered 747s," the airline began a twice-a-week schedule. Cathay Pacific had purchased its first Boeing 747-200B in 1979, and it took over the service to eastern Australia. In 1981, the press began to report Cathay Pacific's HK$400 million order for new Boeing 747-200B aircraft with the latest Rolls-Royce engines, along with spares, to fly nonstop from Hong Kong to Vancouver. The Boeing 707, which had engineered Cathay Pacific's reentry into Sydney in 1974, was phased out between 1982 and 1983. By 1983, the airline was operating a fleet of wide-body airliners comprising eight Boeing 747-200Bs and nine Lockheed L-1011 Super TriStars. Passengers on any of the three service classes were promised to "be enchantingly attended by the grace and beauty of nine Asian lands" during their flight. The inaugural Vancouver service was indeed a festive occasion. The Asian "beauties" who comprised Cathay Pacific's flight and ground staff donned the new uniform designed by Hermès (see chapter 3). With the new service to Vancouver, Cathay Pacific also launched a new advertising campaign that featured the famed Venetian traveler Marco Polo.[50]

48. TNA, BT 245/1853; TNA, BT 245/1854; TNA, FCO 40/1081.

49. *SCMP*, February 13, 1982, 1.

50. Swire HK Archive, Cathay Pacific Airways Limited, Chairman's Statement, April 25, 1978, 3–4; Swire HK Archive, Cathay Pacific Airways Limited Annual Report 1978, 6; Swire HK Archive, Cathay Pacific Airways Limited, Chairman's Statement, April 17, 1980, 1; Swire HK Archive, Cathay Pacific Airways Limited Annual Report 1982, 5; Swire HK Archive, Cathay Pacific Airways Limited Annual Report 1983, 24; *SCMP*, May 27, 1981, 27; *TKP*, July 18, 1982, 5; *WKYP*, July 26, 1982, 5; *SCMP*, April 10, 1983, 87; *SCMP*, April 19, 1983, 6; *SCMP*, May 1, 1983, 5; *SCMP*, May 2, 1983, 14; *WKYP*, May 2, 1983, 6, 12.

FIGURES 5.3 a and b. "Today non-stop to Vancouver." Sources: *SCMP*, May 1, 1983, 5; *Ming Pao*, May 1, 1983, 9.

In the airline's full-page advertisement on the day of the inauguration flight (figs. 5.3 a and b), the nonstop service between Hong Kong and Vancouver was accentuated by a thick-lined schematic arrow that arced the northern rim of the Pacific to connect the two cities. "Today, for the first time ever," advertised Cathay Pacific, "you can fly non-stop to Vancouver. The fastest way to Canada." The advertisement also stressed the possibility of onward journeys from Vancouver, Cathay Pacific's gateway to the rest of North America: "Arriving the same day with immediate connections through to Toronto, Montreal, Calgary, Edmonton and other major cities in Canada and the USA." There was a notable difference between the airline's advertisements in the English and Chinese outlets. Projecting its appeal to the English-reading audience, Cathay Pacific underscored the fact that its nonstop service from Hong Kong to Vancouver was not just unprecedented ("for the first time ever") but also unmatched ("Only on Cathay Pacific") (fig. 5.3a).[51] For the airline's Chinese-reading clientele, emphasis on efficient travel arrangements appeared to suffice. In its English ad, Cathay Pacific appeared keenly aware of competition from the Canadian carrier, which it was eager to dislodge from its entrenched franchise over the trans-Pacific route. The new route marked Cathay Pacific's "entry into the highly competitive trans-Pacific market," noted the airline's newsletter. In conjunction with the new service, Cathay Pacific also established an office in Vancouver.[52]

Cathay Pacific's entry into the market did indeed exert pressure on Canadian Pacific, which had previously enjoyed a monopoly over the Hong Kong–Vancouver route, and ATLA's optimistic forecast for traffic growth did indeed materialize. According to Hong Kong government data, overall passenger traffic between Hong Kong and Vancouver grew 72 percent in the 1983–1984 period, during which Cathay Pacific introduced its new Vancouver services.[53] So successful was the route that Cathay Pacific announced an additional weekly nonstop flight before the year was over.[54] In the first year of the service, Cathay Pacific expanded the market dramatically and readily surpassed its Canadian rival in both directions.[55]

The airline would come to celebrate its Hong Kong–Vancouver service not merely as its entry into the trans-Pacific market but also as a "new concept in long-haul travel." As Cathay Pacific launched its pioneering nonstop service between Hong Kong and Vancouver, it simultaneously reduced flight times and cut out intermediary stops from other routes, thereby offering greater convenience to passengers, and to business travelers in particular.[56] This tremendous feat was made possible not only by support from air services diplomacy in London and Hong Kong but also

51. *SCMP*, May 1, 1983, 5; *Ming Pao*, May 1, 1983, 9.

52. Swire HK Archive, CPA/7/4/1/1/171 Newsletter [November 1982].

53. *HKDCA*, 1982–1983, 26; *HKDCA*, 1983–1984, 29.

54. Swire HK Archive, CPA/7/4/1/1/172 Newsletter [November 1983]; *SCMP*, December 5, 1983, 40.

55. *HKDCA*, 1982–1983, 26; *HKDCA*, 1983–1984, 29.

56. Swire HK Archive, CPA/7/4/1/1/175 Cathay News No. 15 [August 1986]; Swire HK Archive, Cathay Pacific Annual Reports 1982, 1983.

by a major overhaul of the Cathay Pacific fleet. Since the mid-1970s, Cathay Pacific had maintained a fleet that numbered in the mid- to high teens. Phasing out the Boeing 707, by the early 1980s Cathay Pacific had transformed its entire fleet into wide-bodied aircraft, all powered by Rolls-Royce engines. In addition to increased performance and reduced fuel consumption, Cathay Pacific's fleet boasted cutting-edge technology capable of greater range, making it possible for the airline to offer ultra-long-haul nonstop flights.[57] The opening up of the skies amid the new regulatory environment coincided with a wave of technological upgrades, which Cathay Pacific rode just as Hong Kong's economic takeoff was generating sufficient traffic growth to fuel the airline's expansion.

Getting to the Heart of the Imperial Network

The resumption of its service to eastern Australia and the opening up of the trans-Pacific market were indeed triumphant moments for Cathay Pacific. What made these accomplishments all the more impressive was that they unfolded just as the Hong Kong carrier was fighting its way into the imperial air hub: London.

Hong Kong officials had initiated conversations with London about additional services between the two cities in the 1970s. Hong Kong Secretary for Economic Services D. G. Jeaffreson wrote to R. J. T. McLaren at the Hong Kong and General Department of the Foreign and Commonwealth Office in London on March 23, 1979, "to initiate a more formal dialogue" on the possibility of an additional scheduled service between Hong Kong and London. Officials in Hong Kong had voiced their dissatisfaction with the performance of British Airways, especially with regard to punctuality, and there were also voices of discontent in the media. The self-styled "Frequent Traveller," a contributor to the *South China Morning Post*, reproached British Airways for its continued loss of "money and its reputation through strikes, poor maintenance and aloof in-flight

---

57. Swire HK Archive, CPA/7/4/1/1/175 Cathay News No. 15 [August 1968]; Swire HK Archive, CPA/7/4/1/1/184 Cathay News No. 46 [January 1990].

service." Comparing the British state-owned airline to Hong Kong's Cathay Pacific, the contributor judged that British Airways "wouldn't stand a chance" in the face of "competition [from] an efficiently run airline offering a world-beating standard of in-flight service." In another contribution to the paper, a "Hongkong Belonger" wrote that it was time for "our flag carrier, Cathay Pacific" "to carry the flag to London." The Legislative Council also registered repeated complaints about British Airways. The improvements that British Airways subsequently made proved insufficient to placate the Hong Kong public. Competition should be introduced on the Hong Kong–London route to "shak[e] BA into greater efficiency," Jeaffreson declared. To combat the issues of industrial action (workers' strikes) and weather, to which British Airways attributed its deficiencies, "an airline that was not subject to the problems associated with being based in the United Kingdom, in other words by an airline based in Hong Kong," he argued, would provide the most effective competition. Writing again in May 1979, this time to the Maritime, Aviation and Environment Department of the Foreign and Commonwealth Office, the Economic Services Branch of the Hong Kong Government Secretariat substantiated its request for a new service with detailed analysis, declaring that officials had "come to the conclusion that the possibility of another scheduled carrier coming onto the route ought to be seriously pursued." Lending a British voice to support for Cathay Pacific, the airline's backer Adrian Swire wrote to the Foreign and Commonwealth Office stressing that if Cathay Pacific was to win the license to the route, it would fly "with Rolls Royce engined 747s."[58]

In correspondence with his London colleagues in the Department of Transport on June 13, 1979, Deputy Under-Secretary of State Hugh Cortazzi of the Foreign and Commonwealth Office supported the request of the Hong Kong government, which deemed it appropriate for Cathay Pacific to be "given the chance to compete on the London route" and

58. TNA, FCO40/1080; *SCMP*, February 22, 1978, 12; *SCMP*, March 8, 1978, 14. Cathay Pacific's choice of aircraft, and in particular, whether the aircraft used a Rolls-Royce engine, had consumed British authorities since the early 1970s (TNA, BT245/1723). On the term "Hong Kong belonger," see Ku, "Immigration Policies"; Ku and Pun, "Introduction." Introduced in the local immigration law of the colony in 1971, "Hong Kong belonger" developed into an identity that reflected the exclusionary immigration and citizenship policy of the British government.

considered it "unjustifiable" that British Airways should continue its monopoly. In light of British Airways' persistent unsatisfactory performance "a competitor or competitors would likely provide a better service and at the same time spur BA into making improvements." Understanding that other British operators had also expressed interest in the route, the Foreign and Commonwealth Office noted, "If we take our responsibilities towards Hong Kong seriously we must however recognise that CPA has *a natural right* to the route" (emphasis added). This recognition of Cathay Pacific's status in Hong Kong acknowledged the parallel, but at times conflicting, British interests in London and Hong Kong. The position of the Foreign and Commonwealth Office also allowed reciprocal treatment for a Hong Kong entity in the business of connecting the colony to the metropole through the airways. "The Hong Kong travelling public are entitled to an efficient and punctual service," which British Airways had failed to deliver. Cortazzi advised that "it would be unnecessarily damaging to our relations with Hong Kong for us to continue to deny CPA the opportunity to show they can compete effectively." In the spirit of competition and efficiency, British Airways should not be allowed to hold onto its monopoly, and Cathay Pacific should be allowed to enter the market in question. "The removal of this particular grievance," Cortazzi continued, "would also make it easier for Hong Kong to accept any sacrifices they may have to make in the contents of air agreements with Malaysia and China, and indeed in future." It was a matter for the Department of Trade. Cortazzi further warned against any action that would "lead to a confrontation with Hong Kong or deny them a fair opportunity to make their case." It was paramount that the Hong Kong government be reassured that Cathay Pacific "would be given the same consideration as any British airline."[59]

Three companies filed applications with Britain's Civil Aviation Authority (CAA) to enter the market between Hong Kong and London: British Caledonian, Laker Airways (which also applied for the transPacific route from Hong Kong), and Cathay Pacific.[60] Cortazzi was allegedly in agreement with his British Airways contact that "the worst sit-

59. TNA, FCO40/1080.
60. TNA, FCO40/1074 (duplicated as HKRPO, HKMS189-2-326); *SCMP*, July 22, 1979, 11.

Economic Liberalization and Geopolitical Transformations 213

uation would be if Cathay were excluded . . . but either B Cal or Laker were allowed in." To Cortazzi, "it would be best if the two carriers were British Airways and Cathay."[61] However, other London officials disagreed. Claiming that he did not want to interfere in the business of the CAA, Secretary of State for Trade John Nott refused to direct that authority to issue a license to "an airline whose principal place of business was not in the UK."[62]

As the tension escalated, the Hong Kong Executive Council received a briefing on "Hong Kong's Air Links with Britain" in October 1979. The briefing noted that it was unusual for a single carrier to provide scheduled services on an intercontinental air route but that for reasons of "Hong Kong's constitutional relationship with Britain," British Airways held a de facto monopoly on the Hong Kong–London connection. To be permitted to operate scheduled services between the two cities, the three applicants filed applications with both the CAA in London and ATLA in Hong Kong. The two licensing authorities operated "under different statutory criteria."[63] Hong Kong regulations required that operators of all scheduled air services into or out of Hong Kong (with the exception of British Airways on routes specified by the governor) hold a license issued by ATLA. The potential tension with the CAA notwithstanding, this arrangement had "worked reasonably satisfactorily," according to the colonial government secretariat. Until 1979, the licensing body had not needed to hold any formal hearings because it had not received any objection. However, when Cathay Pacific, British Caledonian, and Laker Airways applied for licenses to fly between Hong Kong and London, issues arose. All three applications drew objections from the other applicants and from British Airways.[64]

Reminded that the Hong Kong government's attempt to persuade the British secretary of state for trade to intervene on Cathay Pacific's behalf had failed, members of the Executive Council were invited to note the different hearings of the CAA and ATLA and to "advise whether the Hong Kong Government should advocate the introduction of a second

61. TNA, FCO40/1075 (duplicated as HKPRO, HKMS189-2-327).
62. TNA, FCO40/1080.
63. TNA, FCO40/1081.
64. TNA, BT245/1922.

airline on the Hong Kong/London route."[65] The British government did not need to heed the colonial government's call to aid Cathay Pacific in its application. Nor did the Hong Kong authorities have to look favorably upon all three applications. In November 1979, ATLA announced its decision before the CAA's. Its decision "took many by surprise" because ATLA licensed not only Cathay Pacific, as it was generally assumed it would do, but also British Caledonian.[66] The Hong Kong authority played its cards cautiously: while looking out for the local applicant, as expected, it did not shut out the British competitors completely.

On March 17, 1980, the CAA announced its decision to grant a license to British Caledonian to operate a London (Gatwick)–Hong Kong service and deny the applications of Cathay Pacific and Laker Airways. Acknowledging the inadequacies of British Airways' services and the need for competition, the CAA also recognized the Hong Kong government's preference for Cathay Pacific "because of that airline's existing base at Hong Kong and Far East network." The authority was also satisfied that "Cathay Pacific was a British airline" for the purposes of the application and that all three applicants were financially and operationally competent. However, it found that Cathay Pacific, although "a good regional carrier," focused "primarily on the Asian routes" and would rely on British Airways, its 15 percent owner, for support. The CAA further concluded that Cathay Pacific had proposed an inappropriate type of aircraft, the Boeing 747, for the route and that a smaller aircraft such as the DC-10 would be more suitable. Cathay Pacific's proposal to start its London service with three flights per week was considered "uneconomic" because the Boeing 747 "had a very high break even load factor," which the CAA deemed unattainable. Although both British Caledonian and Laker Airways had proposed using the DC-10, the authority considered British Caledonian's proposal to cater for all fare categories more promising than Laker Airways' targeting of the bargain segment of the market. "Taking account of both aircraft type and marketability of service," the CAA found British Caledonian's proposal best suited for the route. The authority entertained the possibility of two additional carriers but

65.   TNA, FCO40/1081.
66.   TNA, FCO40/1083; *HKDCA*, 1979–1980, 15.

Economic Liberalization and Geopolitical Transformations   215

concluded that the market was not sufficiently large for the profitable operation of three carriers.[67]

The CAA decision aroused a strong reaction in the Hong Kong community. The colonial authorities told the Foreign and Commonwealth Office that the governor was "seriously disturbed by the force of public reaction." The Swires not only clearly indicated their "indignation" to the administrative bureaucracy but also wrote to MP Paul Bryan, chairman of the Hong Kong Parliamentary Group, stressing Cathay Pacific's contribution to the British economy through its commitment to Rolls-Royce-powered aircraft.[68] Senior Legislative Councilor Oswald Cheung concurred. Citing Cathay Pacific's purchase of Rolls-Royce engines worth tens of millions of pounds sterling, Cheung questioned in his March 27 speech to the council how the CAA in London could have "forgotten about reciprocity." After thirty years of British Airways' monopoly, "instead of reciprocating by granting a licence to Cathay Pacific," the CAA had "added a second British based carrier to the route." Cheung rejected the CAA's reasoning, which to him did "not bear examination for one minute." Cathay Pacific's proposed service with the Boeing 747 would indeed fit the traffic pattern ("peaks of both sides of the week end [sic] and dips in mid week . . . very high seasonal demand"), which the CAA had "completely ignored." He suggested that "the decision was political." London needed to rectify the situation, argued Cheung. Hong Kong had "consciously fostered the sentiment of buying British," but that sentiment was "a delicate flower." In spite of Hong Kong's belief in "the good faith of Britain," the CAA's decision cast doubt on whether the United Kingdom practiced what it preached. Cheung could "only hope that wiser counsel will now prevail, lest sentiment withers."[69]

Many others joined the chorus. The Hongkong and Shanghai Banking Corporation, which owned 25 percent of Cathay Pacific (for more on shareholding, see chapter 6), wrote to the governor expressing "dismay" at the CAA decision, calling its reasoning "ill-founded and wholly unacceptable." The bank questioned at a more fundamental level how the CAA

67. TNA, BT384/108. See also *HKDCA*, 1979–1980, 7, 15. See chapter 6 for a detailed discussion of British Airways' ownership of Cathay Pacific.

68. TNA, FCO40/1183.

69. LegCo, March 27, 1980, 673–74; *KSMN*, March 28, 1980, 8.

could vary from the "internationally-recognised principle of reciprocity" in granting air traffic rights. In a letter addressing Nott, the Hong Kong General Chamber of Commerce described Cathay Pacific's record as a Hong Kong–based airline serving a wide international network "second to none" and, in light of the CAA decision, expressed concerns about "the development of British interests in Hong Kong and around Asia." In another letter addressed to Nott, the Trade Development Council called attention to "the special relationship" between Hong Kong and Britain and asked Nott to support "Hong Kong's rightful claim."[70]

The Chinese community in Hong Kong was no less vocal in its complaints about the CAA's treatment of Cathay Pacific. The Chinese Manufacturers' Association of Hong Kong called the decision "lamentable." The association found it regrettable that the British government would not allow a Hong Kong–based airline to serve the Hong Kong–London route, especially after the tremendous efforts on Hong Kong's part to improve the trade situation with Britain. The British government had failed to consider Hong Kong's interests due to "political pressure at home and self-interest." The CAA's rejection of Cathay Pacific's application called into question "how much Hong Kong's interest actually weighed in overall British policy." The Federation of Hong Kong Industries similarly voiced its "grave concern" over what its members considered Britain's "unreasonable treatment" of the colony. The Chinese Executive Club was "shocked" that Cathay Pacific had been turned down in favor of British Caledonian. The Hong Kong Management Association called the British CAA decision a serious "affront," especially as it overrode the colony's ATLA decision. The Hong Kong Plastics Manufacturers' Association condemned the British decision as displaying "total disregard of normal trade reciprocity" and "complete neglect" of the colony's trade relationship with the United Kingdom. Garment manufacturers, as well as the Exchange Banks' Association, also launched complaints.[71]

---

70. TNA, FCO40/1183. See chapter 6 for a detailed discussion of HSBC's ownership of Cathay Pacific.

71. TNA, FCO40/1183; *KSEN*, March 29, 1980, 1; *KSEN*, April 2, 1980, 2; *TKP*, March 29, 1980, 4; *TKP*, April 2, 1980, 5; *WKYP*, March 29, 1980, 22; *WKYP*, April 2, 1980, 5.

Economic Liberalization and Geopolitical Transformations 217

The media was no more generous in its assessment. A Cantonese phone-in radio program logged callers' complaints about the decision, which featured accusations of "exploitation," "discrimination," "greed," "second-rate citizenship," "suppression," and "Britain's sell-out." All of the editorials in sixteen Chinese and four English daily newspapers, with a combined circulation of over one million, condemned the CAA decision. The *South China Morning Post* remarked that "London has displayed a shameless abuse of imperial privilege by shutting out the local airline and dividing the route between two British airlines." The Chinese *Wah Kiu Yat Po* commented that the decision, which "did not make sense," would "ruin the harmonious relation between Hong Kong and Britain." Referring to the removal of Hong Kong from the list of colonies as defined by the United Nations in 1972 (as requested by the PRC), *Ming Pao* pointed out the contradiction of a decision predicated on the subjugation of Hong Kong as a colony, raising the discrimination to the political level. The *Oriental Daily News* called it more "evidence of Britain's imperialistic attitude in ignoring the interests of Hong Kong." The *Kung Sheung Morning News* went further: "In the eyes of the British people, Hong Kong is only a hen that lays golden eggs. As long as it continues to lay eggs, other matters are not important." The pro-Beijing *Wen Wei Pao* noted that the only possible explanation for the CAA decision was "partiality." Governor MacLehose noted the "surprising and unprecedented" protest at the CAA's "failure to license Cathay Pacific (after ATLA had licensed both Cathay and BCal)" and the contagious spread in the community.[72]

The governor reported that many in the city were interpreting the CAA decision as "an example of Britain taking advantage of its constitutional relationship with Hong Kong." He saw grievances emerging from "official and commercial circles" charging that Britain derived benefits in air services agreements worldwide from "the Hong Kong card." Cathay Pacific, as "Hong Kong's airline," should be entitled to some reciprocity. To the public, the CAA decision underscored "the more anomalous aspects of colonial status." The opinion in Hong Kong was especially bitter due to "the contrast with Singapore, the other Chinese city state,

72. TNA, FCO40/1183; *KSMN*, March 23, 1980, 2; *SCMP*, March 18, 1980, 2; *WKYP*, March 21, 1980, 2.

218 *Catapulting Hong Kong*

whose airline has a frequent service to London." The governor warned that if the decision was upheld, "the tide of reaction could become stronger than any similar movement in his experience."[73]

A contentious but prompt appeal process ensued, with both Cathay Pacific and Laker Airways contesting the CAA decision. To support Cathay Pacific's case, the Hong Kong government cited the public reaction, which "came from a wide cross-section of the community (not just the 'taipans' and HK establishment)." The people of Hong Kong resented the "colonialist" attitude manifested in the denial of reciprocal rights to fly to London, "rights which are available to other countries." This public outcry, the Hong Kong government suggested, would make it difficult "to tip decisions on important contracts in favour of the UK." Highlighting the "intangible benefits of licensing a Hong Kong carrier susceptible to pressure from the Hong Kong community," the Hong Kong government acknowledged the advantage in fiscal and economic terms to Hong Kong. With its operational experience in Hong Kong and connections in the Far East, Cathay Pacific, "the Hong Kong flag carrier," could funnel traffic onto the Hong Kong–London route. The Boeing 747 would also be able to handle varying traffic demands with minimum stress to the overextended capacity at Gatwick and Kai Tak. Laker Airways' proposition to target the bargain market appealed to the free-economy mentality of Hong Kong, but the government questioned Laker Airways' cost and market forecasts. Therefore, the Hong Kong government considered it appropriate to grant Cathay Pacific a license with unlimited frequency

---

73. TNA, FCO40/1183. Singapore began to assert its autonomy from the United Kingdom in negotiations of air traffic rights long before its independence (TNA, FO371/127676). For its budding industry, British officials had brokered Singapore's connections with Southeast Asia and Australia (TNA, FCO141/15127). Ironically, BOAC and Australia's Qantas, both of which had provided instrumental assistance in getting Singapore Airlines' predecessor off the ground, ultimately came to block the airline's route expansion (Hickson, *Mr. Sia*, 82). Malaysia-Singapore Airlines managed to launch its service from Singapore to London in 1971 nonetheless (*Straits Times*, June 4, 1971, 1). Its service to Sydney began even earlier, in 1967, on chartered jets (*Straits Times*, April 6, 1967, 15). Although the Singapore carrier was seven years and nine years ahead of Cathay Pacific in its inaugural services to Sydney and London, respectively, the two carriers penetrated the North American market at the same time in 1983, thanks to the introduction of the Boeing 747 (*Straits Times*, May 7, 1983, 12).

Economic Liberalization and Geopolitical Transformations 219

and to reject both British Caledonian and Laker Airways: "the only solution which was right on the merits and met the statutory requirements."[74]

Three months after the CAA's initial decision, the British government announced a change of heart. On June 17, 1980, Nott directed the CAA to issue licenses to Cathay Pacific and Laker Airways, primarily on the basis of competition and the low fares proposed by the latter. Notifying the British prime minister on June 12, Nott called the route between London and Hong Kong "unique in the modern aviation scene in that it runs between two British points . . . and is therefore reserved for British airlines." Nott explained that the Hong Kong government had urged him to direct the CAA to grant a license to Cathay Pacific to quell the deep resentment in Hong Kong about the initial decision. Nott had decided not to leave out Laker Airways, whose proposal, he believed, would open the "untapped market at the bottom end of the price range." Hence, he ruled in favor of all four carriers: British Airways and British Caledonian, which had already been granted their licenses, and Cathay Pacific and Laker Airways on the strength of their appeals.[75]

Nott chose to make his decision public at the Dragonboat dinner held at the Hong Kong Association in London. He addressed what he deemed to be a pragmatic audience for whom "the romance of travel is less significant than the need for efficient services which arrive on time and give the passenger value for money." Claiming that he was not satisfied with the tight regulations on international civil aviation, he suggested that "a more liberal market environment" would better serve consumers and the government alike. In his public statement, Nott expressed conviction that his decision would be welcomed by air travelers in general, as well as "in Hong Kong where freedom to compete is one of the cornerstones on which the economic success of Hong Kong has been built."[76]

Also in public, Governor MacLehose exalted, "how delighted I am, how delighted Hong Kong is, that *our own airline* has at last been licensed to fly to London" (emphasis added). The *Sing Tao Evening News* found the reversal of the British decision significant for helping to wipe out the discord between London and Hong Kong, allowing closer cooperation

74. TNA, BT384/108.
75. TNA, FCO40/1184.
76. TNA, BT384/108.

between the two sides. Although the *Ming Pao Evening News* was not sanguine about the major benefits in terms of local air services, its editorial was particularly optimistic about what the revised decision meant for relations between Hong Kong and Britain: "We are no longer the 'chips' at the negotiating table where we are always the losers and Britain the winner." The *Oriental Daily* appeared cautious about the inevitably fierce competition among the four approved carriers but expressed hope that Britain would be smart enough to compare the economic benefits derived from Hong Kong and the New Hebrides, referring to the politically troubled archipelago in the South Pacific that had declared its independence from Britain and France in 1980.[77] Judging from the Chinese newspapers' response, the fight for "Hong Kong's airline" symbolized the battle for the metropole's regard for Hong Kong. After all, Hong Kong had grown from a reliant colony to a force to be reckoned with in its own right and was no longer content to be treated as a second-rate participant in its dealings with the British government.

The Hong Kong community felt vindicated in securing for Cathay Pacific a license for the London route. After the long and intense battle, Cathay Pacific and British Caledonian finally launched their scheduled services between Hong Kong and London in July and August 1980, respectively, thereby ending the monopoly that British Airways had held over that route for more than thirty years. Eager to celebrate its momentous victory, Cathay Pacific marshaled its resources and launched its service on July 16, 1980, just one day shy of a month since being notified of the success of its appeal against the original licensing decision.[78]

"Thank you for waiting," read the full-page advertisement Cathay Pacific placed in the *South China Morning Post* and the local Chinese newspaper *Wah Kiu Yat Po* on July 16, 1980 (figs. 5.4 a and b). "Today is a great day. We fly to London," proclaimed the proud Hong Kong–based airline. The advertisement in the Chinese newspaper also added that Chinese staff members of Cathay Pacific would be available at Gatwick Airport to help passengers with entry procedures, thereby affording them the convenience of a language-barrier-free entry experience upon landing.[79]

---

77. TNA, FCO40/1184.
78. Swire HK Archive, Cathay Pacific Airways Limited Annual Report 1980, 5.
79. *SCMP*, July 16, 1980, 4; *WKYP*, July 16, 1980, 13.

Economic Liberalization and Geopolitical Transformations   221

FIGURES 5.4 a and b. "Thank you for waiting. Today is a great day. We fly to London."
Sources: *SCMP*, July 16, 1980, 4; *WKYP*, July 16, 1980, 13.

The headline of an article in another Chinese newspaper underscored the significance of the inaugural flight in marking the end of a "successful battle for air traffic rights, breaking a monopolistic arrangement." Marking what the airline newsletter called "the most historic moment for the airline in recent years," Cathay Pacific popped open champagne to toast the full load of passengers on the inaugural flight. In a message to passengers in the airline's in-flight magazine, Managing Director H. M. P. (Michael) Miles recounted the struggle the airline had endured and expressed gratitude for "the tremendous support from the people of Hong Kong and our many friends elsewhere in the world," which had resulted in "Hong Kong's airline being granted the licence." As Cathay Pacific "moved beyond its traditional Asia-Australia-Middle East operating area,"

222 *Catapulting Hong Kong*

Miles pledged to maintain and enhance the reputation of the airline "of which Hong Kong is justly proud."[80]

The new service stimulated substantial traffic growth, resulting in high load factors. Competition led to improvements in standards of service and punctuality. In his annual remarks, the Hong Kong director of civil aviation called the new service "perhaps the most significant development during the year from the point of view of public interest." He reported "an almost bewildering variety of low fares," which had helped to develop the Hong Kong–London route "more rapidly than most people anticipated."[81] Cathay Pacific increased the frequency of its service on the route rapidly in response to the notable uptick in demand. By December 1980, the airline had increased its weekly service from three to four flights; by January, it had risen to five; and by June 30, 1981, it was offering a daily service.[82] Competition resulted in tremendous traffic growth. Passenger traffic between Hong Kong and London spiked a phenomenal 89 percent in a half-year period between 1980 and 1981 in which British Caledonian and Cathay Pacific began serving the route. This strong showing was followed by sustained growth of 35 percent in the subsequent year. With the end of its monopoly over the route, British Airways' share of the traffic declined to 63 percent from 1980 to 1981. Within a year, Cathay Pacific had grown its business to 78 percent of British Airways', claiming the number two spot at 36 percent of market share in 1981–1982 after British Airways' share, which had dropped further to 47 percent.[83] When results were tallied for the year 1982, Cathay Pacific had overtaken British Airways in passenger count in both directions.[84]

---

80. Swire HK Archive, CPA/7/4/1/1/168 Newsletter [January 1981]; *KSMN*, July 16, 1980, 5; *TKP*, July 15, 1980, 4; *TKP*, July 16, 1980, 4; *TKP*, July 17, 1980, 5; *WKYP*, July 16, 1980, 5.

81. *HKDCA*, 1980–1981, 18, 21.

82. Swire HK Archive, Cathay Pacific Airways Limited Annual Report 1980, 5.

83. *HKDCA*, 1979–1980, Appendix VI; *HKDCA*, 1980–1981, 18, 21, 24–25; *HKDCA*, 1981–1982, 24–25.

84. Exhibit CX-1, Cathay Pacific Airways' application to the Hong Kong ATLA dated May 14, 1983, for license to operate a B747 passenger aircraft at a frequency of one flight per week over the route Hong Kong/Bahrain/London (Gatwick) v.v. with Bahrain as an optional traffic stop (JSS, 13/2/12/2 London Route Correspondence).

Economic Liberalization and Geopolitical Transformations   223

One issue remained: Laker Airways now had the go-ahead from the British CAA, but the Hong Kong ATLA had denied Laker Airways' application in November 1979. In spite of the commendations of the director of civil aviation over the new low fares, Laker Airways, the proponent of targeting "the forgotten men and women at the bottom end of the market," continued to be shut out of the Hong Kong–London route. ATLA stood firm in its rejection of Laker Airways' application on the grounds that market demand did not justify more than three services. With Nott's reversal of the CAA's original decision, the Hong Kong government had supported Laker Airways' bid to fly the route, but ATLA exerted its independence before a Hong Kong judge, and there was no provision for an appeal or for directions to ATLA by the Hong Kong governor.[85]

Laker Airways' situation highlighted the potentially conflicting positions of the CAA in Britain and ATLA in Hong Kong. In the process of reviewing Laker Airways' application, the Hong Kong authorities altered their relationship with the British government and sought more symmetrical treatment. Noting the dual licensing procedure, the secretary of state remarked to the prime minister that Hong Kong had "somewhat archaic air transport regulations" but warned that "a price would be paid for . . . a change" in Laker Airways' favor. He further relayed to the prime minister the Hong Kong government's message that "the Executive Council in Hong Kong could not be expected to agree to a revision of the regulations to provide for appeals . . . unless British Airways lose their present exemption from Hong Kong Licensing procedures," an exemption that the Hong Kong government had granted in 1964. The topic was a sensitive one in this period. Although still a state-owned company, British Airways was preparing for privatization as part of Margaret Thatcher's larger deregulation campaign. The diplomatic machinery in London was well aware of the risk of British Airways losing the privileged position the Hong Kong government had granted it just as the airline was ceasing to be a state-owned company. The prime minister was advised "not [to] sacrifice the BA exemption to bring Laker onto the route."[86]

85.   TNA, BT 245/1922; TNA, BT 384/108; TNA, FCO 40/1184.
86.   TNA, BT 245/1854; TNA, BT 245/1922; TNA, BT 245/1923; TNA, PREM 19/1414.

224 *Catapulting Hong Kong*

Writing to Laker Airways founder Freddie Laker on July 17, 1981, Thatcher reiterated the British government's commitment to "greater competition between the airlines and a strong independent sector in the British industry." She allowed that the British government could bring Laker Airways onto the Hong Kong–London route, albeit at the price of "put[ting] at serious risk British Airways' routes out of Hong Kong." In particular, she noted Cathay Pacific's interest in taking over British Airways' routes, especially the Hong Kong–Johannesburg service. Thatcher told Laker that she was "not so sanguine as [Laker] that the interests of an incumbent airline" would be protected. She admitted that "the Hong Kong regulations, if they mirrored the system here, would contain a bias in favour of any Hong Kong-based airline just as the Civil Aviation Acts, except in respect of the London-Hong Kong route, contain a bias in favour of UK-based airlines." In the end, Thatcher supported her minister's conclusion that "the price was not worth paying." In a second letter to Laker Airways, acknowledging again the bias in the British rules, Thatcher conceded that "as long as that is so, it is difficult to object to the reverse bias in the Hong Kong licensing procedures."[87]

Both Cathay Pacific and Laker Airways had sought to expand their networks as deregulation of the airline industry was opening up the skies to competition. However, as the hard-fought battle over the Hong Kong–London route demonstrates, deregulation came with severe limitations. Licensing authorities remained guarded in their approval of new carriers and continued to uphold nationalist agendas. In the case of Hong Kong and Cathay Pacific, the proceedings pitted the colonial authorities against their counterparts in the metropole. In the liberalizing climate of deregulation, Thatcher and her government noted the asymmetry of certain preexisting arrangements and acquiesced to the colony's assertion of greater control.

As Cathay Pacific and British Caledonian entered the fray in the Hong Kong–London market, Laker Airways proceeded to apply to the CAA for a license to commence service from London, but only to Sharjah in the United Arab Emirates, its proposed connection point en route to Hong Kong. British Airways launched further complaints about the midpoint rights of its competitors on the Hong Kong–London route

---

87. TNA, BT 245/1923; TNA, PREM 19/1414.

(Bahrain for Cathay Pacific and Abu Dhabi for British Caledonian). The battle continued for the survivors, but for Laker Airways, which had also applied for a trans-Pacific license to fly to Canada and the West Coast of the United States, the fight was over with the airline succumbing to financial difficulties in early 1982.[88]

Perpetuating the China Dream

Cathay Pacific's entry into the London market reflected the airline's successful lobbying efforts in both Hong Kong and London. However, the airline's accomplishment in this market led to complications in its other venture. Just as the saga over the Hong Kong–London route was unfolding, the PRC began to open airways that had remained largely closed to foreign players since the early 1950s. The initial overtures assumed the humble form of nonscheduled flights toward the end of 1977 from Shanghai to Hong Kong to carry perishable foodstuffs—in particular, freshwater crabs.[89] In the ensuing months, British officials approached their Chinese counterpart to make more formal arrangements.

At that time, while the mainland Chinese market appeared promising, British officials in London and Hong Kong had to handle the situation delicately in order to safeguard another market they treasured—Taiwan. Taiwanese routes were generating tremendous profits for commercial airlines. For Cathay Pacific, earnings from the Taiwan route totaled US$80 million, or 27 percent of the airline's total earnings, for the year ending March 31, 1978. Cathay Pacific faced significant financial consequences if it could not fly to Taiwan.[90]

---

88. TNA, BT 245/1854; TNA, BT 384/113; TNA, PREM 19/1414; *HKDCA*, 1980–1981, 18. Cathay Pacific had offered a service to Bahrain via Bangkok as early as November 1976 (*SCMP*, August 10, 1977, 21; *SCMP*, August 22, 1977, 36).

89. TNA, FCO 40/791.

90. TNA, FCO 40/984 (duplicated as HKPRO, HKMS189-2-260); TNA, FCO 40/985 (duplicated as HKPRO, HKMS189-2-261). Whether Cathay Pacific could serve both Taiwan and the mainland Chinese markets had worried its management and the British authorities since 1965 (TNA, BT 245/1060).

226　　　*Catapulting Hong Kong*

The mainland Chinese government held firm. It insisted on the British government making a formal public statement that affirmed, among other things, that the agreement between the PRC and Britain was an agreement between states whereas Hong Kong–Taiwan flights were "nongovernmental regional air traffic." The mainland government also insisted that the British government not recognize the signage on the Taiwan-controlled China Airlines aircraft as "national insignia," or China Airlines as "an airline representing a state." The mainland Chinese authorities also demanded other provisions such as priorities at Kai Tak for its own airline. British officials were apprehensive of denying China Airlines national status as similar concessions on the part of Japan had led to the cessation of air traffic to Taiwan for over a year between April 1974 and September 1975. In the end, the Taiwanese and Japanese authorities issued operating permits to the airlines as private companies in lieu of any formal air service agreements between the governments, and Japan Airlines opened a dummy company, the wholly owned subsidiary and one-destination carrier Japan Asia Airways, to operate the Japan–Taiwan route.[91]

As British and mainland Chinese diplomats strove to secure flight routes for their airlines, they agreed on a clear articulation of hierarchy that differentiated the Beijing–London connection from the Hong Kong–Taiwan connection: an air services agreement between two governments for the former vis-à-vis an agreement between airlines for the latter; state airlines operating the trunk route between the PRC and the United Kingdom vis-à-vis private airlines flying between Hong Kong and Taiwan. This arrangement paved the way for further negotiations. What was at stake for the British negotiators was more than merely the point-to-point connections. They were eager to gain Beijing's permission for British Airways to fly over China for its connections to Hong Kong, a convenience that was estimated to result in a savings of £2 million per year on the pattern of service in 1978. The subsequent negotiations proved en-

---

91.　TNA, FCO 40/985; JAL Group News, "Planned Integration"; Hsiao, *Foreign Trade of China*, 68. In the early 1990s, Qantas also set up a subsidiary, Australia Asia Airlines, to service the Taiwan route (*Canberra Times*, October 12, 1991, 16). Similarly, British Airways set up British Asia Airways to operate flights to Taipei (*The Independent*, April 23, 1993; *The Independent*, September 14, 1996).

during. By April 1979, the two sides had agreed to the wording of the announcement:

> The People's Republic of China and the United Kingdom of Great Britain and Northern Ireland have handled the matter of regular air traffic between the two countries in accordance with the principles laid down in the 1972 Agreement between them on raising the level of their diplomatic relations to that of Ambassadors. The Air Services Agreement between the People's Republic of China and the United Kingdom of Great Britain and Northern Ireland is an agreement between the two states, while the airline between Hong Kong and Taiwan is operated under non-governmental arrangements.

Eventually, after a ten-week marathon session in Beijing, the British and Chinese negotiators arrived at a UK/China Air Services Agreement, which Mao Zedong's designated successor and the PRC's titular leader Hua Guofeng signed during his visit to London in November 1979.[92]

The air service agreement did not mention Hong Kong at all as the two sides handled delicately the issue of sovereignty, a critical concern in the context of air service. If Hong Kong were under Chinese sovereignty, Chinese airlines could freely operate all services between Hong Kong and the rest of China simply with the blessing of the Chinese government. The Chinese negotiators questioned references to "territory" in the agreement on the regional routes. They would not allow any implication that Hong Kong was British territory. The two sides finally agreed on the understanding that "references to 'territory' . . . will in no way prejudice the position of either Government with regard to the status of Hong Kong." As the negotiations continued, the Chinese side had even proposed the spelling of "Xianggang" instead of "Hong Kong" in the English text of the agreement. In the end, "Hong Kong" prevailed. The discussion was pregnant with territorial assertions. Through the articulation of aerial connection rights, the Chinese authorities were eager to declare their territorial claims over the points connected.[93]

---

92. TNA, FCO 21/1712; TNA, FCO 40/985; TNA, FCO 40/1077; TNA, FCO 40/1079; TNA, FCO 76/2274; *SCMP*, July 27, 1979, 1.

93. TNA, FCO 40/1078; TNA, FCO 40/1079.

228   *Catapulting Hong Kong*

Having acknowledged (and sidestepped) each other's claims to sovereignty, the two sides reached a practical compromise. They prepared two confidential memoranda of understanding (CMUs) to accompany the air service agreement. One dealt with the London–Beijing trunk route and the other, the Hong Kong–China regional routes. The British government was not concerned with whether the regional routes were covered in the air services agreement as much as that officials from London, not Hong Kong, retained control of the negotiations because air traffic rights about Hong Kong were one of the "most powerful bargaining counters in general air service negotiations."[94]

For the regional route, Beijing allowed a "Hong Kong-based airline" one-third of the Hong Kong–Shanghai traffic, as well as the promise of an undefined share of the Hong Kong–Beijing traffic once British Airways ceased to serve the London–Beijing route by way of Hong Kong. In exchange the PRC-controlled Civil Aviation Administration of China (CAAC) commanded traffic rights between Beijing, Shanghai, any four other cities (later expanded to seven), and Hong Kong.[95]

Even after the negotiations on these agreements had concluded, implementation was only partial and acrimonious. The CMU on London–Beijing services allowed each side to operate twice a week, but in reality market demand could barely justify the one service each side started offering in November 1980. The situation of the regional services devolved into a lopsided arrangement in which CAAC routes proliferated and frequency increased while Cathay Pacific's services remained confined to a few weekly flights between Hong Kong and Shanghai. Although the airline commenced its services between Hong Kong and Shanghai in February 1981, it was restricted to two weekly trips with its Boeing 707. There was also a significant, and unforeseen, complication: the CMU governing the regional routes stipulated that the Hong Kong–based airline the British government designated to operate between Hong Kong and Shanghai could not also fly between Hong Kong and London. The Chi-

---

94. TNA, FCO 40/1078.

95. TNA, FCO 40/1078; TNA, FCO 76/2273; TNA, FCO 76/2274. Cathay Pacific's chairman reported in his annual statement of April 17, 1980, that the airline had been "officially designated for the Hong Kong/Shanghai route" (Swire HK Archive, Cathay Pacific Airways Limited, Chairman's Statement, April 17, 1980, 3).

nese government had insisted on this provision lest a British airline provide a direct service between Shanghai and London that the CAAC could not match.[96]

Herein lay Cathay Pacific's problem. The British government had duly designated Cathay Pacific as the British Hong Kong–based airline for the Hong Kong–Shanghai route, but Cathay Pacific had subsequently received approval for the Hong Kong–London route, which it started offering in July 1980. Becoming suspicious of Cathay Pacific even before it began to operate the Hong Kong–London services, the Chinese authorities questioned its network expansion and its relationship with British Airways, especially their shareholdings in each other. One month after Cathay Pacific's inaugural flight from Hong Kong to London, Chinese authorities grounded the airline's services to Shanghai. Paid customers received refunds for their unused tickets. To come up with an arrangement to placate Beijing, Cathay Pacific had proposed reviving its wholly owned subsidiary, Hong Kong Airways, which had not operated for some time (see chapter 2). This proposal stipulated that Hong Kong Airways could lease aircraft from Cathay Pacific and fly between Hong Kong and Shanghai. However, the Chinese government did not find this arrangement acceptable. To provide Beijing with further assurance, Cathay Pacific pledged not to sell through tickets from London to Shanghai.[97]

The issue of Cathay Pacific's service between Hong Kong and London remained unresolved and introduced another layer of complication in the ongoing attempts by the British government to achieve a more equitable share of air traffic between mainland China and Hong Kong. As the discussions dragged on, the Chinese government refused to formally accept the British designation of Cathay Pacific for the Hong Kong–Shanghai services but allowed the Hong Kong–based carrier to continue to operate on that route on an interim basis. Although the British government found that arrangement unsatisfactory, they expected it to be

---

96. TNA, FCO 40/1182; TNA, FCO 76/2273; TNA, FCO 76/2274; Swire HK Archive, Cathay Pacific Airways Limited, Chairman's Statement, March 25, 1982, Schedule C.

97. TNA, FCO 40/1182; TNA, FCO 76/2273; TNA, FCO 76/2274; *KSMN*, August 17, 1980, 8; *WKYP*, August 17, 1980, 5.

230 *Catapulting Hong Kong*

extended.[98] However, the continuation of Cathay Pacific's service entailed further aggravation of this traffic imbalance. In January 1981, CAAC informed the Hong Kong authorities that the PRC carrier planned to operate "no less than 242 return services," including 108 scheduled services, 123 extra sections, and 11 chartered flights between points in the mainland and Hong Kong. In contrast, Cathay Pacific had only applied to operate four chartered services in February. Evidently, there was gross asymmetry in the traffic pattern. The situation was so grave that MacLehose wrote to the Foreign and Commonwealth Office, suggesting that the British reconsider the extension of permits if the PRC authorities remained "totally unresponsive about ways of implementing . . . the regional CMU."[99]

By the time the year was over, the British tallied that in 1981, CAAC offered forty-two scheduled services per week from seven cities, along with one thousand and seven hundred extra flights that year, generating US$35.5 million worth of earnings for the PRC carrier. By comparison, Cathay Pacific's two scheduled services per week between Hong Kong and Shanghai produced only US$3 million worth of earnings. Worse still, CAAC extended their flights linking Hong Kong to Shanghai and Beijing on to Nanjing and Tianjin with four-hour stopovers in these cities, prompting the British government's complaints in 1982 about a breach of conditions.[100] As MacLehose explained in November 1981, "Hong Kong's primary objective is to achieve reasonable, although not necessarily precise, balance" for Cathay Pacific and CAAC on both the Shanghai and Beijing routes. Why was it such an important concern to the British that Cathay Pacific should have a fair share of the Hong Kong–Shanghai and Hong Kong–Beijing traffic? The business might have appeared limited, but that market was growing fast. In fact, the situation appeared so grave to certain British negotiators that they considered serving the Beijing government their one-year notice of withdrawal from the CMU governing regional services if not the entire air services agreement itself.[101]

98. TNA, FCO 76/2275.
99. TNA, FCO 76/2033.
100. TNA, FCO 40/1478.
101. TNA, FCO 21/2154; TNA, FCO 21/2155; TNA, FCO 40/1478; TNA, FCO 76/2034; TNA, FCO 76/2274.

Economic Liberalization and Geopolitical Transformations 231

Such a threat or consideration of the British tough negotiation tactics notwithstanding, the British government and the colonial authorities in Hong Kong were preoccupied at the same time with a more far-reaching issue—the future of Hong Kong. Some called for caution so as not to "risk the coincidence of a showdown with the PRC on air services with negotiations on the highly delicate question of the future of Hong Kong."[102] In November 1982, the secretary of state wrote to the Foreign and Commonwealth Office, copying the prime minister, requesting instructions on air services discussions so as to avoid "risking damage to the wider negotiations which the Prime Minister set in train."[103] Others advised that Britain not allow the Chinese to believe that they could be bullied: "It would be a bad omen for the future of Hong Kong if the UK was seen to be already incapable of supporting Hong Kong's aviation interests." The two issues were intricately related. "The root of the difficulty (and CAAC's obstructionism)," a British diplomat in Beijing wrote to the Foreign and Commonwealth Office in March 1982, "is the question of the status of Hong Kong (which CAAC say is part of China, and that therefore the principle of mutual benefit does not apply to the 'regional' services!)." If the British government failed to resist "a perpetuation of CAAC's unique advantages" and seek reciprocity for a Hong Kong airline, "the implications for 1997 are ominous," the diplomat warned.[104]

Governor MacLehose appeared more guarded. In April 1982, he wrote to remind his colleagues in London that they had "never asserted a claim to normal reciprocity on regional services" and that Cathay Pacific's introduction of the London route fogged the issue. Yet, he suggested that the British negotiators could point out that the Chinese government's heavy-handed tactics in its commercial relations with Hong Kong could undermine confidence. The overarching issue was clear: "To commit ourselves to a show-down with CAAC on an issue which the M[inistry] of F[oreign] A[ffairs] may view as one of sovereignty, at that precise juncture, is obviously risky." "I do not agree," MacLehose concluded, "that to avoid taking issue at this time would be prejudicial to the wider negotiations, though it would be to CPA." Therefore, he advocated not giving

102.   TNA, FCO 76/2272.
103.   TNA, FCO 40/1478.
104.   TNA, FCO 21/2154.

the Chinese an ultimatum in hopes of enhancing Cathay Pacific's share of air traffic: "in the final analysis," the wider negotiations were more critical for "both for the Hong Kong Government and Swires."[105]

Characterizing the problem of Hong Kong–China air services, Alan Donald, the assistant undersecretary of state (for Asia and the Pacific) attributed it to "the oddity of Hong Kong's position" and the advantage the British commanded because of "China's need for Hong Kong" at that time. However, it necessitated in diplomatic discussions "some careful footwork" so that neither side needed to confront "the delicate question of sovereignty." The Chinese had already scored air services points by claiming sovereignty, but the British negotiator needed to avoid that spilling over to sour the atmosphere for the wider negotiations over 1997, he cautioned. Political considerations would continue to restrain their tough posture in air services discussions "as long as we are haggling over the future." Thatcher's visit in September 1982 would not conclude the discussions, he forewarned, "so uncertainty could last for months or even years."[106]

That prediction came to pass in the following years. The asymmetry of traffic volume between Cathay Pacific and CAAC persisted. In the year ending March 1983, connecting Hong Kong with numerous mainland Chinese cities, CAAC embarked some quarter million passengers, and disembarked a similar number at Kai Tak. In the same period, Cathay Pacific handled some seventeen thousand passengers in either direction on their Shanghai route, their only connection between Hong Kong and the mainland.[107] The situation was similar in the following year, although traffic to Shanghai had grown for Cathay Pacific.[108]

In December 1984, the British and Chinese governments signed the Joint Declaration, which resolved the political issue of Hong Kong beyond 1997. Section IX of Annex I expressed the intention to maintain Hong Kong as a center of international and regional aviation and specified that "airlines incorporated and having their principal place of business in Hong Kong and civil aviation related businesses may continue to

105. TNA, FCO 21/2154.
106. TNA, FCO 21/2155.
107. *HKDCA*, 1982–1983, 26.
108. *HKDCA*, 1983–1984, 29.

operate." The provision granted Hong Kong autonomy in the negotiation of air service agreements, but the city's air connections with other parts of the PRC were to be decided by the Beijing regime in consultation with the Hong Kong government. Such connections between the mainland and Hong Kong would be provided by "airlines incorporated and having their principal place of business in the Hong Kong Special Administrative Region and other airlines of the People's Republic of China."[109] London officials understood that they needed to "disentangle . . . Air Services Agreements which . . . include routes to and from Hong Kong as well as routes to and from London so that Hong Kong has a set of its own free standing agreements." The unusual provision in the Joint Declaration was that Hong Kong traffic rights were to be available to airlines that called Hong Kong their "principal place of business," rather than those meeting the traditional requirements of substantial ownership and effective control. This accommodation allowed the continued operation of Cathay Pacific, which could not claim that its ultimate control rested in Hong Kong at that time.[110]

With this new definition of qualified players, competition heated up in Hong Kong. Jardine Matheson, which had competed with Swire through Hong Kong Airways, had resurfaced in the late 1970s, sending delegations to Beijing to discuss CAAC services.[111] What appeared in the mid-1980s was a new cast of players. In July 1985, the colonial government reported that in addition to the city's longstanding carrier, up to three Hong Kong–based airlines had been established in hopes of operating services to mainland China.[112] Among them, Dragon Airlines obtained a license from the Hong Kong ATLA in December 1985 to operate scheduled services between Hong Kong and eight Chinese cities. This newly established carrier had obtained its air operator's certificate in July 1985 and immediately offered nonscheduled services to Xiamen.[113]

109. TNA, FO 93/23/75.
110. TNA, BT 245/1968.
111. TNA, FCO 40/983; TNA, FCO 40/984; TNA, FCO 40/1077; TNA, FCO 40/1078; TNA, FCO 40/1079.
112. TNA, BT 245/2048.
113. *HKDCA*, 1985–1986, 7, 23, 26.

At the same time, the promise of traffic growth finally materialized. The director of civil aviation reported a significant increase in passenger travel between Hong Kong and mainland China. In April 1985, Cathay Pacific finally began their third weekly service between Hong Kong and Shanghai. Between July and October 1985, they could even fly a weekly extra service on the same route. The long-awaited Cathay Pacific flight between Hong Kong and Beijing also came to fruition, albeit as non-scheduled services between July 1985 and March 1986. Not to be left behind, CAAC added capacity to their scheduled services with enhanced frequencies and larger aircraft. The PRC operator also registered a 150 percent increase in nonscheduled passenger services in the year ending March 1986.[114]

China held tremendous growth potential for commercial aviation not only in the region but on a global scale. Hong Kong was in a prime position to leverage this opportunity. Nonetheless, after years of struggle, it appeared even more difficult to penetrate this emerging market than to extend Cathay Pacific's reach to the established locations in Australia, North America, and Europe.

Hong Kong came into its own as a commercial aviation hub at this crucial juncture. With its catchment area of Southeast Asia and long-range connections to Australia, Europe, and North America, Hong Kong's geographic location continued to play a significant role. However, only with favorable political circumstances could the hub's own carrier develop its full potential. Decolonization animated geopolitical developments in Southeast Asia, but the forces pushing for autonomy were not inexorable. Not only did Hong Kong remain a British colony, but the colonial authorities also managed to wrest more power from the metropole, at least in commercial aviation, in the 1970s. The overlapping route discussions discussed herein reveal the multifaceted relations between London and the British territory of Hong Kong in the era. While continuing to underwrite the Hong Kong carrier's reentry into Australia and its expansion to North America, the British government strove to protect UK-based British interests.

---

114. *HKDCA*, 1985–1986, 23.

Economic Liberalization and Geopolitical Transformations   235

Economic takeoff in Hong Kong and elsewhere in the region resulted in more determined local assertions in the commercial aviation arena. At the same time, what constituted "British" or "local" had become more fluid and dynamic. As local forces (or more precisely local British forces in Hong Kong) took hold, power at the metropole eroded. The most notable demonstrations of that phenomenon in the commercial aviation arena were British Airways' yielding of the Sydney route to Cathay Pacific and the British designation of Cathay Pacific for the Hong Kong–Vancouver route. The most iconic was, of course, the inauguration of Cathay Pacific's flight to London, ending British Airways' monopoly over the Hong Kong–London route. As monumental as these specific routes may have been, their realization took a rather gradual course. In the early 1960s, BOAC's share of air passenger and air cargo traffic through Hong Kong hovered in the low double digits. The corresponding percentages had dipped into the single digits by the early 1970s. By 1982, British Airways commanded a meagre 3 percent of passenger and cargo traffic through Kai Tak.[115]

British Airways' monopoly over the Hong Kong–London route gave way to competition not simply because of the economic rise of the British outpost but also thanks to a political climate favoring deregulation and privatization. Deregulation and privatization had softened the titans, particularly in Britain and the United States. Industry deregulation led to a bilateral liberalization of air traffic rights between countries. The success of Cathay Pacific in extending its route map to London cannot be easily construed as a bilateral arrangement in the usual sense of air traffic negotiations but more aptly as underscoring the long-overdue recognition of reciprocity between the colony and the metropole within the British zone.

Nor would Cathay Pacific's introduction of long-haul flights have been possible without the advent of long-range jumbo jets. This new technology extended the airline's reach and offered expanded capacity, two factors paramount to the elimination of the diplomatically problematic issues of en route stops and service frequency. An environment of relaxed regulations, coupled with technological breakthroughs, allowed Hongkongers to exert control of the skyways.

115.   *HKDCA*, 1960–1961; *HKDCA*, 1970–1971; *HKDCA*, 1981–1982.

236 *Catapulting Hong Kong*

As the PRC reopened its airways, the reengagement of Hong Kong, though challenging from the standpoint of negotiation and implementation, reaffirmed Hong Kong's pivotal role. The industry had come full circle. Once again, the market had come to be animated by the same forces that propelled its development in the 1930s and 1940s—the allure of the mainland Chinese market and the prime position of Hong Kong as a nexus to connect not just the regional traffic but flights from as far afield as Australia, North America, and Europe.[116] Nevertheless, the intervening decades had witnessed seismic shifts in geopolitical dynamics. Concomitant with the reemergence of the mainland Chinese market was the PRC's insistence on crafting the rules of engagement, especially as the future of Hong Kong loomed large on the horizon. Various factors—technology, diplomacy, and reengagement with the PRC— would propel the growth of Hong Kong's commercial aviation to another level and at the same time lead to a rebalance of power among the industry participants.

As Hong Kong came to enjoy an elevated economic standing, British politicians and corporate executives in both the colony and the metropole negotiated its flagship carrier's penetration of international aviation networks. Through Cathay Pacific's extending reach, Hong Kong translated its economic success into an expanded presence in the world of commercial aviation. As the colonial government channeled Hong Kong's burgeoning financial prowess to fund an infrastructure upgrade, the colony's budding airline capitalized on the commercial availability of jumbo jets to leapfrog into the long-haul market. Such groundwork primed Hong Kong to take advantage of the opening skies as deregulation transformed the airline industry. As the colony's economy flourished, Cathay Pacific broke free from its regional configuration and arrived at

---

116. The allure of the mainland market had long captivated business operators (see Varg, "Myth of the China Market," for a discussion of an earlier period). Through the lens of commercial aviation in Hong Kong, foreign operators were especially vulnerable to such allure at strategic locations from which they planned to penetrate the mainland Chinese market. In the case of Cathay Pacific, the significance of the mainland market is real. One in ten of the airline's passengers came from the mainland in 1994. In the same year, 70 percent of the airline's cargo originated in the mainland (Clifford, "Mainland Bounty").

Economic Liberalization and Geopolitical Transformations   237

faraway ports in Australia, North America, and Europe. Cathay Pacific's expansion from regional business into the long-haul market underscores not only an airline's commercial accomplishment but also Hong Kong's coming into its own. The Hong Kong carrier's extended reach was but the material manifestation of the city's economic takeoff and growth into a global metropolis.

CHAPTER 6

Recasting Hong Kong

The Making of a "Hong Kong" Airline

Taking on a more Chinese identity is only prudent for Cathay as its home territory, Hong Kong, prepares to be handed back to China in 1997. But analysts and even some Cathay executives acknowledge that shedding a colonial image will require a lot more than a few buckets of paint.

—"Going Native: Cathay Strives to Shake Off Vestiges of Colonial Times," *Asian Wall Street Journal*, August 19, 1993[1]

Mounting tension among the world's great powers has directed scholarly attention to exploring how corporations mitigated political risks in response to shifting geopolitics.[2] The nationality of investors could arouse suspicion, especially in times of regime change. Although the condition of national ownership of key businesses is prevalent in numerous countries in different eras, the case of post–World War II Hong Kong posed a particularly tumultuous situation as the city experienced a rapid succession of political regimes in a few short decades. Regarded as Hong Kong's airline, Cathay Pacific, with a controlling stake in the hands of British investors, continues to face pressure from the PRC regime, which facilitates the routing of many of its flights. Indeed, this situation is not unique to Cathay Pacific or the airline industry. As financial capital wields power in directing commercial flows, political regimes insist on specific investor profiles in industries they consider strategic. The state could challenge the nationality of investors in such industries, especially those with a controlling stake. In the international nexus of Hong

---

1. *Asian Wall Street Journal*, August 19, 1993, 1.
2. For an introduction to this literature, see Casson, "International Rivalry."

The Making of a "Hong Kong" Airline   239

Kong, the strategic industry of commercial aviation has remained in private hands. Focusing on Cathay Pacific, this chapter examines how the privately owned airline fashioned its corporate nationality in a bid to negotiate with political forces that affected its business development.

International businesses and multinationals often convey their "corporate nationality" to accommodate the needs of their host country. Rather than a formal calibration, "corporate nationality" is a cultural construct that management projects to suit their business exigencies.[3] This issue is especially acute in the age of decolonization. Cathay Pacific had ridden the wave of resuming colonial power as British forces returned to Hong Kong in the aftermath of World War II. In an about-face decades later, as the city prepared for a transfer of sovereignty, Cathay Pacific faced the challenge of a new state on which the airline had to rely for continued operational success. Rather than moving from a colonial setting to a national formation, Cathay Pacific's home base was transformed from a British colony to a Special Administrative Region (SAR) in the PRC. In other words, the airline had to prepare to operate under the auspices of an incoming political regime that was to underwrite its business prospects. Not only was the PRC the state in whose name routes were to be negotiated and under whose airspace Cathay Pacific had to fly, but the business opportunities that China in reform afforded also compelled the Hong Kong carrier to signal its commitment. The resulting relationship between Cathay Pacific and the PRC was one that acknowledged mutual benefits. More than the appearance of a local identity, the incoming regime held Cathay Pacific answerable to a gradual recalibration of its shareholding. In response, Cathay Pacific constructed its corporate nationality as a local Hong Kong, if not entirely Chinese, airline.

Throughout its history, Cathay Pacific reshaped its corporate nationality repeatedly, less to respond to the local conditions of its host city than to meet the demands of faraway state powers who were to bankroll the airline's success. After a short stint as an enterprising American-Australian venture, the airline came under the ownership of a British Commonwealth conglomerate (see chapter 2). This chapter focuses on the airline's later campaign to project an image of local Hong Kong to respond to the shifting political climate of Hong Kong. Contrary to the

---

3.   Gehlen, Marx, and Reckendrees, "Ambivalences of Nationality."

conventional view of Cathay Pacific as a quintessential British company, the changing investor profile of the airline reveals a pragmatic business approach. In addition to engineering a shareholding, the airline also altered the face of the cockpit to project a more local image. Paralleling the flexible practice of citizenship that Hongkongers mastered,[4] the airline showcased its agility in constructing its corporate nationality during a period of swift geopolitical shifts that both engendered business opportunities and presented challenges of commitment to Cathay Pacific as well as Hong Kong.

Geopolitical Jockeying

Often celebrated as a laissez-faire economy, Hong Kong was touted as the prime example of economic success with minimal government intervention. Perhaps this narrative is most often cited in Milton Friedman's assessment of Hong Kong as an economy operating mostly on the basis of market dynamics.[5] Other scholars who have echoed Friedman's observations and analyses of Hong Kong's development have also privileged economic freedom as a facilitator of growth.[6] Such narratives of government nonintervention in Hong Kong have been challenged given state involvement in specific activities. Early studies of industrial development in Hong Kong have underscored the government's role in the process.[7] Similarly, scholars have analyzed government intervention in economic activities that involved social issues such as housing in Hong Kong.[8] Recent studies seem to focus on the government's role in regulating the financial markets.[9] Situating the analysis in a comparative framework, some scholars have analyzed the different paths taken by the Little Drag-

---

4. Ong, *Flexible Citizenship*.
5. Friedman and Friedman, *Free to Choose*, 34.
6. Li, *Economic Freedom*.
7. England and Rear, *Industrial Relations*.
8. Keung, *Government Intervention*.
9. Donald, *Financial Centre*.

ons.[10] Yet these studies often emphasized the function of government in solving coordination problems and overcoming market imperfections.[11]

National ownership has attracted scholarly attention for different reasons. Some have studied the role of national ownership in the process of economic reconstruction or reconfiguration against a shifting geopolitical context.[12] Others have assessed the impact of state ownership on economic performance.[13] National ownership of companies in sensitive industries is but one mechanism through which the state can exercise regulatory control over economic activities it deems critical.[14] In the process of reversing state ownership, privatization in different forms and to varying degrees can also affect business structures.[15]

Besides the dichotomy of state-owned enterprises and privately held businesses, corporations have also assumed various shareholding profiles to suit specific purposes. Scholars have studied how a big corporation could organize a subsidiary to broaden ownership and develop political and cultural identities among small investors.[16] Broadening the discussion beyond shareholding to other stakeholders' involvement, companies employ business strategies such as corporate social responsibility to enhance negotiating and bargaining positions while maintaining control of their assets.[17] Building on existing scholarship, this chapter explores the instrumentality of crafting "corporate nationality" through shareholding in allaying political risks and managing business relationships with the state.

Deemed vital to the ruling regime, the airline industry relies on diplomatic intervention to grow and often operates under state ownership,

---

10. Li, *Capitalist Development*.

11. Aoki, Kim, and Okuno-Fujiwara, *Role of Government*.

12. See, for example, Van Hook, "From Socialization to Co-Determination"; Vahtra, Liuhto, and Lorentz, "Privatisation or Re-nationalisation?"; Mitchell and Fazi, "We Have a (Central) Plan."

13. See, for example, Doh, Teegen, and Mudambi, "Balancing Private and State Ownership"; Cohen, "Divergent Paths."

14. See, for example, Pearson, "Business of Governing Business."

15. Lee and Jin, "Origins of Business Groups."

16. Collier, Chandar, and Miranti, "Marketing Shareholder Democracy."

17. Abdelrehim, Maltby, and Toms, "Corporate Social Responsibility."

242 *Recasting Hong Kong*

at least partially. However, on the issue of nationality in the airline industry, scholarly coverage has been sporadic. Derek Levine has explored government involvement in the aviation industry in mainland China, but his analysis addresses mostly the hardware issues of aircraft construction.[18] Gordon Pirie has studied British imperial aviation, and Ken Hickson, as well as Loizos Heracleous, Jochen Wirtz, and Nitin Pangarkar, have published celebratory accounts of Singapore Airlines.[19] In the case of Hong Kong, Cliff Dunnaway has examined the history of the city's aviation industry and has argued that the success of Cathay Pacific caused Hong Kong to become a world city and the gateway to China.[20] State involvement has received scant attention in the study by Lau Chi-pang, who detailed the collective memory of Hong Kong airports, and James Ng, who has chronicled the development of aviation in Hong Kong.[21]

Rather than accepting the ownership of airline companies by the state (for example, British Airways) or in private hands (for example, Pan Am) as a simple dichotomy in business strategies, this analysis investigates Cathay Pacific's changing shareholding as the result of an active negotiation process between the airline and the state over the issue of corporate nationality. Ownership and control of Hong Kong's flag carrier, Cathay Pacific, remained in private hands. Under the auspices of its longtime parent Swire, the airline crafted its shareholding at critical junctures to respond to shifting geopolitics and exercised control over itself to accommodate the needs of various political regimes. Impressions of corporate nationality bore significant implications not only for the accrual of economic profit but also for the state's effort in steering air traffic to specific participants.

This analysis builds on the studies of businesses operating in a decolonizing context in which enterprising operators construct legitimacy against the backdrop of changing political configurations.[22] The politi-

18. Levine, *Dragon Takes Flight.*

19. Pirie, *Cultures and Caricatures*; Hickson, *Mr. SIA*; Heracleous, Wirtz, and Pangarkar, *Flying High.*

20. Dunnaway, *Hong Kong High.*

21. Lau, Wong, and Chin, *Tiankongxia de chuanqi*; Ng, *Xianggang hangkong 125 nian.*

22. Abdelrehim et al., "Ambiguous Decolonisation"; Decker, "Africanization in British Multinationals"; Smith, "Winds of Change."

The Making of a "Hong Kong" Airline 243

cal reconfiguration of Hong Kong did not precisely follow a decolonizing pattern. Cathay Pacific performed "geopolitical jockeying" nonetheless.[23] In the early years, the airline jockeyed for British representation as the returning British regime returned to Hong Kong. In the 1980s and 1990s, as the 1997 handover neared, an emerging rival did pose a challenge to Cathay Pacific's legitimacy as an authentic local Hong Kong concern. However, the overriding threat stemmed from an unusual state issue—the incoming political regime. In a manner that paralleled the strategy of other British concerns in the waning days of colonialism,[24] the carrier did reinforce its local identity by increasing Hong Kong shareholding. Nevertheless, the challenges Cathay Pacific faced extended beyond the city of Hong Kong. Focusing on the evolving shareholding of Cathay Pacific, this chapter explores the airline's strategy in crafting its corporate nationality for its "license to operate"—legitimacy and state sponsorship—during a period of swift geopolitical shifts.[25]

From Commonwealth Holding to British Hong Kong

Under the leadership of Swire, a Commonwealth conglomerate facilitated the early growth of Cathay Pacific in the early years (see chapter 2). In particular, ANA, a founding member of that alliance, offered significant operational guidance to the fledgling airline. In the late 1940s, the

23. Lubinski and Wadhwani define "geopolitical jockeying" as "political positioning with the aim of cultivating alliances with host-country stakeholders in ways that delegitimize rival multinationals from other countries" ("Geopolitical Jockeying"). The rivals of British interests in Cathay Pacific were at times not "multinationals" (for example, the duo of American and Australian mavericks they sought to displace) and at times not "from other countries" (for example, British firms BOAC and Jardine Matheson). However, the airline continuously portrayed its profile as congruent with the political goals of the ruling regime.

24. Smith, "Winds of Change."

25. Decker used the term "license to operate" in a metaphorical sense. In the case of Cathay Pacific, its operating license was literally at stake ("Africanization in British Multinationals").

244 *Recasting Hong Kong*

Australian carrier seconded two of its experts with extensive experience in the industry to "get [Cathay Pacific] established on the right line." They returned to Australia only when the Swire-led team was deemed to have acquired sufficient experience on their own.[26] The tutelage of ANA extended to cabin service as well. In 1950, Cathay Pacific dispatched its newly appointed superintendent hostess to Australia for training by ANA so as to equip her with the skills on her return to Hong Kong to "undertake the training of locally engaged hostesses."[27] In these early years, ANA even lent their senior personnel to serve as Cathay Pacific's check pilots from time to time.[28]

The special relationship with ANA even endured the tumultuous industry conditions in Australia and ANA's takeover by Ansett in 1957.[29] The situation only began to change early in the ensuing decade with further developments in the aviation industry in Australia. Qantas's introduction of the Boeing 707 on the Hong Kong–Sydney route in 1961 rendered Cathay Pacific uncompetitive. Cathay Pacific exited the Australian market and left BOAC as the only operator designated by the British government for the British share of this route.[30]

The Commonwealth conglomerate that controlled Cathay Pacific (table 2.1) gradually gave way to a concentration of British ownership centered in the metropole and the colony. In November 1970, Cathay Pacific recalibrated its share count.[31] Shortly thereafter, in April 1971, the airline welcomed the Hongkong and Shanghai Banking Corporation (HSBC). Assuming a 25 percent stake in Cathay Pacific, HSBC, a prominent British bank in Hong Kong, surpassed BOAC's 15 percent owner-

26. Swire HK Archive, Cathay Pacific Airways Limited Board Minutes, June 25, 1951.

27. Swire HK Archive, Cathay Pacific Airways Limited Board Minutes, November 27, 1950.

28. Swire HK Archive, Cathay Pacific Airways Limited Board Minutes, February 8, 1954.

29. Swire HK Archive, Cathay Pacific Airways Limited Board Minutes, March 5, 1958; *SCMP*, January 8, 1959, 1.

30. Swire HK Archive, Cathay Pacific Airways Limited Board Minutes, September 23, 1961; Swire HK Archive, Annual General Meeting, January 30, 1962. See chapter 5 for discussion of Cathay Pacific's reentry into this market.

31. Swire HK Archive, Cathay Pacific Airways Limited Board Minutes, November 20, 1970.

The Making of a "Hong Kong" Airline   245

ship. With the colony's de facto central bank becoming a significant shareholder, the airline could tap into a wide array of financing opportunities more effectively—a significant advantage especially in the phase of fleet expansion then underway. The public reacted positively to the deal, which "undoubtedly adds considerably to the strength of Cathay Pacific by broadening its capital base and backing for the future."[32] "With expansion, it has become apparent that we should be more broadly based financially," the airline explained. "As noted at the last Board Meeting, The Hongkong and Shanghai Banking Corporation have indicated their willingness to take up a 25% share of the equity of the Company, while BOAC (AC) wish to maintain a 15% position." The remaining 60 percent stayed in the hands of the early investors, but by then, ANA/Ansett had dropped off the list. The Borneo Company and Jardine Matheson had also sold their small holdings.[33] Swire's interest stood at 22 percent, held in the name of two affiliates. Its subsidiary, China Navigation, held an additional 19 percent. P&O also held 19 percent.[34]

The consolidation continued when Swire took over the shares held by China Navigation (then 50 percent owned by Swire) on May 29, 1975, and "Cathay Pacific effectively came under the control of Swire Pacific Ltd. from that date."[35] Soon after, P&O expressed their interest in disposing of their holdings "subject to their receiving similar price to that paid recently to C.N.Co." "The Hong Kong Bank" indicated that it did not wish to increase their holding. BOAC, by then renamed British Airways, expressed some interest in increasing their stake to 20 percent, "but

---

32.   *SCMP*, April 21, 1971, 1. The Chinese newspapers the *Kung Sheung Morning News* ("國泰航空公司新股 二千一百萬元 由匯豐銀行買入," 12) and the *Kung Sheung Evening News* ("滙豐以二千一百萬投資國泰航空公司," 1) carried the story on the same day (April 21, 1971) and emphasized the importance of HSBC's investment to Cathay Pacific's capital base for future expansion.

33.   Swire HK Archive, Cathay Pacific Airways Limited Board Minutes, March 31, 1971; Swire HK Archive, Proposal to Increase the Capital of the Company and to Amend the Articles of Association, March 27, 1971; Swire HK Archive, Extraordinary General Meeting, April 14, 1971.

34.   Swire HK Archive, Cathay Pacific Airways Limited Board Minutes, October 14, 1971.

35.   *SCMP*, April 30, 1975, 25; Swire HK Archive, Cathay Pacific Airways Limited Board Minutes, October 28, 1975. The *Kung Sheung Evening News* had reported Swire's intention to consolidate control of the airline (*KSEN*, April 30, 1975, 7).

246 *Recasting Hong Kong*

Table 6.1 Percentage shareholding of
Cathay Pacific in 1978

	Shareholding
Swire	60.0%
HSBC	25.0%
British Airways	15.0%

Source: Swire HK Archive, CPA/7/4/1/1/162
Newsletter [August 1978].

this was a matter for their Board decision."[36] In the end, British Airways and HSBC maintained their percentage ownership while Swire acquired 87.5 percent of the remaining 60 percent (effectively a majority of 52.5 of Cathay Pacific's equity) with P&O retaining a small stake. This distribution of shares lasted only until July 3, 1978, when Swire Pacific announced that it would purchase all of P&O's remaining shares. As a result, Swire increased its shareholding in Cathay Pacific to 60 percent. The remaining 40 percent remained unaltered, with HSBC holding 25 percent and British Airways 15 percent (table 6.1).[37]

Turning Local Hong Kong

Although the tripartite of Swire, HSBC, and British Airways appeared to present a clear representation of commercial interests in the aviation industry in the British colony of Hong Kong, an undercurrent threatened to disturb the equilibrium. The British government in London had recognized a potential conflict as early as 1972. As they initiated air service negotiations with Beijing, the authorities in London acknowledged that "the interests of Hong Kong based *British* [inserted in handwriting] air-

---

36. Swire HK Archive, Cathay Pacific Airways Limited Board Minutes, December 8, 1975.

37. Swire HK Archive, CPA/7/4/1/1/162 Newsletter [August 1978].

lines (eg Cathay Pacific) diverge from those of BOAC."[38] However, the issue did not come to a head until 1979. Submitted to authorities in both London and Hong Kong, Cathay Pacific's application to operate scheduled passenger and cargo services between Hong Kong and London in 1979 pitted the colony-based airline against the British flag carrier. Cathay Pacific's chairman issued a press release in which he expressed the airline's hope that for the Hong Kong–London route, "a reciprocal service to British Airways should in all equity be operated by *Hong Kong's airline*" (emphasis added).[39] Although a British Airways officer had attended in August 1979 Cathay Pacific's board meeting as a director when the airline discussed the London route application,[40] he declared a conflict of interest in the next meeting and excused himself when the board discussed the application.[41] The divergence of the two airlines' interests resulted in British Airways notifying Cathay Pacific of its intention to sell its shares in 1980. In the end, British Airways sold its 15 percent stake, held since 1959, for £6.3 million.[42] Their respective purchases from British Airways made Swire and HSBC holders of 70.6 percent and 29.4 percent of Cathay Pacific's shares.[43] Rights offering and other share adjustments resulted in a clean 70/30 split in Cathay Pacific's equity ownership by Swire and HSBC, respectively, by 1983.[44] Thus concluded the process by which the airline consolidated its ownership in the hands

38.   TNA, FO 21/995.

39.   Swire HK Archive, Cathay Pacific Airways Limited Board Minutes, July 19, 1979.

40.   Swire HK Archive, Cathay Pacific Airways Limited Board Minutes, August 3, 1979.

41.   Swire HK Archive, Cathay Pacific Airways Limited Board Minutes, October 15, 1979.

42.   Swire HK Archive, Cathay Pacific Airways Limited Board Minutes, December 9, 1980.

43.   Swire HK Archive, Cathay Pacific Airways Limited Board Minutes, December 11, 1980; Swire HK Archive, CPA/7/4/1/1/170 Newsletter [November 1981]; *SCMP*, December 12, 1980, 41.

44.   Swire HK Archive, Cathay Pacific Airways Limited Board Minutes, December 17, 1980; Swire HK Archive, Cathay Pacific Airways Limited Board Minutes, May 25, 1983; Swire HK Archive, Cathay Pacific Airways Limited Board Minutes, June 6, 1983, issue of share certificates.

of local British Hong Kong firms. Throughout this process, Swire maintained its control of and provided management support to Cathay Pacific.[45]

The powerful duo of Swire and HSBC worked effectively as local British interests in Hong Kong in resolving conflicts with the metropole. However, even as they represented the local presence of British power, the sustainability of such British interests came to be challenged in the 1980s as political uncertainty over the future of Hong Kong mounted. In this turbulent period, the management of Cathay Pacific launched a series of restructuring measures. Management transformed the ownership of this Hong Kong airline to reflect a new definition of local identity and to embrace a new alliance in accordance with shifting geopolitical realities.

In the negotiations leading up to the signing of the Sino-British Joint Declaration on December 19, 1984, which prescribed the return of Hong Kong sovereignty to the PRC in 1997, the market postulated that mainland Chinese investors would take up the majority ownership of the Hong Kong carrier.[46] The carrier's chairman acknowledged that in light of the 1997 issue, Beijing would determine traffic rights for the Hong Kong–London route. "It might also be assumed that, in due course, some part of the share capital would pass to Beijing."[47] At the same time, Cathay Pacific strove to reaffirm its commitment to Hong Kong. Featured in the airline's newsletter in September 1984, Managing Director Peter Sutch said that "come 1997 and well into the 21st century," he expected "the familiar red and blue flag of the Swire Group [to fly] over Hong Kong." He dismissed rumors that Cathay Pacific would move from its home base of Hong Kong and pledged that the airline would "stay up to and well beyond 1997." Remarkable in Sutch's statement is the emphasis on Swire's "house flag" at the expense of the Union Jack. International businesses created in the heyday of colonialism had learned to be flexible in con-

---

45. "Since 1949, Cathay Pacific has had agreements with Swire for the provision of management support services" (Swire HK Archive, Cathay Pacific Airways Offer for Sale, April 22, 1986, 26).

46. *WKYP*, September 15, 1984, 6.

47. Swire HK Archive, Cathay Pacific Airways Limited Board Minutes, September 19, 1984.

The Making of a "Hong Kong" Airline 249

structing their corporate nationality. "We believe we still have a major contribution to make to the future." He anticipated that Cathay Pacific would affiliate more closely with China and displace British Airways on the Hong Kong–Beijing route. "Why should BA enjoy the rights to the Hong Kong–Beijing route? Shouldn't it be served by *Hong Kong's own carrier*?" (emphasis added).[48] Cathay Pacific was not alone in its claim to be the city's carrier. Within months of the signing of the Joint Declaration, Dragonair arose as its home base rival as it applied to operate charter flights from Hong Kong to China.[49]

Without any time to waste, Cathay Pacific made a dramatic gesture to reinforce its roots in Hong Kong. On November 28, 1985, the airline announced that it would seek to list its shares in Hong Kong "with a view to introducing direct public participation in the Company." Although Swire would remain the majority shareholder with the two existing shareholders retaining their existing share proportions, a public offering was to be made in the first half of 1986 to "achieve the maximum widespread distribution among individual Hong Kong investors," though this would give preferential treatment to the airline's staff.[50]

There had been considerable speculation that Swire would list its airline subsidiary separately on the stock exchange.[51] In fact, Cathay Pacific had studied the possibility for a year with input from Swire's London management. Although Hong Kong presented its own political issues, Cathay Pacific was to be in the good company of then-privatizing airlines (Singapore Airlines, Malayan Airways, and British Airways). As airlines flew under traffic rights granted by governments, it was politically wise to allow "nationals . . . to participate directly in the airlines." However, Michael Miles, chairman of Swire and Cathay Pacific, insisted that management was "not just following fashion. We [have been] doing this for our own very good reasons, and have had it under active consideration for more than a year."[52] Up to 25 percent of the airline's share was

48. Swire HK Archive, CPA/7/4/1/1/173 Newsletter [September 1984].

49. Swire HK Archive, Cathay Pacific Airways Limited Board Minutes, July 16, 1985.

50. Swire HK Archive, Cathay Pacific Airways Limited Board Minutes, November 28, 1985.

51. *SCMP*, November 20, 1985, 30; *SCMP*, November 21, 1985, 36.

52. *SCMP*, November 29, 1985, 29.

250 *Recasting Hong Kong*

to be offered to the public, and HSBC had given its consent. Management stressed the importance of indicating to the public that "this exercise was not in any way diminishing Swire's commitment to Hong Kong." Rather, going public was considered the airline's increasing "commitment to the people of Hong Kong."[53] Commentators were quick to call the offer "a political rather than a commercial move" as Swire had no immediate plans for the proceeds from the offering and must have recognized "the need to rectify the image of Cathay as a British airline operating out of Hongkong."[54] Management's statements confirmed such suspicions. While refuting rumors that Cathay Pacific's flotation was a response to Dragonair's attempt to undermine Cathay Pacific's local identity, former Chairman of Cathay Pacific Duncan Bluck asserted, "Cathay Pacific is Hongkong's airline and has been Hongkong's airline for many years. If anybody questions this, we have to make sure we satisfy the question."[55]

Political motivations of the airline's management notwithstanding, Cathay Pacific's flotation generated extraordinary enthusiasm among Hong Kong investors (fig. 6.1).[56] "Turnover Soars as Cathay Takes Off," read a *South China Morning Post* headline the day after the offering. The stock opened at HK$5.10 and registered a high quote of HK$5.35. The stock closed at HK$5.15 on the first day of trading, "a hefty 33 per cent premium" on the initial offer price of HK$3.88.[57] Pro-Beijing newspaper *Ta Kung Pao* noted the enthusiastic reception of Cathay Pacific's shares in an otherwise lackluster day in the Hong Kong stock market, lauding the performance of the pathbreaking listing for its support of trading volume in Hong Kong.[58]

The airline was eager to promote the offering in its own newsletter. "CX Share Offer Breaks All Records," read the headline of *Cathay News* in August 1986. Called "a resounding success," the public share offering

53. Swire HK Archive, Cathay Pacific Airways Limited Board Minutes, November 28, 1985.

54. *SCMP*, November 29, 1985, 29.

55. *SCMP*, December 19, 1985, 35.

56. *SCMP*, April 23, 1986, 1; *SCMP*, April 23, 1986, 29; *SCMP*, May 1, 1986, 25; *SCMP*, May 2, 1986, 29; *SCMP*, May 6, 1986, 27.

57. *SCMP*, May 16, 1986, 33.

58. *TKP*, May 16, 1986, 13.

The Making of a "Hong Kong" Airline

FIGURE 6.1. "People scramble for Cathay shares" at the headquarters of HSBC, *SCMP*, April 23, 1986. Photo credit: David Wong, *South China Morning Post*.

was said to have broken financial records on the Hong Kong Stock Exchange. The offering was 32.6 times oversubscribed. Excluding the shares allotted to the airline's staff (10 percent of the total) and to "certain Hong Kong institutions," the offering was subscribed "a staggering 56.4 times." Not merely an indication of public interest in aviation shares internationally, the results underscored "the Hong Kong public's growing identification with Cathay Pacific as Hong Kong's airline." The offering attracted over HK$51 billion in applications, which "amounts to almost HK$10,000 for every man, woman, and child in the territory."[59] The airline noted that the amount generated was "significantly higher than Hong Kong's money supply" and exceeded by a considerable margin the size of any corporate transaction in the territory's history. Sutch called the success of the public offering "a huge vote of confidence in the airline."[60] In its travel magazine, the *Discovery*, the airline also emphasized that the public offering "shattered records in Hong Kong," highlighting that "a total of

---

59. Swire HK Archive, CPA/7/4/1/1/175 Newsletter [August 1986].
60. Swire HK Archive, CPA/7/4/1/1/175 Newsletter [August 1986].

22.5 per cent of Cathay Pacific shares are now in the hands of Hong Kong residents, businesses or institutions."[61]

Interestingly, in this calculation, the list of "Hong Kong residents, businesses or institutions" did not include Swire or HSBC. In conjunction with the shares offered to the general public, Swire and HSBC agreed to sell 5 percent and 2.5 percent of the airline, in proportion to their existing holdings, to a group equally owned by Cheung Kong (Holdings) Limited and Hutchison Whampoa Limited (both under the control of Hong Kong magnate Li Ka-shing), and to Hysan Development Company Limited (of a prominent Lee family in Hong Kong), respectively. The two purchasers, who bought the shares at the same price as the offer price for the general public, had expressed their intention to hold the shares as a long-term investment.[62] The market had speculated earlier that shipping tycoon Sir Yue-kong Pao would take a substantial stake in the airline to accentuate its local character and to distance the airline from "the British identity that some observers believe may already have been a disadvantage."[63] Underscoring the enduring connection between maritime transport and commercial aviation, Pao was a member of the board of Cathay Pacific and chair of Dragonair, Cathay Pacific's budding rival. That he did not belong to the final lineup of local investors in Cathay Pacific was not surprising—he had resigned from Cathay Pacific's board in March, ahead of the flotation.[64]

The conspicuous inclusion of local Chinese investors in this round of shareholding restructuring reflected more than the impact of the recently executed Sino-British Joint Declaration. Ethnic Chinese entrepreneurs had risen in profile in Hong Kong, most notably in Li Ka-shing's assuming control of the British conglomerate Hutchison Whampoa in 1979 and taking over its chairmanship in 1981. As part of this profound change in the distribution of power in the world of commerce in Hong Kong, Cathay Pacific arrived late to the party in its restructuring of the airline's shareholding.[65] Moreover, the restructuring did not result in a

61. Swire HK Archive, *Discovery* [September 1986].
62. Swire HK Archive, Cathay Pacific Airways Offer for Sale, April 22, 1986, 8.
63. *SCMP*, November 29, 1985, 29.
64. *SCMP*, March 29, 1986, 21.
65. *New York Times*, January 14, 1981, D3.

Table 6.2 Percentage ownership of Cathay Pacific before and after the initial public offering, April 22, 1986

	Before the offering	After the offering
Swire Pacific	70.00%	54.25%
HSBC	30.00%	23.25%
Cheung Kong/Hutchison Whampoa		5.00%
Hysan		2.50%
Public/float		15.00%

Source: Swire HK Archive, "Cathay Pacific Airways Offer for Sale," April 22, 1986.

yielding of British control.[66] After the public offering, Swire Pacific continued to be the largest shareholder with a 54.25 percent stake, followed by HSBC's 23.25 percent (table 6.2). The airline's chairman, managing director, three of its six executive directors, and seven members of its senior management team were employees of Swire.[67] In the August 1986 issue of the travel magazine, the editor underscored the airline's new tagline, "Hong Kong is the home base of Cathay Pacific Airways."[68]

The airline could not readily dispose of its British ties, but at the same time it needed to develop a claim to serve the Hong Kong Special Administrative Region that was to be formed after the 1997 handover. It had to continue to operate under bilateral air service agreements and the designation of airlines "effected by the British government acting with the advice of the Hong Kong government."[69] The Joint Declaration specified that it was the PRC government's policy to "maintain the status of Hong Kong as a centre of international and regional aviation" and that Cathay Pacific, as well as "airlines incorporated and having their principal place of business in Hong Kong . . . may continue to operate."[70] After the public offering, the airline took on more local Hong Kong, non-British

66. Cathay Pacific's restructuring of its shareholding parallels the gradual erosion of British control in other settings (see, for example, Smith, "Winds of Change").

67. Swire HK Archive, Cathay Pacific Airways Offer for Sale, April 22, 1986, 26.

68. Swire HK Archive, *Discovery* vol. 14, no. 8 [August 1986].

69. Swire HK Archive, Cathay Pacific Airways Offer for Sale, April 22, 1986, 23.

70. Swire HK Archive, Cathay Pacific Airways Offer for Sale, April 22, 1986, 24, citing Section IX of Annex I to the Joint Declaration.

254   *Recasting Hong Kong*

investors, and to echo the requirements of the Joint Declaration, it prominently promoted Hong Kong as its home base.

Embracing Mainland Chinese Investments

Unfortunately for Cathay Pacific, these provisions did not include services that connected Hong Kong to, from, or through mainland China,[71] for which Dragonair remained a formidable home rival. Formed on April 1, 1985, Dragonair enjoyed significant mainland Chinese backing. Along with Yu-kong Pao, Chao Kuang-pui, who founded one of the largest textile companies in Asia, was instrumental in the establishment of the new Hong Kong–based airline.[72] The Air Transport Licensing Authority had approved the license application for Dragonair to fly to eight cities in China starting on Christmas Eve in 1985, giving "official recognition to Cathay Pacific's first domestic competition," which Cathay Pacific acknowledged in its newsletter. Dragonair still needed to apply to British and Chinese governments for traffic rights. To fend off unnecessary competition, Cathay Pacific appealed to the public by citing the "40 years of hard work and substantial investment to reach its position as one of the world's leading carriers." The carrier asked all involved to remember, "CX has been providing Hong Kong with a sound, competitive, and high quality of air services for some considerable time" and so "perhaps . . . deserve[s] at least some credit for that." Sutch asserted, "The arrival of Dragonair has been a sober reminder that we have no God-given right to earn a profit and that we have really got to live up to our promises of helping people arrive in better shape if we are going to *continue to be Hong Kong's airline*—which I can assure everybody is very much our intention" (emphasis added). He considered Dragonair more of an "irritant" than a threat "on the realization—albeit somewhat misguided—that there is a greater need for air service between Hong Kong and China." Claiming that Cathay Pacific was not monopolistic, Sutch maintained that the airline would "probably have as much if not more competition to face than

71.   Swire HK Archive, Cathay Pacific Airways Offer for Sale, April 22, 1986, 25.
72.   Davies, *Airlines of Asia*, 274.

most airlines in our home base."[73] Even after Cathay Pacific's public offering, Dragonair, which one op-ed in the local English-language newspaper called "Hongkong's unrelenting new airline," continued to challenge Cathay Pacific, "the territory's long-standing sole carrier," by purchasing additional aircraft, planning to "operate to 12 points in China" by fall of 1987, and expressing its determination to offer scheduled services.[74]

In its defense against Dragonair, Cathay Pacific needed to be mindful of Dragonair's investor base, which included not only Pao and other Hong Kong shareholders but also "mainland Chinese heavyweights."[75] Cathay Pacific's public offering might have rendered the airline more local to Hong Kong as measured by its shareholder base, but it had yet to facilitate mainland Chinese ties for the airline. The market soon speculated that mainland state-run China International Trust and Investment Corporation (CITIC) would buy a stake in Cathay Pacific and that such a politically inspired bargain would allow the airline to win mainland Chinese support in its bid to thwart Dragonair's mounting challenge to its position in Hong Kong.[76]

The news came on January 28, 1987: "Beijing Buys $2B Stake in Cathay." CITIC acquired 12.5 percent of the shares of Cathay Pacific "whose aircraft still carry the Union Jack on their tails."[77] CITIC bought 212,186,040 new shares for HK$5 per share. The market reacted favorably, as the airline's stock price rose steadily. In addition, CITIC acquired 145,877,902 shares from HSBC at HK$6 per share. A joint announcement explained that the shares HSBC sold would accrue dividends for the year 1986, unlike the new shares sold at the lower price.[78]

The transaction reduced Swire's ownership from 54.25 percent to 50.23 percent, and HSBC's from 23.25 percent to 16.43 percent. Spokespeople for Cathay Pacific explained that ownership must remain

---

73. Swire HK Archive, CPA/7/4/1/1/174 Newsletter [February 1986].

74. *SCMP*, January 1, 1987, 24; *SCMP*, January 5, 1987, 31.

75. *SCMP*, February 1, 1987, 17.

76. *SCMP*, January 27, 1987, 21.

77. *SCMP*, January 28, 1987, 1.

78. *SCMP*, January 28, 1987, 20. Trading of the stocks of Swire and Cathay Pacific was suspended the afternoon before the announcement of the deal. Both the English and Chinese media had anticipated CITIC's acquisition of Cathay Pacific's shares (*SCMP*, January 27, 1987, 1; *SCMP*, January 27, 1987, 21; *WKYP*, January 27, 1987, 5).

256   *Recasting Hong Kong*

in the hands of British owners because Cathay Pacific was still "a British airline for the purposes of international air transport licence consideration." Chairman Michael Miles also said that a dilution of British ownership to below 50 percent would be harmful to the airline's interest.[79] The airline was quick to point out that Swire and HSBC "still hold 66.66 per cent of the airline."[80] That the two biggest shareholders owned two-thirds of the airline was intentional, said the chairman.[81]

In a press conference, Miles underscored Swire's long-term intention to maintain its majority shareholding in the airline. He added that "the addition of a minority mainland Chinese shareholding was felt by the Board of Directors of Cathay Pacific, Swire Pacific, and the HSBC to be beneficial to the long term future of the airline" and that "CITIC had fully accepted the existing policies of the airline, particularly with regard to staff." "We are very conscious of the necessity of building relations with the People's Republic of China in a sensible balanced way," Miles said. "In 1997, sovereignty over Hong Kong reverts to China and it will be very necessary to have the active support of Beijing. A HK$2 billion investment by the PRC in Cathay Pacific makes it very very clear that we do indeed have Beijing support," Miles added.[82]

The airline explained that the local and international media generally considered the transaction "as a logical step for the airline in the long term, and a positive one for Hong Kong as a whole." CITIC's purchase "lift[ed] the threat of political favour from clouding the aviation scene in Hong Kong and allow[ed] that sector to continue to develop with renewed confidence." CITIC officials issued a press statement that stated the development of the airline industry was "of great importance to Hong Kong's stability and prosperity" and that the investment demonstrated CITIC's "full confidence in the bright future of Hong Kong." As part of the investment, the vice chairman of CITIC joined the board of Cathay Pacific.[83]

---

79. *SCMP*, January 28, 1987, 1.

80. Swire HK Archive, CPA/7/4/1/1/178 Newsletter [March 1987].

81. *SCMP*, January 28, 1987, 1.

82. Swire HK Archive, CPA/7/4/1/1/178 Newsletter [March 1987].

83. Swire HK Archive, CPA/7/4/1/1/178 Newsletter [March 1987]; *SCMP*, January 28, 1987, 1; *WKYP*, January 28, 1987, 5; *TKP*, January 28, 1987, 9.

The deal was said to have "largely laid to rest Dragonair's claims that it is the only genuine Hong Kong airline in terms of ownership."[84] Cathay Pacific claimed both of its major stakeholders, Swire and HSBC, to be "substantially owned by Hong Kong shareholders." At the same time, their being "British owned and controlled" also allowed Cathay Pacific to continue to operate from Hong Kong and have their traffic rights negotiated by the British government. The airline hastened to add that Cathay Pacific had "one of the largest registers of local shareholders of any public company in Hong Kong" and in that regard was "more substantially a Hong Kong airline than Dragonair."[85]

Sutch asserted that "as a group"—Cathay Pacific and its parent company Swire—they had always been sensitive to the environment in which they operated and "if political or economic changes in Hong Kong called for a change in the shareholding structure, such might take place." The airline's embrace of CITIC as a 12.5 percent shareholder was thus "totally consistent with that stated policy," and for Cathay Pacific, it was "business as usual."[86]

CITIC's investment in Cathay Pacific might have engineered a détente between the two rival Hong Kong carriers. However, the financial performance of Dragonair remained lackluster. It was reported that the fledgling carrier remained a loss maker and that the June Fourth crisis in Beijing in 1989 dimmed hopes just as its performance approached break-even.[87] It was from that context that the unlikely partnership between the two rival carriers emerged. On January 17, 1990, Cathay Pacific and Swire Pacific announced their acquisition of 30 percent and 5 percent of Dragonair for HK$294 million and HK$49 million, respectively.[88] The press release stated the two airlines' intention to cooperate, "with Dragonair initially concentrating on the development of routes between Hong Kong and the People's Republic of China." As part of the deal, Cathay Pacific also entered into an agreement to provide management, technical, and

---

84. Swire HK Archive, CPA/7/4/1/1/178 Newsletter [March 1987].
85. Swire HK Archive, CPA/7/4/1/1/178 Newsletter [March 1987].
86. Swire HK Archive, CPA/7/4/1/1/178 Newsletter [March 1987].
87. *SCMP*, January 12, 1990, 48. The abrupt and violent end to the democratic movement in Beijing on June 4, 1989, heightened international tension.
88. *SCMP*, January 18, 1990, 39, 41.

administrative services to Dragonair—a classic Swire arrangement.[89] At the same time, CITIC, through its subsidiaries, increased its Dragonair ownership to 38 percent from 16.6 percent.[90] Although CITIC's 12.5 percent investment in Cathay Pacific was public knowledge, the mainland Chinese concern had not previously confirmed its investment in Dragonair. The Chao family, a founder of Dragonair, which had recently purchased Pao's shares, trimmed its holdings in the carrier to 22 percent. After the deal, businesspeople "from Hongkong and Macau with strong links to Beijing" held the remaining 5 percent of Dragonair.[91] Although reports circulated that CITIC had pressed Cathay Pacific into the transaction,[92] Cathay Pacific's spokesperson insisted that the move was purely "commercial."[93] Similarly, CITIC general manager Wei Mingyi underscored the commercial benefits of the airlines' alliance.[94]

"Commercial" or "political," the deal made sense as an international airline requires the support of the governing political body to negotiate for it and grant it traffic rights. A newspaper report called such a relationship the "nationality" or "citizenship" of an airline—"the country or territory in which it is owned and based."[95] The preponderance of the consideration pivoted not on the local situation in Hong Kong but on the insistence of state administrators from afar. The alliance of CITIC-Swire in Cathay Pacific and Dragonair constituted a Sino-British joint venture that enabled the transition of air service rights negotiation and assignment in the years leading up to the handover. Aviation sources believed that Cathay Pacific's coalition with Dragonair would "pave the way for the bigger carrier's smooth operation after 1997."[96]

Shortly after the Dragonair deal in 1990, mainland Chinese shareholding in Cathay Pacific came to eclipse that of its major local Hong

89. *SCMP*, January 18, 1990, 41. The arrangement extended when Cathay Pacific became Dragonair's sales agent the following month (*TKP*, February 17, 1990, 17; *WKYP*, February 18, 1990, 21).

90. *SCMP*, January 18, 1990, 39, 41.

91. *SCMP*, January 18, 1990, 39.

92. *SCMP*, January 11, 1990, 33.

93. *SCMP*, January 18, 1990, 39.

94. *TKP*, January 20, 1990, 1.

95. *SCMP*, January 12, 1990, 48.

96. *SCMP*, January 18, 1990, 50.

The Making of a "Hong Kong" Airline

Kong investors. The HSBC had sold large chunks of its holdings, most recently in 1991, to raise some HK$1.7 billion. This lowered the bank's holdings to a 13.78 percent stake in the airline.[97] Gone was the two-thirds majority that Swire and HSBC had crafted in 1987 as they admitted CITIC into Cathay Pacific. By 1992, Li Ka-shing's companies and Hysan also did not register significant holdings in Cathay Pacific.[98] This should probably come as no surprise. Even though these investors, who received their allocation in the initial flotation, had claimed to be long-term holders of Cathay Pacific shares, according to a newspaper report, Hysan's managing director, H. C. Lee, had conceded shortly after the allocation that "in Hongkong six months could be construed as long term."[99]

The year 1992 witnessed the loss of one of Cathay Pacific's longest-standing shareholders. As the HSBC Holdings chairman asserted that "aviation is clearly not a core business for a financial services group," the bank ended its twenty-one-year-old equity in Cathay Pacific. HSBC "shed the last vestige" of its stake in Cathay Pacific, reported the local English newspaper. This sale of the remaining shares of Cathay Pacific held by HSBC amounted to 10 percent of the airline's total. The bank sold half of its holdings to China National Aviation Corporation (CNAC, a reincarnation of the enterprise by the same name in the prewar Republican era) and half to China Travel Service (CTS). These two entities, CNAC, a subsidiary of the Civil Aviation Administration of China that regulated mainland China's airlines, and CTS, the local arm of mainland China's official travel bureau, tremendously bolstered Cathay Pacific's ties to the PRC. Mainland Chinese holdings in Cathay Pacific increased from the 12.5 percent already held by CITIC to a total of 22.5 percent for the three mainland-controlled entities.[100]

Cathay Pacific Chairman Peter Sutch said that the airline was "sorry" to lose HSBC but was "pleased to have CNAC and CTS as shareholders"

---

97. *SCMP*, April 10, 1991, 33; *Asian Wall Street Journal*, July 14, 1992, 1; *SCMP*, July 14, 1992, 51.

98. Swire HK Archive, Cathay Pacific Annual Report 1991.

99. *SCMP*, September 2, 1986, 29.

100. *Asian Wall Street Journal*, July 14, 1992, 1; *SCMP*, July 14, 1992, 51; *Wall Street Journal*, July 14, 1992, A12.

and believed their involvement reflected "increasing co-operation between airlines in Hongkong and aviation and tourism in the People's Republic of China." As part of the transaction, the chairpersons of the two new investors joined the board of Cathay Pacific. "Participation by CNAC and CTS as shareholders in Cathay Pacific will also help to assure one objective of the Joint Declaration: that the Hongkong Special Administrative Region should remain—and indeed be further developed—as a centre for international and regional aviation," read a joint statement issued by Cathay Pacific and its two new investors. The statement called this deal a positive development both for the airline and for the future of Hong Kong. The vice director of the Hong Kong branch of the New China News Agency echoed the positive sentiments, calling the deal "good for the stability and prosperity of Hongkong."[101] At the conclusion of this transaction, Swire barely retained its majority shareholding of 51.8 percent.[102]

Swire did hold onto its majority stake in Cathay Pacific for a few more years, but even under its watch, the airline needed to mute its British shareholding ahead of the handover. "Going Native: Cathay Strives to Shake Off Vestiges of Colonial Times," read the headline of an *Asian Wall Street Journal* article in 1993, which reported that the airline had erased the Union Jack from its livery. "We're taking a good, hard look at ourselves and seeing we're no longer a British airline," said a senior executive at Cathay Pacific. "We're a Hong Kong airline. There's really no point in having a British flag." Shedding the airline's colonial image, the article noted, would "require a lot more than a few buckets of paint." The article cited an Asian aviation veteran and former chief executive of Dragonair, who declared that there would be "no alternative to Chinese shareholders acquiring the major controlling interests at the very latest by 1 July 1997."[103]

Cathay Pacific strove to dispel rumors of Swire reducing its stake. News articles on May 5, 1995, reported Swire's intention to maintain its majority holding in the airline. The carrier's managing director was quoted

---

101. *SCMP*, July 14, 1992, 51; Swire HK Archive, *Cathay News* No. 76 [August 1992].

102. Swire HK Archive, Cathay Pacific Annual Report 1992.

103. *Asian Wall Street Journal*, August 19, 1993, 1.

The Making of a "Hong Kong" Airline   261

as saying, "We have three mainland shareholders who hold a 22.5% stake. I think it's about right."[104] In August, however, the Swire chairman admitted that CNAC was in negotiations with Swire and Cathay Pacific to buy a stake in Dragonair.[105] The partial sale of Swire's and Cathay Pacific's interests in Dragonair would come to fruition the following year. In April 1996, CNAC announced its purchase of 35.9 percent of Dragonair from Swire, Cathay Pacific, and CITIC. The Dragonair deal was but a component of a major reshuffling of mainland Chinese interests in the airlines of Hong Kong. Along with CNAC becoming Dragonair's largest shareholder, CITIC bought new shares of Cathay Pacific, increasing its stake to 25 percent. As a result, although it retained all its shares, Swire saw its holding diluted to 43.9 percent,[106] below a simple majority.

It was a reasonable deal for Cathay Pacific. CNAC had announced plans to set up its own airline in Hong Kong to compete with Cathay Pacific, but as part of this transaction, it agreed to drop those plans and would instead pursue its interests in Hong Kong through Dragonair. To finance their purchase, CNAC would sell some of its shares in Cathay Pacific.[107] CITIC's stepping up of its interests in Cathay Pacific also represented a reversal in its strategy. Earlier, the CITIC chairman had left the board of Cathay Pacific, causing concerns that ties between the two companies had fractured. Late in 1995, CITIC had even sold 2 percent of Cathay Pacific. CITIC's increased stake signaled a resolution of any issue. The transaction entitled CITIC to two more board seats (which had included twenty directors before the transaction) and two representatives to the airline's executive committee (which had eight members).[108] The handover of 1997 might have seemed smooth, but it was only possible through a calibrated shareholding realignment and deliberate boardroom maneuvers.

A securities analyst commented that the deal afforded Swire and Cathay Pacific "a political guarantee that their future is much more

104. *Asian Wall Street Journal*, May 4, 1995, 3; *SCMP*, May 4, 1995, 41.
105. *SCMP*, August 18, 1995, 1.
106. *Wall Street Journal*, April 30, 1996, A14.
107. *SCMP*, June 12, 1996, 37.
108. *SCMP*, May 20, 1996, 36; *SCMP*, May 20, 1996, 37.

262 *Recasting Hong Kong*

safeguarded after 1997." The *Asian Wall Street Journal* went as far as to say that "Swire Pacific Ltd. is swapping a big chunk of its airline empire for post-1997 political insurance."[109] A *Wall Street Journal* headline reported, "Hong Kong Airline Protects Its Turf with Sale to China."[110] Although a headline in London's newspaper *The Independent* lamented that "Swire Loosens Grip on Cathay as Chinese Move In,"[111] the same newspaper carried a report a few days later that celebrated "Happy Landings for Hong Kong: The Swires Have Shown That It Pays to Kowtow to China's Capitalists."[112] Hong Kong's own *South China Morning Post* celebrated "Cathay on Course to Brighter Future as Share Accord Clears Political Clouds."[113] Swire sold control of the airline "on the cheap, but this is the price of political accommodation with the 1997 hand-over looming," said another local newspaper article.[114] "Airline Pact Seen as a Neat Solution" read a headline as yet another journalist reported on the "historic shakeup."[115]

The Hong Kong government readily accepted the deal. A spokesperson confirmed that both Cathay Pacific and Dragonair would continue to be designated as Hong Kong carriers even though British ownership in them would fall below 50 percent. The clause that Hong Kong's airlines must be "substantially owned and effectively controlled by British nationals" would not have to apply after the handover as the territory had negotiated with all major countries.[116] Hong Kong Governor Chris Patten called the decision "commercial" and highlighted Cathay Pacific's investment in the new Chek Lap Kok airport as an indication of its determination to remain a forceful player in Hong Kong's aviation industry.[117] In Beijing, the deal won the accolades of Qian Qichen, the vice premier, who praised Swire for "striking the historic deal that marked the end of British domination of the local aviation industry."[118]

109. *Asian Wall Street Journal*, April 30, 1996, 1.
110. *Wall Street Journal*, April 30, 1996, A14.
111. *The Independent*, April 30, 1996, 20.
112. *The Independent*, May 5, 1996, 8.
113. *SCMP*, April 30, 1996, 37.
114. *SCMP*, April 30, 1996, 56.
115. *SCMP*, April 30, 1996, 35.
116. *SCMP*, May 2, 1996, 37.
117. Swire HK Archive, *The Weekly*, Issue 82 [May 3, 1996].
118. *SCMP*, July 10, 1996, 1.

In preparation for the handover, Cathay Pacific changed its shareholding: Swire with 43.9 percent, CITIC with 25.0 percent, and other shareholders with (including CNAC's 4.2 percent) 31.1 percent. Dragonair's shareholding was as follows: CNAC owned 35.86 percent, CITIC 28.50 percent, Cathay Pacific 17.79 percent, Swire 7.71 percent, the Chao family 5.02 percent, and others 5.12 percent.[119] "We believe that the [Cathay Pacific] share placement and the Dragonair transaction are in the best interests of the shareholders as well as the employees of Swire and Cathay Pacific," said Sutch.[120] Swire's simple majority shareholding in Cathay Pacific was finally eclipsed a year before the 1997 handover.

Change of Heart: Putting More Local Representation in the Cockpit

Cathay Pacific did not merely refashion its investor base for a more local Hong Kong profile in the 1980s. Not coincidentally, as the airline scaled back the British colonial profile of its shareholding during the period of Sino-British negotiations over the future of Hong Kong, it also revamped its policy on cockpit hiring to shed its previous image of a British Commonwealth crew.

Cathay Pacific had touted its recruitment of English-speaking flight hostesses from the various Asian countries it served (see chapter 3). Although the Hong Kong contingent did not constitute a majority and lagged behind the number of hires from the airline's large Japanese pool, the composition registered respectable home representation in the cabin crew.

The faces in the cockpit told a different story. Although Cathay Pacific had moved on from its early tagline calling itself a "British Airline with British Pilots,"[121] in the cockpit, what mattered was its British Commonwealth pilots' experience traversing "the Orient." An article in a 1970 issue of the publication offered details on the airline's cockpit crew,

---

119. *SCMP*, May 20, 1996, 36; *SCMP*, May 20, 1996, 37; *SCMP*, June 12, 1996, 37.
120. Swire HK Archive, *The Weekly*, Issue 88 [June 14, 1996].
121. Swire HK Archive, CPA7/4/1/1/1 Newsletter [1958].

264   *Recasting Hong Kong*

which by April 1970, included forty-eight captains, forty-seven first officers, and forty-nine flight engineers. "Our ideal man would be air force or navy trained, with five to eight years in the service flying jets, and then would have spent two years with a commercial airline," explained Cathay Pacific's deputy operations manager, Captain Alec Wales. The licensing authority in the British colony allowed ready conversion of first-class British and Australian air transport pilot licenses "straight to the Hong Kong license." Captain Wales believed that flying crew found Cathay Pacific's rate of expansion appealing. "This is faster than the largest airlines. Therefore, promotion is faster. Salaries and working conditions are good, and Cathay Pacific enjoys a good reputation in the aviation world." Most British pilots joined Cathay Pacific from BOAC and its affiliate BEA or a charter company. Australian recruits came from Qantas, Ansett, or TAA. Cathay Pacific's pilots ranged in age from thirty to fifty-five, the retirement age. The airline's cockpit crew flew an average of sixty hours a month, enjoyed six weeks of annual leave, and received pay that reportedly "compare[d] favourably with bigger airlines."[122]

A tally of Cathay Pacific's flight crew confirmed the dominance of British, Australian, and New Zealand influence. By 1971, the airline's flight crew had grown to fifty-six captains, fifty-eight first officers, and fifty-eight flight engineers. Of this total, more than half (eighty-seven) came from Australia. The British contingent (seventy-two) followed at 41.86 percent of the crew. The rest were New Zealanders (ten) and holders of United States passports (three). Broken down by rank, the representation was similar: thirty-one of the fifty-six captains were Australian, and twenty-two were British; thirty-five of the first officers came from Britain, and seventeen, from Australia.[123] This composition persisted into the early years of the 1980s. As Cathay Pacific touted its services provided by "cabin attendants selected from 9 Asian countries, who combine[d] warm and friendly Asian femininity with first class service," it remained steadfast in guarding its "flight deck crews recruited from around the world, in particular—Australia, Britain and New Zealand."[124]

---

122.   Swire HK Archive, CPA/7/4/6/38 *Cathay News* No. 46 [March—April 1970].

123.   Swire HK Archive, CPA7/4/1/1/124 Newsletter [May 1971].

124.   Swire HK Archive, CPA7/4/1/1/170 Newsletter [November 1981]. In 1989, Singapore Airlines reportedly advertised for more foreign pilots because of insufficient

The Making of a "Hong Kong" Airline

Cathay Pacific finally relaxed this exclusive entry into the cockpit, for decades open only to nonlocal pilots. The airline began its recruitment of "Chinese flight crew" in 1987. In the same issue of the newsletter that announced its 12.5 percent share sale to the mainland-controlled CITIC, Cathay Pacific announced, "Chinese flight crew recruitment nears realisation. . . . A step which has long been on the cards in the development of Cathay Pacific, *particularly in its role as Hong Kong's airline*, came a step nearer realisation late last year when it was confirmed that the airline is to instigate an 'ab initio' [from the beginning] training scheme for *Hong Kong Chinese* pilots" (emphasis added). In explaining its previous policy of employing only expatriate flight crew, Cathay Pacific cited "the lack of local resources." "Hong Kong has no air force, which is the prime source of pilots for most national airlines, nor any formal pilot training establishment." Until that point, the airline did not find that an issue because its "high flying standards and conditions of employment" had enabled Cathay Pacific to attract "the cream of the world's pilots, who [had] come complete with extensive flying experience." There was no operational or economic reason for the airline to establish an in-house training school.[125]

Attributing its change of mind to the "heavy" crew that ultra–long haul nonstop flights required, Cathay Pacific "look[ed] seriously at training a corps of Hong Kong Chinese pilots." With these ultra-long-haul flights, the airline instituted the position of a second officer (new to Cathay Pacific but not to the industry). "Acting as relief crew, the Hong Kong pilots will be able to chalk up long hours of hands-on training, observing flight procedures and techniques. By the time they qualify to move to the right hand, or First Officer's seat, these pilots will have attained a level of experience similar to that of junior pilots joining Cathay Pacific from outside Hong Kong."[126] The local recruits, of Chinese descent, would join the cockpit at the bottom of the hierarchy. Their subordinate

---

numbers locally. At that time, the airline boasted "one of the most cosmopolitan mix of pilots—they come from more than 30 countries" (*Straits Times*, April 23, 1989, 18). The city-state's airline had hired both local and overseas pilots and had incorporated pilots from the city's air force into the cockpit (*Straits Times*, August 24, 1977, 25; September 22, 1977, 7; October 18, 1978, 1).

125. Swire HK Archive, CPA/7/4/1/1/178 Newsletter [March 1987].
126. Swire HK Archive, CPA/7/4/1/1/178 Newsletter [March 1987].

266   *Recasting Hong Kong*

position notwithstanding, this change in recruitment policy in 1987 presented the first opportunity for local entry to the command center of the aircraft.

The training would remain in British hands. Recruits were to be subjected to "an intensive training course, probably at one of Britain's top pilot training schools." When the airline inaugurated this program, its schedule called for recruitment to begin in 1987, recruits' enrollment in pilot training school in 1988, and their first appearance as Hong Kong pilots on the flight deck by 1989, reaching first officer status "in the early to mid 1990s," a transition meticulously choreographed to synchronize with the city's handover in 1997.[127]

Cathay Pacific made it clear to the public that the program, which targeted Hong Kong residents for commercial airline training, was costly to the airline. "By the time we get [the trainees] back to the airline," said the airline's director of flight operations, "they will have cost us about a million or very close." Management asserted that the program was intended to respond to "a need for a back-up for the first officer" for the airline's recently introduced nonstop transcontinental flights. Since the last tally in 1971, not only had Cathay Pacific expanded its route network, but its flight deck crew had also more than tripled in size. Before launching the cadet program in 1987, the airline employed one hundred and eighty captains and two hundred first officers, and their representation had not changed: "mostly from Britain, Australia and New Zealand," and "all hired as experienced pilots."[128]

"The career opportunity of a lifetime: be a pilot with Cathay Pacific," Cathay Pacific advertised in the *South China Morning Post* in October 1987. Open to "permanent local resident[s] of Hong Kong" who fulfilled certain criteria, the highly selective program entailed an "intensive 70 week training course," after which the cadet would become "a qualified commercial pilot and . . . return to Hong Kong in the rank of Second Officer with Cathay Pacific." As second officers, the graduates would work in the cockpits of Cathay Pacific's Boeing 747s and Lockheed Tri-Stars. Cathay Pacific offered a starting salary of HK$10,000 per month, rising to over HK$15,700 per month, along with travel concessions,

127.   Swire HK Archive, CPA/7/4/1/1/178 Newsletter [March 1987].
128.   *SCMP*, August 5, 1987, 25.

The Making of a "Hong Kong" Airline

medical benefits, retirement, and leave benefits (the median monthly income in the British colony was HK$2,573 in 1986 and HK$5,170 in 1991). Promotion to first officer could happen after six years with sufficient flying experience and successful completion of additional training. "Those who continue to meet our extremely high standards can expect to reach the ultimate career goal of Aircraft Captain." The advertisement urged potential applicants not to miss "this career opportunity of a lifetime," encouraging those who were not old enough or did not qualify in 1987 to apply in the future.[129] Indeed, this career prospect offered attractive terms. What was not discussed was the compensation gap that remained between local hires and expatriate pilots.

The airline recorded on-time performance with its first cohort of cadet pilots. From a pool of over four-hundred applicants, Cathay Pacific selected "eleven ambitious young Hongkong men," noted the *South China Morning Post* on its front page.[130] Most of them graduated from the program on time in 1989. "Cloud nine!" read the headline of Cathay Pacific's November 1989 newsletter. Nine cadets, "all born in Hong Kong" with Chinese names romanized according to Hong Kong conventions, finished the program at the British Aerospace Flying College at Prestwick, Scotland, and received their "CX wings" from Governor David Wilson on October 24, 1989. Upon their completion of the seventy-week training program at Prestwick, which included two thousand hours of flying by day and night, they received UK commercial pilot's licenses and instrument ratings and joined Cathay Pacific as second officers. In Hong Kong, their training continued with further technical and simulator instruction before they joined the Boeing 747-400 fleet, "the first step on the career ladder to becoming a CX Captain."[131]

In their choice of the right candidates, "flying aptitude skills are essential," explained Cadet Pilot and Administration Manager Peter Baxter. Along with good motivation, fluency in English, and other

---

129. *SCMP*, October 21, 1987, 19; *SCMP*, October 28, 1987, 13; "Summary Findings of the 1991 Population Census," *Hong Kong Monthly Digest of Statistics,* November 1991, 114.

130. *SCMP*, April 27, 1988, 1. The newspaper continued to follow keenly their development (see *SCMP*, November 8, 1988, 4).

131. Swire HK Archive, CPA/7/4/1/1/183 Newsletter [November 1989].

268   *Recasting Hong Kong*

educational qualifications, the airline looked for hand-eye coordination. "In the sort of degree that we're looking for, these are skills that only a minority of people have." Even if language proficiency ("fluent English"), educational background ("maths and physics at Advanced Level"), and motor coordination knew no ethnic boundary, many Hong Kong applicants found themselves handicapped in one crucial criterion. Cathay Pacific required "20/20 vision (uncorrected)" and did not accept any applicant who needed corrective lenses. "Many Hong Kong people wear spectacles," and "the medical test 'knocks out' a lot of applicants."[132]

Although Cathay Pacific aimed to present a more local profile for its cockpit crew, the airline had to start with its British rulebook. They borrowed the RAF interviewing format, modified to meet its own requirements. The airline did not choose the training school in Prestwick "simply out of fondness for the Scottish weather," quipped the airline's newsletter. After all, Prestwick was "the first establishment specifically designed to produce airline pilots," and Cathay Pacific cadets only required conversion training upon their return to Hong Kong. British Airways was the major customer at Prestwick. The courses enrolled half their students from Cathay Pacific and half from British Airways. "The idea is that some of the culture will rub off in each direction. Our lads are really good at applying themselves to the books, while the Westerners are relatively more outgoing. The system works very well, especially when it comes to improving English skills." Acculturation to the international flying world, where English was the lingua franca, required that the Hong Kong Chinese cadets learn from their British counterparts.[133]

Cathay Pacific refined its selection process so that by 1990, applicants at an advanced stage of selection went on a flying grading course. For that assessment, Cathay Pacific sent the applicants to the Jandacot Flying College near Perth. The airline could have opted to conduct all the training in Australia, but the Australian license was less acceptable to the Hong Kong Civil Aviation Department than its UK equivalent. Australian pilots needed to sit for extra examinations to satisfy Hong Kong's requirements.[134] In the flying world, qualifications were not created equal. While

132.   Swire HK Archive, CPA/7/4/1/1/186 Newsletter [March 1990].
133.   Swire HK Archive, CPA/7/4/1/1/186 Newsletter [March 1990].
134.   Swire HK Archive, CPA/7/4/1/1/186 Newsletter [March 1990].

The Making of a "Hong Kong" Airline 269

there was a clear preference for credentials in the Commonwealth, documentation in the United Kingdom counted for more.

The graduates appreciated the rigorous training and the career opportunity the program afforded them. "There's a lot to learn," Pioneer Cadet Richard Wong Kwok Ho said. "Prestwick is tough," said Wong, "The Cathay way." "Quite mad about airplane[s]," Wong held a degree in aeronautical engineering from Queen Mary's, London, although he did not feel that those without such qualifications should feel disadvantaged. Some of Wong's fellow cadets "had never been in an airplane before joining CX, *not even as a passenger*" (emphasis added). For those who persevered, the future was bright. "They will be on exactly the same pay scale as a new joining expatriate," the Cadet Pilot and Administration Manager Baxter explained, "though without the expat allowances, of course."[135]

In a 1990 review of the cadet pilot scheme, Cathay Pacific traced the plan's origin to 1987 when the airline "foresaw a global shortage of pilots" and the impending impossibility of recruiting "the cream of the world's aircrews." It was "the first time" Cathay Pacific could "afford the expense of having pilots trained from scratch." The review hastened to add, however, that "it seemed only right that *Hong Kong's airline* should give *local* young people the opportunity of a career as a pilot, which had never been available before" (emphasis added).[136]

Cathay Pacific intended to make this cadet scheme an ongoing program. In March 1990, after the first batch of graduates, the program had produced nine qualified second officers with another twenty-seven undergoing training (some sixteen of whom were expected to graduate shortly).[137] Five more cadets received their wings in July and became second officers. The airline was careful to note that all graduates in the second batch were born in Hong Kong, "with the exception of Dennis Liang Bun, who was born in China."[138] The airline planned to start three more groups of eight cadets each in 1990, with recruitment continuing at the rate of twenty-four per year. "Ultimately, we'll only be able to say that this scheme has been a success," Baxter said, "when we

135. Swire HK Archive, CPA/7/4/1/1/186 Newsletter [March 1990].
136. Swire HK Archive, CPA/7/4/1/1/186 Newsletter [March 1990].
137. Swire HK Archive, CPA/7/4/1/1/186 Newsletter [March 1990].
138. Swire HK Archive, CPA/7/4/1/1/190 Newsletter [July 1990].

270   *Recasting Hong Kong*

can look back in 10 or 12 years' time and say that these people have become good captains."[139]

The Impetus: Politically Inspired Competition to Be the Locally Operated Airline

The decades of the 1980s and 1990s challenged Cathay Pacific to redefine itself. The changing political tides and the competition that Dragonair presented made it imperative for Cathay Pacific to adjust its governmental ties and refashion its brand. As the airline sold equity to mainland Chinese interests and hired local pilots, it also launched an advertising campaign that its manager of passenger marketing explained had sprung from an instinct: "We knew we were the flag carrier of the Asia/Pacific rim and the airline for the new world of international travel."[140] During those years of uncertainty around Hong Kong's future, the rejection of a rigid political alignment was understandable, and Cathay Pacific adroitly claimed to be the bearer of a larger standard that projected the airline to a more extensive market. Similarly, the airline meticulously crafted its position statements and rephrased its reference to the multiethnic cabin crew. Instead of calling them "flight hostesses . . . from nine Asian *countries*" (emphasis added) as it did in the early 1980s, Cathay Pacific referred to them as "flight attendants from 10 Asian *lands*" (emphasis added) in 1990,[141] a move that represented a reversion to the 1960s.[142]

This move to a nondefinition in politics did not suffice in Cathay Pacific's protection of its home turf in Hong Kong. Dragonair, its budding rival, strove to establish its true homegrown status as Hong Kong's carrier. In fact, Dragonair's recruitment of local pilots predated Cathay Pacific's, and the rivalry of the two in this regard, at least in publicity,

139.   Swire HK Archive, CPA/7/4/1/1/186 Newsletter [March 1990].
140.   Swire HK Archive, CPA/7/4/1/1/195 Newsletter [December 1990].
141.   Swire HK Archive, CPA/7/4/1/1/168 Newsletter [1981?]; Swire HK Archive, CPA/7/4/1/1/170 [November 1981]; Swire HK Archive, CPA/7/4/1/1/191; Swire HK Archive, CPA/7/4/1/1/193; Swire HK Archive, CPA/7/4/1/1/195 Newsletter [1990].
142.   Swire HK Archive, CPA/7/4/1/1/40 [July 1962].

The Making of a "Hong Kong" Airline

271

transpired in the open stage of Hong Kong. In its nascent days, Dragonair management had stressed that the airline "run by the people of Hong Kong" was to follow a localization policy in its hiring of pilots. The budding airline announced in 1985 that it had hired Hong Kong's first cadet pilot and boasted a completely local cabin crew.[143] Hong Kong Chinese newspapers across the political spectrum featured this pilot, Bancho Kwan, a Hong Kong native who had received flying instruction, and his commercial license in 1981 in the United States.[144] In the first issue of its inflight magazine in 1987, Dragonair boasted "Hong Kong's first airline pilots": "Two young Hong Kong men . . . are putting themselves, and Dragonair, in the record books. They are the first locally-born airline pilots to be employed by a Hong Kong airline, the first graduates of the airline's flight crew training scheme established in 1985." Just like the Cathay Pacific cadets who would come after them, Dragonair's two local pilots received their training in Scotland (at a different school). Two more Dragonair trainees were also undergoing flight training there.[145] The two Hong Kong Chinese trainee pilots joined thirty-year-old Chris Thatcher, previously a flying instructor at the Aviation Club of Hong Kong. Dragonair was quick to note that Thatcher was "the first ever locally trained pilot to join a Hong Kong based airline."[146]

A 1988 newspaper article that reported the appointment of Bancho Kwan as the "first Hongkong-born Chinese pilot" underscored "the airline's policy of employing and training Hongkong citizens as commercial pilots." As Dragonair's managing director claimed, the airline had maintained a "long-term plan to increase the proportion of Chinese to expatriate pilots."[147] In mid-1988, Dragonair forecasted that in one year, 20 percent of its pilot workforce would be Hong Kong born.[148] Although they remained a minority even for an airline startup whose flight crew totaled only twenty at that moment, these local pilots pioneered Hong

143. *TKP*, July 5, 1985, 7.
144. *TKP*, July 5, 1985, 7; *WKYP*, July 6, 1985, 13.
145. Swire HK Archive, Dragonair *Golden Dragon* Vol. 1 No. 1 [June–July 1987].
146. Swire HK Archive, Dragonair *Golden Dragon* Vol. 2 No. 3 [June–July 1988].
147. *SCMP*, August 27, 1988, 2. The pioneering appointment of Kwan was also reported in *WKYP*, August 27, 1988, 3.
148. Swire HK Archive, Dragonair *Golden Dragon* Vol. 2 No. 3 [June–July 1988].

Kong representation in the cockpit, a remarkably late development for an aviation hub.

Dragonair paved the way for more pioneering achievements by Hong Kong locals. In 1988, the airline employed Rosa Chak, "the first female trainee pilot," who upon completion of her training was to become "Hong Kong's first female commercial airline pilot."[149] A second female trainee pilot had also joined Dragonair.[150] The following year, Dragonair celebrated Chak's promotion to the rank of first officer. Upon completion of intensive training in Scotland and certification in the United Kingdom, she returned to Hong Kong and received her Hong Kong commercial license in June 1989. Her appointment as first officer afforded her the status of being "the only Hongkong-born Chinese woman flying for a commercial airline."[151] When she compared her pioneering achievement to female participation in the industry in Hong Kong, China, and the West, Chak noted, "I understand Cathay Pacific changed their flight manuals recently to read he/she when referring to the pilot."[152]

It was against this backdrop of hometown competition that Cathay Pacific inaugurated its cadet program in 1987. In fact, the airline had endured criticism for years for the absence of local Chinese representation in its upper echelons. In January 1985, the editor of *Cargo-lines Asia* magazine, David Slough, chastised the dominance of expatriates "who fill[ed] most of the airline's senior posts." At Cathay Pacific, "there isn't a single ethnic Chinese Hongkong pilot on the staff," Slough emphasized.[153]

Cathay Pacific's cadet program formed an integral part of the airline's effort in formulating a Hong Kong identity for itself. Indeed, the battle for local recruitment extended beyond the cockpit. In 1991, Dragonair touted that 95 percent of its trainees for cabin services came from Hong Kong.[154] An article in the December 1992 issue of the Cathay Pacific employee newsletter that sought to raise the airline's profile in Hong Kong

---

149. Swire HK Archive, Dragonair *Golden Dragon* Vol. 2 No. 3 [June–July 1988].

150. Swire HK Archive, Dragonair *Golden Dragon* Vol. 2 No. 6 [December 1988–January 1989].

151. *SCMP*, September 9, 1989, 3; Swire HK Archive, Dragonair *Golden Dragon* Vol. 3 No. 5 [October–November 1989].

152. *SCMP*, March 31, 1990, 94.

153. *SCMP*, January 5, 1985, 22.

154. Swire HK Archive, Dragonair *Dragon News* Vol. 2 No. 9 [September 1991].

simply stated that "CX and the community cannot be considered as separate." Cathay Pacific called itself "one of Hong Kong's biggest private-sector educators," providing the local population with "training in a wide variety of skills—flying, information technology, English, management and so on." Attributing the airline's success to "the territory's geographic location" and declaring its commitment to the city in the process of political transformation, Cathay Pacific expressed its desire for "the people of Hong Kong to be proud of CX's achievements and to consider the airline's accomplishments as a reflection of their own."[155]

To incorporate local faces into Cathay Pacific's headcount, especially at higher levels, was paramount to this enterprise of engineering local pride and ownership. The airline's local recruitment expanded beyond the commanding positions in the cockpit and extended to executive levels in the airline. Featuring three employees whose names in English followed local romanization, a February 1992 advertisement in the *South China Morning Post* carried the tagline "become a Cathay Pacific executive and explore new territories." The advertisement appealed to Hong Kong readers who might aspire to be one of the airline's "high caliber managers to sustain its rapid growth and excellent reputation."[156]

Forecasting continued growth, Cathay Pacific targeted Hong Kong applicants in its recruitment: "If you have the ability and inclination, an acceptable level of numeracy, and can speak fluent English *and Cantonese*, then the sky is really the limit with Cathay Pacific" (emphasis added).[157] The requirement of fluency in English was the continuation of the airline's longstanding policy, but the addition of Cantonese marked a decidedly local turn. At a job fair at the Hong Kong Convention and Exhibition Center in 1993, Cathay Pacific strove to demonstrate its commitment to "promoting the airline as the preferred employer in labour-short Hong Kong." During the event, the airline "distributed 14,000 brochures and persuaded more than 100 people to apply right away (several started work the following week)."[158]

---

155. Swire HK Archive, *Cathay News* No. 80 [December 1992].
156. *SCMP*, February 20, 1992, 42.
157. *SCMP*, February 20, 1992, 39.
158. Swire HK Archive, *Cathay News* No. 84 [April 1993].

The race to become Hong Kong's "local" airline proved intense during the period of the city's political transformation. In a *South China Morning Post* interview, Managing Director Rod Eddington dispelled rumors of an identity crisis for Cathay Pacific (not "quite sure whether it's an Asian airline or a European one"). He offered a resounding answer: "It's Hong Kong's airline." Its cockpit crew was emblematic of the identity overhaul of its entire personnel profile. On the issue of "locals and expats," the airline maintained, "we have good local managers and good overseas managers, and we have an ambitious local management development programme." Vowing to offer "the same opportunities within the organization," Cathay Pacific asserted that "what matters most is your ability to contribute and deliver, not where you came from—we need a cosmopolitan management team with people committed to Hong Kong."[159] The airline's newfound interest in local talent indicated that the airline was exerting a conscious effort to define cosmopolitanism by incorporating more local Hong Kong elements, at least in its payroll, at all levels.

Also in 1993, the airline advertised prominently, "Wanted: women cadets." Five years after the establishment of its cadet pilot scheme and five years after Dragonair had admitted its first female pilot trainee, Cathay Pacific expanded the program "to encourage the development of flying as a career option for women in Hong Kong." The airline's director of flight operations explained that the expanded program would ensure "equal opportunity in what had traditionally been a male-dominated employment area among most Asian airlines."[160]

"The competition among experienced qualified pilots wanting to work for Cathay is so great that to date we have not been in a position to accept any of the few women pilots who have applied to join the company," reasoned Captain Gerry Clemmow, director of flight operations of Cathay Pacific. "Working through the cadet pilot scheme, with all applicants starting on the same footing, we will be in a better position to encourage young women in Hongkong to consider flying as a career," he continued. The *South China Morning Post* article that reported this development at Cathay Pacific was quick to draw a comparison with Drag-

159. Swire HK Archive, *Cathay News* No. 89 [September 1993].
160. Swire HK Archive, *Cathay News* No. 87 [July 1993].

The Making of a "Hong Kong" Airline

onair, which by that point was hoping to employ two other women pilots besides Rosa Chak. The article quoted the general secretary of the Hongkong Aircrew Officers' Association as saying, "We are delighted that Cathay Pacific Airways is now moving with the times."[161]

The cadet pilot program of Cathay Pacific had allowed local men in Hong Kong to enter an occupation once reserved for foreign men. The first five years of the scheme had yielded seventy graduates—all male—who entered the privileged domain of the cockpit as second officers. Seven of these graduates had received promotions to acting first officer.[162] The progress at Cathay Pacific paled in comparison with Dragonair's, where Jack Ip, who joined as a student pilot in 1986, had become the airline's "first locally qualified Captain" in 1994.[163]

Although lagging behind Dragonair in the promotion of local pilots, Cathay Pacific made an effort to keep pace. The results proved lackluster at first. The *South China Morning Post* reported only "a trickling response." Fielding "tough competition from male cadets," Cathay Pacific accepted no female applicants in the first year the cadet program removed its gender restriction.[164] Candy Wu, an engineering graduate from the University of Hong Kong, became the first woman to join the pilot cadet scheme at Cathay Pacific in 1994. Rejecting any explanation of low female participation on the basis of physical fitness, Wu said, "in the air force they are supposed to have a fit body, to be more aggressive as a fighter pilot. But for civil aviation pilots it's just the same for men and women: there's more emphasis on the thinking side than the physical side."[165]

By 1996, Wu had completed the necessary training and reached the rank of second officer serving as a relief pilot on long-haul flights.[166] In an interview published in the airline's newsletter that year, Wu mentioned how her job at Cathay Pacific smashed her earlier reservations about joining the profession. Having never dreamed of becoming a pilot until three years ago, Wu, who grew up in Tokwawan right next to the Kai

161. *SCMP*, June 18, 1993, 4.
162. Swire HK Archive, *Cathay News* No. 87 [July 1993].
163. Swire HK Archive, Dragonair *Dragon News* No. 28 [September–October 1994].
164. *SCMP*, May 11, 1994, 40.
165. *SCMP*, August 31, 1996, 3; *SCMP*, August 4, 1997, 20.
166. *SCMP*, August 31, 1996.

Tak airport, had "imagined you had to be much older and had to come from a very rich family." She noted, "Yes, most Hong Kong people are very surprised because there are very few female pilots here—but in Europe and Australia there are quite a few." "Hong Kong, too, is changing," concluded that article in the newsletter.[167]

"Life's dandy for Candy," Cathay Pacific reported on its first local female pilot who by 2002 had risen through the ranks to become a senior first officer. Candy relished the honor of her pioneering achievement but admitted that it could be "tough being a woman in what is seen as a man's world." Although she enjoyed the "freedom" of her high-flying job, the work environment was "not particularly kind to body and skin," issues that might otherwise receive scant attention in an article for her male colleagues.[168]

Beyond the issue of gender, Cathay Pacific remained on the defensive in their local recruitment of pilots. In an open letter published in the *South China Morning Post* on July 13, 1994, aircrew Recruitment Manager Stephanie Heron-Webber defended the airline's traditional policy of recruiting its pilots from the United Kingdom, Australia, New Zealand, and Canada, attributing it to "the compatibility of air crew licences amongst the Commonwealth countries." The airline was said to have been "fortunate enough to have received considerable numbers of suitable applications from these countries to fill all available vacancies." The manager pledged that Cathay Pacific would consider applicants outside of their "traditional recruiting areas" and reminded readers of their recruitment of cadet pilots from Hong Kong "as a corporate commitment to Hong Kong," which by then had produced "around 70 *Chinese* Second Officers and Acting First Officers" (emphasis added).[169] Despite these results, Cathay Pacific understood that structural issues stood in the way of reducing its dependency on expatriate pilots. The airline's recruitment efforts were complicated by the failure of half of its applicants' failing the "Hong Kong residency qualification." For the rest, poor eyesight continued to be an obstacle. "The rate of sight deterioration is faster here than elsewhere," noted Heron-Webber on a different occasion. "For this

167. Swire HK Archive, *The Weekly* Issue 99 [August 1996].
168. Swire HK Archive, *CX World* Issue 70 [January 2002].
169. *SCMP*, July 13, 1994, 18.

The Making of a "Hong Kong" Airline

reason we will not accept any cadet who wears any form of eyesight correction."[170]

Cathay Pacific's orchestrated effort to expand local representation in its headcount was evident, but certain hurdles remained. A local recruitment event in 1995 attracted eight hundred hopeful applicants to join its cabin crew. The recruitment controller felt vindicated: "Since Cathay Pacific is a Hong Kong-based airline, there is a need to recruit more local flight attendants." However, structural issues remained. As the controller explained, local applicants for cabin crew jobs faced the main obstacles of "the English exam," and, like applicants for the cockpit jobs, "the eyesight test."[171] Although the airline continued to insist on this vision requirement, it relaxed its restriction of permanent residency for aspiring cadet pilots, allowing men and women of Hong Kong to apply for the job provided they could present "evidence of . . . being part of the community," such as their having "gone to school here or lived here a number of years." However, competition for local aspirants remained intense as Cathay Pacific continued to recruit pilots from other airlines as well as from air forces "in the UK, Canada, Australia, Hong Kong, New Zealand and occasionally from South Africa and Zimbabwe."[172] The predilection for a Commonwealth background and training persisted. It would therefore come as no surprise that when the airline decided in 1995 to relocate its training from Prestwick, Scotland, where close to one hundred CX recruits had received flying lessons, it chose an Australian aviation college near Adelaide.[173] The airline demonstrated an earnest effort to introduce

---

170. *SCMP*, October 12, 1994, 45. Grindrod also compared Cathay Pacific's situation with those of Singapore Airlines (SIA) and Royal Brunei Airlines (RBA). SIA was said to employ more expatriate pilots than any other airline in the region and did not confine its cadet program to locals. Instead, it recruited a large number of Malaysians, in addition to a significant number of British and Australian cadets and "a smattering of other nationalities including Irish, Germans, Norwegians and Americans." RBA began its government-sponsored cadet pilot scheme in 1979 when the airline was five years old. The localization scheme proved difficult for Brunei, which had a population of only 276,000, and many nationals who appeared reluctant to stay long periods away from home.

171. Swire HK Archive, *The Weekly* Issue 20 [January 1995].

172. Swire HK Archive, *The Weekly* Issue 34 [May 1995].

173. Swire HK Archive, *Cathay News* No. 95 [March 1994]; Swire HK Archive, *The Weekly* Issue 22 [February 1995]; Swire HK Archive, *The Weekly* Issue 34 [May 1995]. In

local Hong Kong faces to Cathay Pacific's staff, including those who occupied the commanding seats in the cockpit. At the same time, the airline desired that its crew continue to receive its training and build its knowledge base according to British Commonwealth standards.

Cathay Pacific's localization efforts persisted into the final days of colonial rule. A 1995 article in the *South China Morning Post* continued to advertise job prospects with the airline where "the sky is the limit." The article reiterated, among other requirements, the ability "to converse fluently in both English and Cantonese."[174] Nine years into the program, the total number of graduates surpassed one hundred, but the media would report that "airline training [was] slow to take off." Despite the potential of a HK$100,000 monthly salary for those reaching the top of their profession, "none of the cadets has made it to the rank of captain." More regrettably, many failed to enter the program in the first place because of persistent issues with eyesight, physical fitness, or English proficiency. "A lot do not pass because of potential deterioration of the eyes," said Jennifer Ng Siu-hoi, assistant manager of aircrew recruitment, "or they're not tall enough, or there are heart problems."[175]

As critics took stock of the situation two years after the 1997 handover, a *South China Morning Post* article lamented the slow progress of localization with the headline "Career Prospects up in the Air." The article quoted Steve Miller, the former Dragonair chief who initiated its localization program in 1985: "It's a shame the localisation programmes are going so slowly." The figures in 1999 proved his point. At Cathay Pacific, only one hundred and eight of its one thousand and three hundred strong pilot force were local after more than a decade of its cadet program. At Dragonair, of its one hundred and fourteen pilots, only nine were local. Both airlines had fueled their expansion mostly by hiring experienced pilots, many foreign, rather than recruiting through the much-touted local

---

comparison, China trained its cadets in the United States, Australia, Britain, and New Zealand; Taiwan's China Airlines and Japan's airlines sent theirs to the United States; and Vietnam Airlines' pilots received their training in Australia, France, and Britain (*SCMP*, November 9, 1994, 37).

174. *SCMP*, February 23, 1995, 63.

175. *SCMP*, July 22, 1996, 3.

The Making of a "Hong Kong" Airline

279

cadet programs.[176] Cathay Pacific cited high training costs and a limited pool of eligible applicants as the causes of slow progress. However, the airlines knew of the extensive investments that the training programs entailed, and there had been no initiative to address the known structural issues in applicant selection.[177]

Cathay Pacific's localization efforts paralleled the process whereby the British colonial government transitioned power to local hands ahead of the handover. Before this transition period, the colonial government co-opted non-British elites into the government. Ambrose King has called the strategy "the administrative absorption of politics" whereby the British authorities derived legitimacy.[178] But as Steve Tsang has pointed out, staff localization proceeded very slowly, even after the government adopted a general policy in 1961 "not to appoint expatriate officers to the permanent establishment 'unless there appears to be no possibility of Chinese with the appropriate qualifications being available in the next few years.'" The pace hastened quickly with the countdown to 1997 after the signing of the Joint Declaration in 1984. Starting in 1985, the government recruited only ethnic Chinese as administrative officers. Locally recruited administrative officers also received preferential treatment in promotion as expatriate officers were granted compensation for forced retirement or missed promotion opportunities. Similarly, until 1982 female officers faced institutionalized gender prejudice without full equality.[179] There are many reasons for localization of the commanding positions in the power structures of Hong Kong. Of course, merit considerations such as one's familiarity with the community call for the localization of civil service, but

176. Dragonair hired four First Officers with at least five years' airline experience and basic training at a military or civil school. Its program received enthusiastic response, "mainly from the various Commonwealth countries." Despite its initial publicity of local recruitment for the cockpit, the airline was still in need of foreign hires with qualified backgrounds and selected its applicants with criteria similar to those Cathay Pacific had adopted. (Swire HK Archive, Dragonair *Dragon News* Vol. 3 No. 19 [March 1993]; Swire HK Archive, Dragonair *Dragon News* Vol. 3 No. 20 [May–June 1993]; Swire HK Archive, Dragonair *Dragon News* Vol. 3 No. 22 [September–October 1993]).

177. *SCMP*, October 8, 1999, 21.

178. King, *China's Great Transformation*, chap. 7.

179. Tsang, *Governing Hong Kong*, chap. 7.

280   *Recasting Hong Kong*

the process quickened largely for political reasons.[180] As the possibility of democratization was remote, the British colonial government showed its accountability to the local population partly through the localization of government staff, which also fit Beijing's desire to have Chinese nationals gain control of the administration of Hong Kong.[181]

Although the airline industry in Hong Kong shared the motive of reducing British control with the transitional government in the final days of colonial administration, Cathay Pacific did not have to undergo as complete a transformation as the Hong Kong government. Just as its capital base continued to increase local Hong Kong and mainland Chinese representation beyond the 1997 handover, the localization of Cathay Pacific's cockpit crew was an ongoing process. The persistent representation of expatriate pilots, well past 1997, testifies to the resilience of management and its ingenuity in embracing the requests of the incoming political regime and undertaking sufficient measures to satisfy the requirements of a local Hong Kong airline.

The timing was telling. Cathay Pacific rushed to recast its commanding positions in the cockpit to incorporate local Hong Kong faces just as it began to embrace the infusion of capital from local and mainland Chinese concerns. The airline's pace in local cadet recruitment also reflected the competition Dragonair exerted in their bid to claim recognition as the airline of Hong Kong. The rivalry between Cathay Pacific and Dragonair abated as the former absorbed the latter. Although local recruitment continued for both airlines, the changes in hiring the 1980s and 1990s did not have any transformative effect in the cockpit practices for either airline. Instead, the local recruitment of pilots produced a more of a lasting impact in the public images of the airlines, as well as the cosmopolitan network of a city in the midst of its political realignment.

The varying national/local profile of Cathay Pacific underscores the dynamic interactions between the state and the investment market in an ever-changing geopolitical landscape. As the conclusion of Sino-British negotiations in 1984 sealed the fate of Hong Kong beyond 1997, the airline assumed a local Hong Kong facade. This refashioning of the airline

---

180.   Lee and Huque, "Transition and the Localization."
181.   Cheung, "Rebureaucratization of Politics."

as a local company did not adequately satisfy the requirement of the incoming regime, especially as Dragonair garnered the support of investors with mainland ties and competed with Cathay Pacific for genuine local status. The competition with Dragonair extended beyond the financial markets as the rush to incorporate Hong Kong faces in their workforce (or at least to create the perception thereof) promised to allow local entry into command positions.

In the end, loyalty to locality did not guarantee Beijing's backing. To continue to rule the airways over Hong Kong, Cathay Pacific embraced red capital from mainland-controlled enterprises with direct connections to Beijing. Throughout this process, the carrier's dominant shareholder, Swire, demonstrated considerable agility and resilience against a rapidly shifting geopolitical backdrop. Swire's perceived Britishness had allowed its assumption of control from the airline's Australian-American founders and facilitated investments in the airline by those with ties that London deemed appropriate. Ironically, the same Britishness came to haunt the group's investment in the airline when the political tides turned. Perception of such British roots would not dissipate readily. Even as the airline restructured, its shareholder base could not develop a local Hong Kong identity that Beijing would find palatable. After the shareholding reshuffling in 1987, British shareholders managed to hold onto an aggregate two-third majority. By 1992, the airline's British backer Swire could barely retain a simple majority, and its position proved indefensible in 1996, a year before the 1997 handover. The incremental infusion of mainland Chinese investment in and influence over the Hong Kong carrier, calibrated to reflect a comfortable compromise, enabled a seamless transition over the handover, which entailed a change of sovereignty (and the attendant air service rights) overnight on July 1, 1997—no mean feat in light of the delicate diplomatic sensitivity and vast financial interests involved.

Like local British institutions whose expertise seemed vital to the development of successor institutions in other cases of decolonization,[182] Cathay Pacific proved its sustaining power in Hong Kong through the transition period to Chinese sovereignty. However, in the case of Cathay Pacific, its British owners needed to share power and profits not just with

---

182. Stockwell, "Imperial Liberalism."

the "indigenous elites" in Hong Kong but also with the incoming mainland Chinese power.[183] Not only was the airline instrumental to the success of commercial aviation in Hong Kong, but the operational competence it developed during the colonial period could help the growth of the budding airline industry and airport infrastructure in mainland China. Yet, as in other decolonizing settings, British businesses would experience eroding bargaining power with the local government. Their staying power depends on their entrepreneurial dynamism and their ability to integrate with the local political and economic powers after decolonization.[184]

The dexterity of Cathay Pacific in crafting its ownership structure at different points of its history tested at a very late stage the application of "gentlemanly capitalism" in a transnational setting.[185] The carrier's challenges before the 1980s might have been confined to the greater British circles. However, its transformation in the transition era ahead of 1997 stretched the shareholder base beyond its comfortable social network to encompass investors with a radically different background. Among these new investors are some that represented mainland Chinese state interests critical to the continuing success of the airline. Not unlike the British, the incoming political regime exercised its state influence within the framework of the free market. Just as BOAC, the longstanding investor in Cathay Pacific, was a state-owned enterprise, CITIC, CNAC, and CTS represented the interests of the mainland Chinese state in the Hong Kong carrier.

The airline was equally adroit in designing the composition of its personnel to meet business exigencies. Besides fashioning the female lineup of its cabin crew (see chapter 3), Cathay Pacific initiated its program of hiring pilot and management trainees in Hong Kong as the appearance of localizing its more highly paid command posts was deemed strategically profitable. Nevertheless, structural hindrances (for example, the vision requirement for pilots) kept this initiative limited numerically as the

183.   Louis and Robinson, "Imperialism of Decolonization," 463.

184.   Jones, *Merchants to Multinationals*; White, *British Business*.

185.   For a discussion of the development of "gentlemanly capitalism" in the case of Britain, see Cain and Hopkins, *British Imperialism*.

airline continued to insist on technological and institutional indoctrination through British and British Commonwealth training.

Historical precedents could engender apprehension about state intervention in the marketplace, but state actions vary according to the agenda of the times. The PRC's assumption of sovereignty over Hong Kong inaugurated a major shift in the development of the city's aviation industry. Compared with the experience in the early years of the PRC, the ascending control of mainland Chinese interests in Hong Kong's primary carrier unfolded more subtly and involved none of the open conflict that state seizure entailed in the 1950s.[186] The difference reflected not only the particular arrangements of the handover but also the drastically transformed posture of the PRC government in the reform era. However, the airline industry remains a critical sector for the state, and a carrier's success relies heavily on its ability to garner political support.[187] Unlike its longtime investor HSBC, which chose to transform its profile from the colony's largest bank to a global financial conglomerate,[188] Cathay Pacific deepened its mainland connections in its many rounds of corporate realignment in the transitional years leading up to the 1997 handover.

In colonial Hong Kong, Cathay Pacific was both foreign and local—foreign because its shares were not primarily locally held but also local because its routes radiated from its home base in Hong Kong. That peculiar combination provided the airline legitimacy as a British concern in Hong Kong. This source of legitimacy was becoming obsolete with the looming sovereignty transfer in 1997.[189] In response, Cathay Pacific first

---

186. Previous scholarship on urban transformation during the early decades of Communist rule in China focused on the penetration of the Communist state into local society (Vogel, *Canton under Communism*; Lieberthal, *Revolution and Tradition*; Gao, *Communist Takeover of Hangzhou*). For focused discussions of the experience of capitalists in the early years of the PRC, see Gardner, "Wu-fan Campaign"; White, *Careers in Shanghai*; Krause, *Class Conflict*; Cochran, "Capitalists Choosing Communist China"; and Leighton, "Capitalists, Cadres, and Culture."

187. For a discussion of the general development of the airline industry in the PRC, see Le, "Reforming China's Airline Industry."

188. *Asian Wall Street Journal*, July 14, 1994, 1.

189. Bucheli and Kim, "Political Institutional Change."

accentuated its local profile to minimize its "liability of foreignness" involving certain British or British Commonwealth characteristics.[190] As an increased local Hong Kong profile would not suffice, the airline accepted another source of "foreign" (read: from outside of Hong Kong) ownership and yielded to ascending representation by the incoming political regime among its shareholders.

In the multiple rounds of shareholding reshuffling and local hiring initiatives, Cathay Pacific's preemptive efforts, as well as the persistence of the PRC state in escalating its representation in the Hong Kong carrier, point to an understanding that economic decolonization often lags behind political decolonization.[191] Indeed, British representation remained dominant in Cathay Pacific even after the handover, but British endurance was tolerated for not only its instrumentality in ensuring operational efficiency locally in Hong Kong but also its connections to the budding industry in mainland China. Seen from the perspective of the PRC's national politics, the gradual shifts in Cathay Pacific's corporate profile indicate the willingness of the incoming regime to delay economic nationalization for the sake of its own reform agenda.

Cathay Pacific was not a simple local operator. From its longstanding home base in Hong Kong, the airline commanded a multinational presence that extended first to a regional network encompassing East and Southeast Asia, and in due course, to Europe and North America. This analysis underscores that, in its examination of the transnational setting of such business enterprises as Cathay Pacific, the role of the state is crucial.[192] Firms operating in a transnational setting altered the sociopolitical landscape within which they run their business.[193] Conversely, geopolitical powers also condition and influence business choices and enforce jurisdictional boundaries that endure in spite of globalizing forces. The case of Cathay Pacific's restructuring underscores business exigencies for operating in the crevices between shifting definitions of states. Set against a global backdrop, the business of Cathay Pacific pivots not only on its

190. Bucheli and Salvaj, "Political Connections."
191. Jones, *British Multinational Banking.*
192. Rather than standing in contrast, this observation complements Boon's call to foreground transnational elements ("Business Enterprise and Globalization").
193. Fitzgerald, *Rise of the Global Company.*

resonance with the local (but not quite national) circumstances but also on the agreement of a changing cast of state actors.

The role of the state is all the more important in industries that the state deems strategic and over which the state maintains oversight. The case of Cathay Pacific in the run-up to the handover was unusual: unlike multinationals that leverage their operations in their host and home countries, the airline constructed a local presence in Hong Kong and muted its British roots in a bid to appease its new host in Beijing. In that round of their restructuring, Cathay Pacific adopted "aspirational political practices" to address the requirements of a new governing regime.[194] The airline needed to restructure more fundamentally than simply masking its British roots in order to manage its political risks and earn the trust of the incoming power that was to underwrite its continued success. It does not require warfare or open conflict for a business to experience an existential threat stemming from political risk. In the protracted process through which Britain yielded jurisdictional power of Hong Kong to the PRC, Cathay Pacific responded preemptively first by enhancing its local profile, and then by appealing to economic nationalism of the sovereign state poised to take charge.

Whereas the people of Hong Kong were known for their practice of "flexible citizenship" in their negotiation with nation states,[195] the transformation of Cathay Pacific's investor profile demonstrates that its corporations, too, were equally adept in adjusting their ownership structure to function effectively under different political regimes. Equally importantly, Cathay Pacific's ability to vary its image along a shifting political spectrum attests to the pliability of the notion of "corporate nationality." Scholarly focus on national ownership presumes too much of a binary (state-owned versus private enterprises) to appreciate sufficiently corporate agility, as in the case of Cathay Pacific; unbridled celebration of the free market in Hong Kong also does not recognize the influence of

---

194. Lubinski and Wadhwani discuss the use of "aspirational political practices" to address "the inherently future or goal oriented character of nationalism" ("Geopolitical Jockeying"). Although the handover of Hong Kong did not reflect a nationalistic move within the city, the airline's strategy satisfied the political requirements of the PRC that demanded greater control of the airways above its territories upon asserting jurisdiction over Hong Kong.

195. Ong, *Flexible Citizenship*.

the state (in its different formations) in the economic activities of the city. Instead of examining the mobility of global capital,[196] this chapter focuses on the corporate nationality that Cathay Pacific projected at different points in its history. Although capital as calibrated by denominations could transcend, within limits, divides by foreign exchange in the global marketplace, the presumed identity of the investor, individual or corporate, confers corporate nationality on the business, along with certain implications of political allegiance.

Seen through the lens of Cathay Pacific's changing corporate profile, the free market of shareholding, rather than being a polar opposite of state control, facilitated the negotiation of state claims and enabled the fluid transfer of business interests between political regimes. Augmented by modifications in Cathay Pacific's hiring practices, shifting shareholding allowed the airline to construct its corporate nationality as the Hong Kong–based carrier charted its course through the turbulent waters of geopolitics and mitigated political risks that threatened to destabilize the ascent of its business.

196. Globalization and capital mobility have attracted the attention of many economic historians and political scientists. Some have focused on their converging effect (see, for example, Williamson, "Globalization, Convergence"), whereas others have studied the benefits such trends accrue to particular groups of people (Obstfeld, "Global Capital Market"; Stallings, "Globalization of Capital Flows").

CONCLUSION

What Is Next for Hong Kong?

Through commercial aviation, Hong Kong built on its role as a maritime transport hub and connected with evolving regional and global networks. Political forces governed this prolonged and ongoing process of connections while economic factors animated commercial participation as aviation linkages proliferated and intensified. The expansion of aviation networks that included Hong Kong overhauled the social environment of the city as this new industry generated mobility, both in terms of physical movement and changes in power structure. Accelerated movement through Hong Kong heightened interactions with elements in Southeast Asia and beyond, producing cultural forms that reflected the globality of commercial aviation and the particularity of Hong Kong.

As a new technology, commercial aviation wove connections across the skies as a "space of flows" over which budding infrastructure was to facilitate novel forms of exchange and interaction.[1] Yet the network this new technology engineered was conditioned by preexisting connections, the natural environment, and the political framework in Hong Kong. At the same time, the development of commercial aviation in Hong Kong generated an infrastructure that served as a material substrate integrating disparate elements. The relationship thus fashioned among the new technology of commercial aviation and its attendant network and infrastructure gave

---

1. Castells, *Rise of the Network Society*; Larkin, "Politics and Poetics."

288   *Conclusion*

shape to modern Hong Kong's politics, economy, society, and culture.[2] Institutionally, key players in the logistics business that had dominated previous generations of transportation, in particular shipping, needed to safeguard their commercial interests by asserting their presence on this new technological platform. It should therefore come as no surprise that Swire and Jardine Matheson entered the fray, and the competition between the two over the airways in Hong Kong persisted. In terms of infrastructure, before enabling facilities such as runways became sufficiently mature for technical and commercial viability, flying boats circumvented hardware challenges and got aircraft off the ground for pioneering carriers. Politically, while the new technology compressed distances and threatened to redraw boundaries, territorial jurisdictions claimed their rights over their aerial turf and dictated the rules of engagement over connections in the skies. Keenly aware of the disruptive potential of commercial aviation, entrenched interests shaped the new industry to their advantage and mitigated the risk of seismic impact on the competitive landscape.

The development of commercial aviation in postwar Hong Kong mirrored that of the colony. At critical junctures, favorable conditions in all these aspects came together to cultivate Hong Kong as an aviation hub. At the same time, opportunities for Hong Kong and its readiness to invest in infrastructure reflected the shifting situation of the colony throughout this period of expansion. Furthermore, the construction of aviation networks through the colony transformed the dynamics within Hong Kong as well as its interactions with elements beyond the city. Hong Kong blossomed into a hub only against the backdrop of its advantageous geopolitical setting; in turn, its flourishing commercial aviation industry consolidated its centrality in the networks in which the city was increasingly embedded.

Economic Opportunities in a Politically Liminal Space

It is tempting to attribute the development of commercial aviation in Hong Kong to British sponsorship. After all, the metropole and the im-

---

2.   Jensen and Morita, "Introduction."

What Is Next for Hong Kong? 289

perial airline underwrote the negotiation of air service agreements at crucial moments in the early phases. Yet, commercial aviation in Hong Kong owed its origin not only to Britain but also to the confluence of budding traffic from Europe, North America, and China. Political and economic alliances between US and Republican Chinese interests had initiated traffic from North America to this side of the Pacific. However, routes could not achieve true circumnavigation until trans-Pacific traffic met with airways stemming from Europe, linking up at Hong Kong with European imperial traffic to Southeast Asia and Australia. US aviation interests along with their Chinese partners thus connected with the UK's Imperial Airways at this southern tip of the Chinese landmass before the outbreak of World War II.

The return of Hong Kong to the control of Britain after the war rejuvenated interest in developing the colony into a hub of commercial aviation. Rather than its status as a British colony, it was Hong Kong's proximity to mainland China that proved critical to its potential as a nexus of traffic flows. For the development of the industry in the immediate aftermath of World War II, the appeal of Hong Kong as a hub for commercial aviation pivoted on its location outside of China proper.[3] However, as important as its jurisdictional autonomy from China was, the British colony's territorial attachment to China and the attendant potential to serve as a gateway to China, literally and metaphorically, loomed larger in the calculation of those who steered the design of airways as they constructed a new network of flow. Rather than agonizing about political allegiances as subjects had in war-torn zones,[4] enterprising individuals made full use of Hong Kong's hybridity as a foreign enclave at the tip of mainland China, a precious remnant of Western imperialism at the conclusion of World War II. Hong Kong's peculiar position was its advantage. The city was integral to China's network, prompting mainland and Hong Kong interests to include Hong Kong from without (*baokuo zaiwai* 包括在外) as they co-opted the strategic location of the British enclave to provide China an outlet to the wider world.[5] The dramatic growth of aviation traffic through Hong Kong in the late 1940s attested

3. Carroll, *Edge of Empires*, 57.
4. Fu, *Passivity, Resistance and Collaboration*.
5. I borrow this term from Sinophone Studies. David Wang has used this term to describe "Sinophone writers' strategic appropriation of the Chinese sign in their

to this image of its cruciality. The crossroads at Hong Kong arose not just from its location outside of China but also through its ability to facilitate exchange with China along its periphery.

Unfortunately, expectations of the hub's pivotal potential proved overblown. Communist control of the mainland severed aerial flows and dashed inflated hopes for the budding aviation hub. Ironically, the events in the mainland settled the rivalry between two British camps in Hong Kong, with Swire-backed Cathay Pacific emerging victorious over its competitor backed by Jardine Matheson and BOAC. On the strength of its Southeast Asian network, which inflaming Cold War dynamics reinforced, Cathay Pacific fashioned a regional configuration that fed into the trunk systems of BOAC and other industry titans. Although the aviation industry of Hong Kong suffered a devastating blow as the regime change in the mainland truncated promising Chinese routes from its network, all was not lost as the colony reverted back to depending on its British sponsor for a place in the imperial network.

For Cathay Pacific and Hong Kong, business grew as geopolitics fostered opportunities in the crevices of state powers. As Cathay Pacific proved itself sufficiently British (or British Commonwealth) to the colonial government and authorities in the metropole, it was groomed as the de facto flag carrier for the colony. Although mainland China remained off limits to commercial aviation from Hong Kong, Cathay Pacific's operation on the other side of the bamboo curtain afforded the fledgling carrier ample business opportunities that in time expanded from Southeast Asia to Japan and Taiwan.

The growth of commercial aviation through Hong Kong, which began in the 1950s and kicked into high gear in the 1960s, facilitated and reflected the favorable economic development of the colony. As Cold War dynamics propelled the flow of traffic, the United States initially represented the majority of incoming visitors to Hong Kong (28 percent in 1969). This US lead was soon eclipsed by Japan, which held the largest share of incoming visitors to the British colony throughout the 1970s, reaching a high of 37 percent in 1973. Southeast Asia took the larger share in 1980 only to yield to Japan by the mid-1980s. Taiwan rose to promi-

---

articulation of local sensibilities" (Lin, "Writing beyond Boudoirs," 255, citing Wang's *Wenxue xinglu*).

What Is Next for Hong Kong? 291

nence in the late 1980s and represented one-fifth of the total during the last decade of colonial Hong Kong.[6] Besides benefiting from mounting passenger flow, the British colony also profited from skyrocketing exports, registering double-digit growth in all but four years between 1962 and 1988. During that period, total exports grew at a compound annual growth rate of 17 percent. While exports to the United Kingdom showed respectable increases, growth in total exports was fueled by surging volumes of export to the United States. The British colony's exports to the United States overtook exports to the United Kingdom by the early 1960s, growing to four or five times that of the shipment to the United Kingdom. From 1965 to 1989, Hong Kong's shipments to the United States consistently exceeded 30 percent of the colony's total export, peaking at 44 percent in 1984 and 1985. The loss of business with mainland China dealt Hong Kong a severe blow, but its alignment with the camp on the other side of the bamboo curtain propelled the British colony into a formidable economic power. Concomitantly with this economic takeoff, commercial aviation traffic escalated through Hong Kong (fig. C.1).[7]

Total exports growth waned by the 1990s but not until it had received one last boost from the reopening of the PRC. Exports to the PRC from Hong Kong expanded from 1977 until 1992, rising from a negligible fraction of the total to 30 percent by the handover. The PRC eventually overtook the United States as the most popular destination for Hong Kong exports in the mid-1990s.[8] Similarly, mainland China's share of total visitor arrivals to Hong Kong spiked, reaching 22 percent of the total in the year of the handover.[9] Air traffic volume continued to climb, reflecting the retooling of the city's economy and its acute reorientation toward the mainland. The allure of the mainland market manifested itself

---

6. *Hong Kong Tourist Association Annual Report*, 1970, 21; *A Statistical Review of Tourism*, 1976, Table A12; *A Statistical Review of Tourism*, 1986, Table 1.12; *A Statistical Review of Tourism*, 1995, Table 1.11; *A Statistical Review of Tourism*, 1997, Table 1.13.

7. *Hong Kong Statistics, 1947–1967*; *Hong Kong Annual Digest of Statistics, 1978 Edition*; *Hong Kong Annual Digest of Statistics, 1985 Edition*; *Hong Kong Annual Digest of Statistics, 1994 Edition*.

8. *Hong Kong Annual Digest of Statistics, 1978 Edition*; *Hong Kong Annual Digest of Statistics, 1985 Edition*; *Hong Kong Annual Digest of Statistics, 1994 Edition*; *Hong Kong Annual Digest of Statistics, 1998 Edition*; *Hong Kong Annual Digest of Statistics, 2000 Edition*.

9. *Hong Kong Tourist Association Annual Report*, 1997/98, 18.

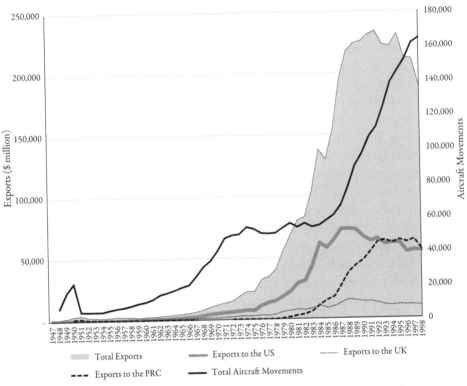

FIGURE C.1. Annual Hong Kong exports and aircraft movements, 1947–98. Sources: *Hong Kong Statistics, 1947–1967, Hong Kong Annual Digest of Statistics,* 1978, 1985, 1994 1998, 2000 editions.

not only in export volumes from Hong Kong but also in competition over Hong Kong and mainland connections in aviation networks.

Commercial aviation is a business that connects places, often transcending political divides. It is also an industry that requires heavy capital investment. Not only did Cathay Pacific need to fund its capital investments in equipment upgrades, but the colonial government also had to develop the financial wherewithal to fund infrastructure construction for the continued growth of commercial aviation through Hong Kong. Against the backdrop of the colony's robust economic growth, Cathay Pacific's revenue increased sevenfold in the decade between 1958 and 1968, and another tenfold to HK$1.8 billion in 1978. Operating profits also rose

What Is Next for Hong Kong? 293

from the respectable level of HK\$2 million in 1968 to a staggering HK\$259 million in 1978, providing the requisite war chest for the airline's expansion strategy.[10] Similarly, the Hong Kong colonial government profited from the city's economic takeoff. From a balanced budget in 1967, the colony grew to amass a sizeable reserve balance as the government recorded a surplus for all but four years in the three decades leading up to the handover.[11] Alongside the city's airline, the colonial government deployed the financial resources the city accumulated toward infrastructure and equipment upgrades, building Hong Kong into a hub for commercial aviation.

The suspension of commercial aviation with the mainland was trying for the industry in the 1950s, but renewed economic growth, generated by Cold War dynamics, invigorated the British colony and underwrote its development into a hub for commercial aviation on one side of the geopolitical divide. In addition to the flow of goods as air cargo, tourist growth also powered aviation traffic through Hong Kong. Just as the Cold War thawed and industrial development in Hong Kong subsided, the reform era in mainland China revitalized the city's economic ties with the awakening giant to the north, and sustained the momentum of the development of Hong Kong and its aviation industry. At decisive moments, Hong Kong interests displayed "liminal innovativeness" and fortified the pivotal function of the city as an aviation hub.[12]

The evolution of Hong Kong into an international hub of commercial aviation proceeded in phases. Leveraging a favorable geopolitical location, Hong Kong managed to extend its role as a maritime hub to the budding world of aviation. Re-anchoring itself in the network of flows enabled by new technologies did not guarantee continued importance, however. Overcoming a seismic shift that severed aerial traffic with mainland China, policymakers and commercial enterprises persevered and developed the requisite infrastructure for the emerging industry of aviation.

---

10. Swire HK Archive, Cathay Pacific Airways Limited Annual Reports, 1962, 1968, 1978.

11. *Hong Kong Annual Digest of Statistics, 1978 Edition*; *Hong Kong Annual Digest of Statistics, 1985 Edition*; *Hong Kong Annual Digest of Statistics, 1994 Edition*; *Hong Kong Annual Digest of Statistics, 1998 Edition*.

12. I borrow the term from Duara, "Hong Kong," 228.

294 Conclusion

Cold War dynamics underwrote the development of Hong Kong as a connection point and connected it to a network that linked Southeast Asia to Europe at one end, and via the Philippines, Taiwan, Japan, and South Korea to North America at the other. In time, forces rooted in Hong Kong extended their control beyond the regional orbit and exerted their influence over a larger footprint. Against a shifting geopolitical backdrop, Hong Kong needed to reinforce its cruciality, differentiate itself from rival hubs, and play an active role in network development.

Post-handover Networking

The planning for the new airport at Chek Lap Kok, which required coordination with mainland authorities, predated the signing of the Joint Declaration in 1984.[13] In spite of Sino-British disputes, the HK$155 billion (US$20 billion) airport, which took seven years to complete, was opened by PRC President Jiang Zemin on July 2, 1998, a year after the handover. Tung Chee-hwa, the first chief executive of the Hong Kong SAR, attributed its completion to "the wealth of the Hong Kong people, their confidence in the future and Beijing's support." Hours after Jiang had boarded his Air China flight back to Beijing, Air Force One touched down in Hong Kong—the first international flight to land at Chek Lap Kok, bringing US President Bill Clinton to Hong Kong for an en route stop at the end of his China trip. The first sitting US president to set foot in Hong Kong, Clinton spoke, at a dinner hosted by Tung at the Government House, of Hong Kong's vital importance "to the future not only of China and Asia, but of the US and the world as well." The significance of the choreographed tandem visits of Jiang and Clinton to Hong Kong on the inaugural day of Chek Lap Kok was unmistakable. The planning and funding of Chek Lap Kok had transcended political divides. The new airport was to reinforce Hong Kong's role as a transport hub and nexus of trade, finance, and tourism, as Tung remarked (fig. C.2).[14] The PRC

13. TNA, FCO 40/1312.
14. *SCMP*, July 3, 1998, 1; *Wen Wei Po*, July 3, 1998, 1; *Ming Pao*, July 3, 1998, 1.

What Is Next for Hong Kong?

FIGURE C.2. Hong Kong newspapers of different political persuasions reported on their front page the tandem visits of the leaders of the PRC and US on the inaugural day of the new airport at Chek Lap Kok. Source: Author.

and US governments had indicated, in 1998, their intention to maintain the criticality of Hong Kong.

Hardware development is but one of the criteria for Hong Kong's centrality as an aviation hub. In concert with the expansion in airport infrastructure which extended to the construction of a purpose-built airport (a project that echoed the plans to abandon Kai Tak in the aftermath of World War II [see chapter 2]), in the post-handover era, the skyways over Hong Kong also saw continued transformations in such areas as the proliferation of flight routes, ownership structure of airlines, and refinement of the Hong Kong brand.

Expanding beyond its initial North American foothold of Vancouver, Cathay Pacific launched on July 1, 1990, its service between Hong Kong and Los Angeles, "the longest scheduled non-stop commercial flight in the world."[15] A year before the handover, the Hong Kong airline also

15. Swire HK Archive, Cathay News No. 54 [September 1990].

296   *Conclusion*

took a "bite out of the Big Apple," introducing on July 1, 1996, service five times a week to New York, its first to the East Coast, via Vancouver.[16] A year after the handover, Cathay Pacific eliminated the en route stop for this service and inaugurated the first nonstop flight between Hong Kong and New York. Dubbed *Polar One* for its route over the North Pole, the Boeing 747-400 touched down in Hong Kong on July 6, 1998, the first commercial flight to land at the new airport in Chek Lap Kok.[17] In 2000, Cathay Pacific extended this transpolar service to its Hong Kong–Toronto connection using an A340-300 aircraft.[18] These technological feats not only extended the reach of Cathay Pacific's network but also literally redrew the map of commercial aviation, making Hong Kong a solid anchor along the skyways.

In the post-handover period, the Hong Kong airline engineered phenomenal growth of its network. In 1998, Cathay Pacific "offered scheduled passenger and cargo services to 48 cities in five continents."[19] Ten years later, the airline counted "116 destinations in 35 countries and territories."[20] Fast-forward another decade: Cathay Pacific "directly connect[ed] Hong Kong to 109 destinations in 35 countries worldwide (232 and 53 respectively with code share agreements)," reported the airline in its 2018 annual report. Those counts included "26 destinations in Mainland China," the airline noted.[21]

The proliferation of its flight routes stemmed in part from its incorporation of Dragonair. In a round of restructuring in 2006, Cathay Pacific took over all of Dragonair and accepted 35 percent of mainland holdings of its own shares split between Air China (the PRC's national flag carrier, which was the majority shareholder of CNAC) and CITIC. At the same time, Cathay Pacific expanded its stake in Air China to 17.3 percent, up from the 10 percent it had acquired in 2004.[22] The air-

---

16.   Swire HK Archive, *The Weekly*, Issue 57 [October 27, 1995].

17.   Swire HK Archive, *CX World* Issue 10 [August 7, 1998]; *SCMP*, July 7, 1998, 3.

18.   Swire HK Archive, *CX World* Issue 51 [June 2000].

19.   Swire HK Archive, Cathay Pacific Airways Limited Annual Report 1998, 1.

20.   Swire HK Archive, Cathay Pacific Airways Limited Annual Report 2008, inside cover.

21.   Swire HK Archive, Cathay Pacific Airways Limited Annual Report 2018, 2.

22.   Swire held onto 39.99 percent of Cathay Pacific's shares. *SCMP*, June 10, 2006, 1; Swire HK Archive, Cathay Pacific Airways Limited Annual Report 2006, 30.

line called this "a historic shareholding realignment" as management promised to "work to optimize the significant commercial opportunity provided by [their] purchase of Dragonair." In particular, management spoke of "synergies and opportunities that arise from linking Cathay Pacific's international network with Dragonair's extensive Mainland China services."[23]

Articulating its expanding flight network and the mounting significance of mainland China in its operations, Cathay Pacific launched a makeover of its flight attendants. In 1998, the Hong Kong airline hand-picked home-grown talent Eddie Lau whose ensemble of "eye-catching colours and striking designs" was to "finally lay to rest" the Nina Ricci outfits introduced during the last years of colonial Hong Kong.[24] Building on his famous collections "East is Red" and "Shanghai Chic,"[25] Lau's design debuted in 1999, featuring an "Asian cut" in "colourful hues of red, blue and purple." Prominent in the design was the stand-up collar that was to "echo a mandarin collar and give the airline a more Asian appearance."[26] The airport's general manager noted that this "nice and elegant" design conveyed what he considered the intended message: "we are the locally-based carrier of Hong Kong, with an international flavour."[27] This characterization harks back to the Chinese-inspired, rose-colored uniform of 1962 (see chapter 3), except that this time, the ensemble projected more confidence in its modern Asian elegance, as one crafted not by an American fashion expert but a local Hong Kong professional. So convinced was Cathay Pacific of this direction of uniform design that when it came time for a refresh, the airline reengaged Lau.[28]

For the industry as a whole, commercial aviation experienced spectacular growth. From 1997 to the peak in 2018, the number of air passengers grew almost 150 percent. Registering a threefold increase during the same period, air cargo growth was even more impressive (figs. C.3 a and b).[29] In 2019, Cathay Pacific was the world's eighth-largest airline in

---

23. Swire HK Archive, Cathay Pacific Airways Limited Annual Report 2006, 3, 21.
24. Swire HK Archive, *CX World* Issue 6 [June 1998].
25. *SCMP*, August 28, 1979, 16; *SCMP*, April 22, 1980, 16.
26. *SCMP*, September 30, 1999, 6.
27. Swire HK Archive, *CX World* Issue 42 [November 5, 1999].
28. Swire HK Archive, *CX World* Issues 103 [October 2004], 104 [November 2004].
29. *HKDCA*; Airport Authority Hong Kong website.

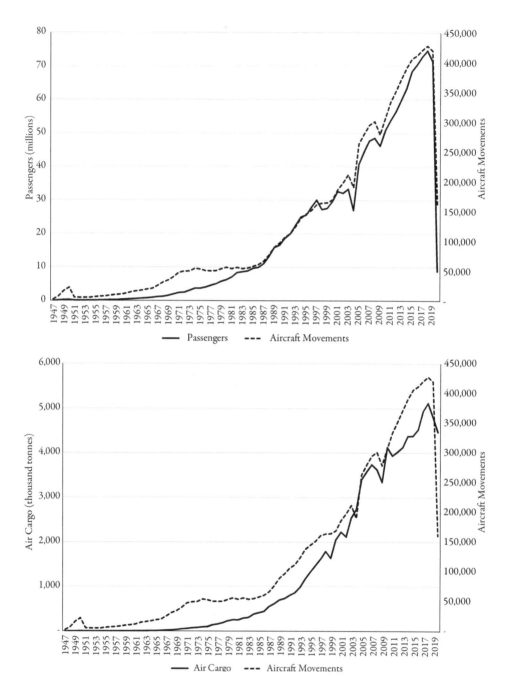

FIGURES C.3 a and b. Annual air traffic statistics at the Hong Kong International Airport, 1947–2020. Sources: *HKDCA* annual reports and the Airport Authority Hong Kong.

terms of passenger traffic, and the third-largest in international air cargo.[30] For a while, strong tailwinds appeared to propel the development of Cathay Pacific and the industry of commercial aviation in Hong Kong, extending the city's global reach, regional presence, and local self-assurance.

Positioning for a City in Flux

Commercial aviation is a global industry and airlines, with the exceptions of those flying only domestically, are multinational enterprises in that their business presence crosses borders and extends beyond the boundaries of their home jurisdictions.[31] Rooted in the local milieu, air carriers confront the political and economic constraints at home as they search for business opportunities in connecting their home base with destinations overseas. In the process, airlines stretch the political confines they encounter in their home jurisdiction as they cast an ever-changing international presence for the country or city they represent. Cathay Pacific and the industry of commercial aviation have done just that for Hong Kong since the late 1940s, thereby anchoring the city in the global network and transforming the city's place in the world economy.

The growth of Hong Kong into an aviation hub entailed the development of not only Cathay Pacific but also the entire commercial aviation industry in the British colony. The definition of the "firm" of Cathay Pacific or its majority shareholder Swire extended beyond the simple confines of their corporate structure. Commercial interests mingled with political forms to produce the necessary superstructure and business environment for Cathay Pacific's business expansion as Hong Kong matured into a nexus for aerial traffic. Although the benefits of this confluence of traffic accrued to all Hongkongers, the drivers in this period of development remained decidedly the Hong Kong British.[32]

---

30. Swire HK Archive, Cathay Pacific Airways Limited Annual Report 2020, 2.

31. Wilkins, "Foreword," xiv. For a summary of recent literature on multinationals, see Wilkins, "History of Multinationals."

32. The connection between the carrier and the colonial government could be quite seamless at times. After serving as the chairman of Cathay Pacific from 1973 to 1980, John Bremridge became the financial secretary of the British colony in 1981, a post

300   *Conclusion*

Operating in the interstices required that Cathay Pacific be agile in response to political demands from various corners. Although its being sufficiently British had helped secure flying rights connecting the colony to destinations brokered by London, the reemergence of the mainland Chinese market altered the game. With the allure of the mainland market and the ongoing negotiation over the future of Hong Kong, Cathay Pacific had to rebrand its corporate nationality. Although the PRC had yet to assume sovereignty over Hong Kong, Beijing authorities wielded power over the city's connections with mainland destinations, and more importantly, traffic rights through the Chinese air space. The Hong Kong British configuration of Cathay Pacific that London had deemed sufficiently British was now considered too British by Beijing. To recast its image, Cathay Pacific infused its Hong Kong British elements with Hong Kong Chinese flavor through hiring and shareholding. Yet, Beijing did not approve of this formulation and insisted on not just Hong Kong Chinese but mainland Chinese constituents in the city's carrier.

International airlines are at once local and multinational. They need to secure the backing of the state in their home jurisdiction as they connect with destinations in foreign domains. The case of Cathay Pacific was made all the more complicated as its pre-handover shareholding structure invited questions about the carrier's home jurisdiction. Was it the city of Hong Kong or Britain? As air traffic resumed with mainland China, was the Hong Kong carrier extending its service to a foreign host or the turf of its incoming power? The ambiguity of Hong Kong's status necessitated the definition of Hong Kong as a politically discrete entity (and Cathay Pacific as a nonnational flag carrier) even when its survival and continued success hinged on the support of state powers beyond the city's limits. The peculiar configuration of Cathay Pacific as Hong Kong's carrier stretched the application of "multi-*national*" to the limits as the airline found itself ill at ease in any national construct.

Businesses that operate across political jurisdictions need to engage in "geopolitical jockeying" to secure their footing as their operating en-

---

he kept until 1986 (Swire HK Archive, Cathay Pacific Airways Limited Annual Reports, 1972–1979; *The Independent*, May 14, 1994).

What Is Next for Hong Kong?   301

vironment is transformed.[33] Although political risks often arise from open warfare, posing at worst existential threats to businesses,[34] sea changes in the geopolitical landscape that entail peaceful power transfer can also jeopardize business prospects, and companies still need to mitigate political risks in peacetime. For the period this book covers, one such sea change was the wave of decolonization. In the case of airlines that had operated in the British Empire, family-backed Tata Air Lines became Air India, in which the independent government held a majority share.[35] Launched in 1947, Singapore-based Malayan Airways received heavy funding from BOAC. Its development adhered to a British Commonwealth formation in the 1950s and 1960s, not unlike Cathay Pacific. After numerous name changes that reflected the reconfiguration of political entities on the Malay Peninsula, the operations splintered in 1972 into what came to be known as Malaysia Airlines and Singapore Airlines.[36] Outside of the dissolving British domain, US penetration and Cold War development influenced the formation of airline companies. Philippine Airlines was funded by Philippine and US interests, including a 28 percent shareholding by TWA. Established as a private concern in 1951, Japan Air Lines received assistance from Northwest Airlines of the United States.[37] Organized by South Korea's Ministry of Transport in 1949, Korean National Airlines operated until its reconstitution as 60 percent state-owned Korean Air Lines in 1962.[38] In contrast, Cathay Pacific navigated the political developments in Hong Kong as it remained a private concern.

The case of Hong Kong took a unique path in that the city returned to British colonial jurisdiction and remained a British colony until it became a Special Administrative Region in the PRC. Unlike businesses repositioning for survival in a decolonizing context, businesses in Hong

---

33. Lubinski and Wadhwani, "Geopolitical Jockeying." See chapter 6 for an extended discussion.

34. Jones and Lubinski, "Managing Political Risk"; Kurosawa, Forbes, and Wubs, "Political Risks and Nationalism."

35. Davies, *History of the World's Airlines*, 389–96.

36. TNA, BT245/1060; British Airways Archives, "O Series," 178, Geographical, 3464–72, 10000–10004; Davies, *History of the World's Airlines*, 411–12.

37. Davies, *History of the World's Airlines*, 406, 412.

38. Davies, *Airlines of Asia since 1920*, 516–19.

302   *Conclusion*

Kong did not have to focus on combating a rising *national* consciousness either in its early years or in the run-up to the 1997 handover. Instead, Cathay Pacific, along with British operators and those not necessarily favored by the incoming regime, had to respond to intensifying *state* demands emanating from beyond the city. The political risk for Hong Kong–based businesses stemmed not from the population in the city but from the state power under whose authority the SAR administration was to govern a political entity formulated under a novel scheme.

Steering Cathay Pacific clear of political turmoil thus required positioning the airline vis-à-vis various political formations in different stages of its development. In its nascent stages of development, the carrier needed to conform to the colonial framework and stay clear of mainland interference. In that period, Cathay Pacific did not pose much of a threat to either British or Chinese interests, thereby protecting the carrier and the city it served from unwanted attention. As the airline grew, Cathay Pacific had found it easier to survive and even prosper on the fringes of a weaker state, be it the PRC in its struggling early decades or the British Empire in the closing decades of decolonization. Conversely, the resurgence of such state powers troubled the Hong Kong–based carrier whose commercial viability depended on political and diplomatic sponsorship.

Among airlines operating in greater China, Cathay Pacific's unique private ownership status required more resourceful coordination of political and commercial interests. In Taiwan, after the collapse of CNAC, which was established as a state-owned concern and later received funding and assistance from Pan Am, the Republican regime organized in 1959 another state-owned carrier, China Airlines. Although the shares of the Taiwanese carrier have been traded on the Taiwan Stock Exchange since 1993, the state retains significant control through its major shareholding.[39] In the mainland, after its industry reform and consolidation, Air China became the PRC's national flag carrier under its majority ownership of state-related entities.[40] The large, private shareholding of Cathay Pacific in the hands of Swire stands in sharp contrast.

---

39.   China Airlines, Ltd., *Financial Statements for the Years Ended December 31, 2019 and 2018 and Independent Auditors' Report*, 11. Established in 1989, privately owned EVA Air is another major carrier in Taiwan.

40.   Air China Limited, 2017 Annual Report, 21.

Recent years have witnessed an erosion in Hong Kong's capacity to define itself in the liminal space. As the city continues to profit from an infrastructure design inherited from the closing years of British rule, the system in Hong Kong, in particular the city's dominant carrier, could be considered too reminiscent of its colonial legacy as economic benefits continue to accrue to parties not appointed by the current political regime. Cathay Pacific became Hong Kong's airline thanks to the collapse of the market of northbound traffic from Hong Kong after the Communist takeover of the mainland decimated the business of its local rival. As the PRC asserts its authority over Hong Kong and the sky around it, Cathay Pacific is experiencing a reversal of fortune. In the post-handover era, the Hong Kong carrier has faced not only challenges from emerging rival hubs in the mainland but also stepped-up pressure to embrace mainland shareholding, as evidenced by its incorporation of Dragonair and the increase in mainland Chinese stakes in its own corporate structure. Just as the British regime had sought to exercise its influence in Cathay Pacific through BOAC's investment in the Hong Kong–based carrier, PRC authorities constructed its formidable presence in the aviation industry in its SAR by weaving together an intricate scheme of cross-ownership.

Like the city it represents, Cathay Pacific connects and transcends the divides between China and the West, North, and South, the local and global, the market and state. Its continued success depends on the delicate balancing of the interests its routes connect. Tilting toward one side along any dimension would distort the shifting equilibrium that its network facilitates, necessitating a rebalancing act on the part of the airline and the polity that diplomatically underwrites its operations. Overreliance on one side would disrupt flows and diminish economic incentives to sustain movement in the system.

Wiring Global Circuitries

Wiring global Hong Kong through commercial aviation entails the proactive positioning of the city in regional and global circuitries of flight routes. Thanks to its British colonial status and its location on the periphery of mainland China, Hong Kong had already occupied a prime location in global networks. The city owed its geographic value first as a connection point for China and the traffic from Europe and North America. In that capacity, Hong Kong completed the circuitry and became the confluence point of aerial traffic from mainland China and emerging commercial aviation networks originating from Britain, the United States, and beyond.

The unrelenting rewiring of Hong Kong accentuates the global dimensions of the city's connections. British colonial control mapped Hong Kong onto the world while its proximity to mainland China provided the persistent appeal of potential aerial connections. The Communist takeover of China severed the British colony from its crucial mainland Chinese linkages. The shrinkage of the world around Hong Kong heightened its regional ties, reorienting the city toward the regional network of Southeast Asia, a configuration refashioned out of Nanyang, a Sinocentric characterization of the maritime cluster of ports located to the south of China. In this region, Hong Kong developed a footprint of commercial aviation, expanding the network to conform to the rules of this new industry of international connections. The city's airline held onto the hope of Cathay, the mythical market of China, while charting its world in the Pacific.[41]

The extending reach of the airline signified the process by which Hong Kong wired itself into a regional formation in Southeast Asia before its forays into global systems. The development of Cathay Pacific's catchment area and Hong Kong's regional command parallels Carolyn Cartier's discussion of regional formation.[42] Through this process, the new industry of commercial aviation reinforces preexisting connections between Hong Kong and Southeast Asian ports while forging new ties with other emerging hubs in the area.[43] The resulting circuitry through Hong Kong thus combines trunk routes that continue to stretch to Europe and North America with a regional network that links cities strug-

---

41. Osterhammel discussed the issue of "metageography"—a spatial schematization of the world and the naming of spaces (*Transformation of the World*, chap. 3).

42. Cartier, "Origins and Evolution."

43. This process echoes Otmazgin's study of the role popular culture played in shaping East Asia as a region ("New Cultural Geography").

gling to recover from wartime damages during their late colonial and postcolonial stages of development.

Sustainability in this regional network was not a foregone conclusion. Hong Kong competed with those with financial wherewithal and political backing not only in continuous infrastructure upgrades, both of airport facilities as well as airplane equipment, but also in service differentiation, accomplished largely by packaging its frontline personnel. By this stage of the process, survival as a node in the circuitry of commercial aviation required that Hong Kong generate the very elements that the network was to transport—goods and people. Along with select hubs, Hong Kong benefited from Cold War developments that routed traffic along the Pacific Rim that formed the bamboo curtain and more importantly, propelled the British colony into economic expansion through export growth.

Eventually, the city's phenomenal economic success allowed Hong Kong to break out of its regional configuration and extend its command over long-haul flight routes. Against the backdrop of privatization and economic liberalization, the metropole yielded certain trans-Pacific routes to its overgrown colony, which finally came to rival the machinery in London over the Hong Kong–London connection. Just then, the mainland Chinese market reopened, resuscitating the old circuit that had lain dormant since 1949. Fueled by its own economic accomplishments and buttressed by global developments, Hong Kong did not merely deliver on the earlier promise of serving as the linchpin of transcontinental and trans-Pacific traffic flows with mainland Chinese connections, but the burgeoning hub also asserted its prowess and assumed control of the conduits of flow.

Ahead of the handover in 1997 and in view of Chinese ascendancy, Hong Kong rewired its flight pattern yet again, this time to acknowledge the incoming power that was to control the switch of this circuitry of commercial aviation through the city. The appeal of the mainland Chinese market drew operators flying into and out of Hong Kong into the budding network in the PRC, which proved guarded in its control of its airspace. The erosion of British commercial interests was inevitable for Hong Kong's industry of commercial aviation. More than two decades after the handover, this transition remains an ongoing process.

Throughout its period of growth in the second half of the twentieth century, Hong Kong was instrumental in globalization. The city served

306 *Conclusion*

as an active agent in these global processes and became an element in the resulting global structure. In the fluid space that underwent continuous transformation, Hong Kong made itself an anchor in the network of flows. In the post-handover period, the city has experienced equally transformative dynamics: the persistent expansion of the Chinese economy, its evolving relation with the PRC leadership, a changing British and American posture toward Hong Kong and China, to name just a few. Against this backdrop, Cathay Pacific has grown its revenues threefold in the two decades after the handover, although its fluctuating operating profits have failed to register similar growth.[44] While many macro factors are inevitably beyond the control of enterprising operators, entrepreneurship should know no ethnic or demographic boundaries. It remains to be seen whether the new crew steering the development of commercial aviation in Hong Kong will prove themselves as adroit and pragmatic as their predecessors in the previous decades.

Although the system has withstood bouts of political turbulence, which impact industry development from time to time, commercial aviation in Hong Kong, along with its global partners, has proved defenseless in face of the COVID-19 pandemic that swept the world in 2020. Already enervated by social unrest and political suppression in recent years, Hong Kong's economy, as well as the city's primary carrier, had to search for a new impetus for its next phase of growth. With the outbreak of COVID-19, commercial aviation came to a grinding halt. As all flights to and from Hong Kong are cross-border by definition, tightened border control has had an outsized impact on commercial aviation in Hong Kong, reducing flight movements by over 60 percent and passenger traffic by nearly 90 percent in 2020.[45] The pandemic devastated Cathay Pacific. The airline's passenger revenues plummeted over 80 percent. Cargo services proved more resilient, registering a 17 percent increase. Previously accounting for one-fifth of the airline's total revenues, cargo services contributed nearly 60 percent of total revenue in 2020.[46] The situation has

---

44. Swire HK Archive, Cathay Pacific Airways Limited Annual Reports, 1997–2019.

45. Hong Kong International Airport, "HKIA 2020 Passenger Volume."

46. Swire HK Archive, Cathay Pacific Airways Limited Annual Report 2020, 23. See figs. C.3 a and b for statistics for all Hong Kong traffic.

only deteriorated in 2021. In the first half of the year, Cathay Pacific had to cut its passenger capacity to 5 percent of its prepandemic level. Despite its reduced capacity, the airline only filled 18.9 percent of its seats in that period. Eighty-nine of its 238 aircraft remained parked.[47] Debilitated even after the infusion of government funds, Cathay Pacific had to reduce its operations to prepare for a prolonged industry shutdown. The carrier discontinued the operations of Cathay Dragon (as Dragonair was rebranded in 2016), and reduced its staff numbers by 24 percent.[48] With rumors circulating that a fledgling mainland carrier was eyeing Cathay Dragon's routes, the Hong Kong government indicated that Cathay Pacific would not necessarily be given the traffic rights of its defunct subsidiary.[49] In this surprising turn of events, Cathay Pacific extinguished the brand of its one-time local rival, which had focused on connections to mainland China, a market that had long fascinated the Hong Kong–based operator.

In a similar series of restructuring moves, Cathay Pacific also announced the closure of its bases in Canada, Germany, Australia, New Zealand, and London.[50] Reportedly the fourth-largest carrier in 2019 in terms of passenger traffic between North America and the Far East, Cathay Pacific has cut or suspended four of its ten North American gateways, reducing its route map to "a shadow of its former self."[51] The pressure on Cathay Pacific is intense. On the issue of runway slots, for which airlines are known to compete fiercely, International Air Transport Association (IATA) Chief Willie Walsh reminded operators of the global practice of granting such prime resources on a "use it or lose it" basis.[52]

How does the Hong Kong–based airline intend to position itself in the web of global aviation? The story will continue to unfold with twists

47. Swire HK Archive, *Cathay Pacific Airways Limited Analyst Briefing: 2021 Interim Results*; Swire HK Archive, *Cathay Pacific Airways Limited Announcement, 2021 Interim Results*.

48. *SCMP*, June 9, 2020; Cathay Pacific Airways, "Cathay Pacific Group Announces"; Swire HK Archive, *Cathay Pacific Airways Limited Annual Report 2020*, 10.

49. Wong, "Cathay Dragon Traffic Rights."

50. Lee, "Cathay Pacific to Close Vancouver"; Richards, "Cathay Confirms"; Lee, "Cathay Pacific's London Pilot Base."

51. Pearson, "10 to 6."

52. Lee, "Hong Kong's Tough Pandemic Measures."

308 *Conclusion*

and turns, keeping the network of commercial flights and Hong Kong's position therein fluid and unpredictable. Hong Kong think tanks remain optimistic about the sustainability of Hong Kong as an international aviation hub. With coordinated development in civil aviation, Hong Kong could see its international status consolidated and enhanced. Air freight would serve a critical function as Hong Kong plays to its advantage in a global supply chain and its strength in high-end logistics. In conjunction with the logistical hub of Zhuhai on the mainland, Hong Kong could serve as a nexus connecting the southwestern regions of the mainland to international airborne traffic. So argue a group of scholars and experts on logistics.[53]

As for Cathay Pacific, the airline's preview into its rebranding campaign offers some clues. In late May 2021, Cathay Pacific unveiled a video clip on its Facebook and YouTube pages, showcasing what appears to be a new corporate logo. The new design retains the airline's signature brushwing but changes the color scheme to sky blue. More tellingly, the word "Pacific" is dropped from the corporate name—it will simply be "Cathay."[54] Gone is the "Pacific" that had pointed to the transoceanic world beyond the mythical "Cathay"—China in Marco Polo's travel account.[55] If a change in corporate name is at least as indicative of new business directions as the airline's redesigns of schematic flight patterns, one may infer from this video preview that the Hong Kong–based airline would double down on its emphasis on China at the expense of connections overseas.

---

53. Feng, Yeh, Enright, and Chang, *Creating Hong Kong's New Advantages*; interview of Chang Ka Mun and Stephanie Li by author at the Fung Group office, Hong Kong, August 13, 2021.

54. Cathay Pacific Airways, "We'll Soon Be Elevating Your Cathay Experience 國泰嶄新體驗・身心全面昇華," Facebook, May 28, 2021, https://www.facebook.com/cathaypacificHK/videos/525886458426380; Cathay Pacific Airways, "We'll Soon Be Elevating Your Cathay Experience 國泰嶄新體驗・身心全面昇華," YouTube, May 27, 2021, https://www.youtube.com/watch?v=WZRcQfecTDo.

55. There is no indication that the airline would change its Chinese name. In the 1950s, corporate paraphernalia indicated that the airline included in its Chinse corporate name "太平洋" to correspond to "Pacific" in Cathay Pacific (see fig. 3.3). Subsequently, this Chinese reference was dropped, but arguably, "泰" already signifies "pacific" as in "tranquil" and "peaceful" ("國泰" could literally be translated as "stability" and "prosperity" in the [unspecified] country).

Commercial aviation in Hong Kong has flown full circle. Aerial connections cultivated since the aftermath of World War II now face the challenge of a worldwide unraveling of transport and trade networks. Although COVID-19 has leveled devastating damages to commercial aviation in Hong Kong and elsewhere, it too shall pass. The skies will reopen, and certain industry players will recover and come back stronger. The restarting of commercial aviation worldwide will usher in another round of rewiring that promises to rewire the circuitries of flows. The conclusion of the pandemic could witness the dissolution of Hong Kong's regional and global ties or the revitalization of Hong Kong's status as a hub. Playing out in the skies over Hong Kong, the outcome pivots on the city's ability to demonstrate yet again its nimble posture and delicate footwork in connecting global powers on the periphery.

Bibliography

Periodicals

Asian Wall Street Journal
Canberra Times
CM. The Chinese Mail 華字日報. Accessed through the Multimedia Information System, Hong Kong Public Libraries.
Flight & Aircraft Engineer
Hongkong Standard
Hong Kong Sunday Herald
Hong Kong Telegraph
The Independent
KSEN. Kung Sheung Evening News 工商晚報. Accessed through the Multimedia Information System, Hong Kong Public Libraries.
KSMN. Kung Sheung Morning News 工商日報. Accessed through the Multimedia Information System, Hong Kong Public Libraries.
Ming Pao 明報
New York Times
North China Herald
SCMP. South China Morning Post
Shenbao 申報
Straits Times
TiKP. Tien Kwong Po 天光報. Accessed through the Multimedia Information System, Hong Kong Public Libraries.
The Times
TKP. Ta Kung Pao 大公報. Accessed through the Multimedia Information System, Hong Kong Public Libraries.
Wall Street Journal
Wen Wei Po 文匯報

WKYP. *Wah Kiu Yat Po* 華僑日報. Accessed through the Multimedia Information System, Hong Kong Public Libraries.

Archival Sources

Airport Authority, Hong Kong website (https://www.hongkongairport.com/en/the-airport/hkia-at-a-glance/fact-figures.page).
British Airways Archives. London, UK.
British Library. Asia and Africa Studies, India Office Records, London, UK.
HKDCA. *Hong Kong Annual Departmental Reports by the Director of Civil Aviation.*
HKPRO. Hong Kong Public Records Office. Government Records Service, Hong Kong.
Hong Kong Heritage Project. Kowloon, Hong Kong.
Hong Kong Tourist Association Annual Reports.
Hoover Institution Archives. Stanford, California.
JSS. John Swire & Sons Ltd. Archive, London, UK.
LegCo. Official Report of Proceedings, Hong Kong Legislative Council, Hong Kong.
National Archives of Australia, Sydney.
National Archives of Singapore. Singapore.
Pan Am. Pan American World Airways Inc. Records. University of Miami Libraries Special Collections. Miami, Florida, US.
Qantas Archives.
Swire HK Archive. Swire Archives, Hong Kong.
TNA. National Archives of the UK. Kew, UK.

Published Sources

Abdelrehim, Neveen, Aparajith Ramnath, Andrew Smith, and Andrew Popp. "Ambiguous Decolonisation: A Postcolonial Reading of the Ihrm Strategy of the Burmah Oil Company." *Business History* 63, no. 1 (2021): 98–126.
Abdelrehim, Neveen, Josephine Maltby, and Steven Toms. "Corporate Social Responsibility and Corporate Control: The Anglo-Iranian Oil Company, 1933–1951." *Enterprise & Society* 12, no. 4 (2011): 824–62.
Administration Reports for the Year 1935. Hong Kong: Government Printer, 1936.
Administration Reports for the Year 1938. Hong Kong: Government Printer, 1939.
Air China Limited. *2017 Annual Report.* http://www.airchina.com.cn/en/investor_relations/images/financial_info_and_roadshow/2018/04/24/F238A48A6F4F31EA6ADD4072A0E210C8.pdf.
Annual Report on Hong Kong for the Year 1946. Hong Kong: Local Printing Press, Ltd., 1947.
Annual Report on Hong Kong for the Year 1947. Hong Kong: Local Printing Press, Ltd., 1948.
Aoki, Masahiko, Hyung-Ki Kim, and Masahiro Okuno-Fujiwara, eds. *The Role of Government in East Asian Economic Development: Comparative Institutional Analysis.* Oxford: Oxford University Press, 1998.

Bibliography

Appadurai, Arjun. "Disjuncture and Difference in the Global Cultural Economy." *Theory Culture Society* 7, no. 2–3 (1990): 295–310.

Appiah, Kwame Anthony. "Cosmopolitan Patriots." In *For Love of Country*, edited by Joshua Cohen, 21–29. Boston, MA: Beacon Press, 1996.

Arnold, Wayne. "For the Singapore Girl, It's Her Time to Shine." *New York Times*, December 31, 1999, C4.

Ashton, Stephen R. "Keeping Change Within Bounds: A Whitehall Reassessment." In *The British Empire in the 1950s: Retreat or Revival?*, edited by Martin Lynn, 32–52. Basingstoke, England; New York: Palgrave Macmillan, 2006.

A Statistical Review of Tourism. Hong Kong: Research Department, Hong Kong Tourist Association, 1976.

A Statistical Review of Tourism. Hong Kong: Research Department, Hong Kong Tourist Association, 1986.

A Statistical Review of Tourism. Hong Kong: Research Department, Hong Kong Tourist Association, 1995.

A Statistical Review of Tourism. Hong Kong: Research Department, Hong Kong Tourist Association, 1997.

Banner, Stuart. *Who Owns the Sky? The Struggle to Control Airspace from the Wright Brothers on*. Cambridge, MA: Harvard University Press, 2008.

Barnes, Victoria, and Lucy Newton. "Women, Uniforms and Brand Identity in Barclays Bank." *Business History* (2020): 1–30.

Barnett, K. M. A. *Population Projections for Hong Kong, 1966–1981*. Hong Kong: Government Printer, 1968.

Barrett, Sean D. "The Implications of the Ireland-UK Airline Deregulation for an EU Internal Market." *Journal of Air Transport Management* 3, no. 2 (1997): 67–73.

Barry, Kathleen. *Femininity in Flight: A History of Flight Attendants*. Durham, NC: Duke University Press, 2007.

Beck, Ulrich, and Edgar Grande. "Varieties of Second Modernity: The Cosmopolitan Turn in Social and Political Theory and Research." *British Journal of Sociology* 61, no. 3 (2010): 409–43.

Bednarek, Janet R. *Airports, Cities, and the Jet Age: US Airports Since 1945*. Cham, Switzerland: Springer International Publishing, 2016.

Bhatia, Nandi. "Fashioning Women in Colonial India." *Fashion Theory* 7, no. 3–4 (2003): 327–44.

Bickers, Robert A. *China Bound: John Swire & Sons and Its World, 1816–1980*. London: Bloomsbury Business, 2020.

———. "Loose Ties That Bound: British Empire, Colonial Autonomy and Hong Kong." In *Negotiating Autonomy in Greater China: Hong Kong and Its Sovereign before and after 1997*, edited by Ray Yep, 29–54. Copenhagen: NIAS Press, 2013.

———. "The Colony's Shifting Position in the British Informal Empire in China." In *Hong Kong's Transitions, 1842–1997*, edited by Judith M. Brown and Rosemary Foot, 33–61. London: Macmillan, 1997.

Black, Prudence. "Lines of Flight: The Female Flight Attendant Uniform." *Fashion Theory—Journal of Dress, Body & Culture* 17, no. 2 (2013): 179–95.

Bibliography

Boon, Marten. "Business Enterprise and Globalization: Towards a Transnational Business History." *Business History Review* 91, no. 3 (2017): 511–35.

Breckenridge, Carol A., Sheldon Pollock, Homi K. Bhabha, and Dipesh Chakrabarty, eds. *Cosmopolitanism*. Durham, NC: Duke University Press, 2002.

Bucheli, Marcelo, and Min-Young Kim. "Political Institutional Change, Obsolescing Legitimacy, and Multinational Corporations." *Management International Review* 52, no. 6 (2012): 847–77.

Bucheli, Marcelo, and Erica Salvaj. "Political Connections, the Liability of Foreignness, and Legitimacy: A Business Historical Analysis of Multinationals' Strategies in Chile." *Global Strategy Journal* 8, no. 3 (2018): 399–420.

Button, Kenneth, ed. *Airline Deregulation: International Experiences*. New York: New York University Press, 1991.

Cain, P. J., and A. G. Hopkins. *British Imperialism, 1688–2000*, 2nd ed. Harlow, UK: Longman, (2002) 1993.

Carroll, John M. *A Concise History of Hong Kong*. Lanham, MD: Rowman & Littlefield, 2007.

———. *Edge of Empires: Chinese Elites and British Colonials in Hong Kong*. Cambridge, MA: Harvard University Press, 2005.

Cartier, Carolyn. "Origins and Evolution of a Geographical Idea: The Macroregion in China." *Modern China* 28, no. 1 (2002): 79–142.

Casson, Mark. "International Rivalry and Global Business Leadership: An Historical Perspective." *Multinational Business Review* 28, no. 4 (2020): 429–46.

Castells, Manuel. *The Rise of the Network Society*. Cambridge, MA: Blackwell, 1996.

Cathay Pacific Airways. "Cathay Pacific Group Announces Corporate Restructuring: Group Will Cease Cathay Dragon Operations, and Reduce Workforce and Passenger Capacity as It Adapts to the New Travel Reality," news release, October 21, 2020. https://news.cathaypacific.com/cathay-pacific-group-announces-corporate -restructuring.

Cathay Pacific Airways. "We'll Soon Be Elevating Your Cathay Experience 國泰嶄新體驗，身心全面昇華," Facebook, May 28, 2021. https://www.facebook.com/cathay pacificHK/videos/525886458426380.

Cathay Pacific Airways. "We'll Soon Be Elevating Your Cathay Experience 國泰嶄新體驗，身心全面昇華," YouTube, May 27, 2021. https://www.youtube.com/watch?v=WZ RcQfecTDo.

Chan Lau Kit-ching. *China, Britain, and Hong Kong, 1895–1945*. Hong Kong: The Chinese University Press, 1990.

Cheah, Pheng, and Bruce Robbins, eds. *Cosmopolitics: Thinking and Feeling beyond the Nation*. Minneapolis: University of Minnesota Press, 1998.

Chen, Philip N. L. *Greatest Cities of the World*. Hong Kong: University of Hong Kong, Centre of Asian Studies, 2010.

Cheung, Anthony B. L. "Rebureaucratization of Politics in Hong Kong: Prospects after 1997." *Asian Survey* 37, no. 8 (August 1997): 720–37.

Cheung, Gary Ka-wai. *Hong Kong's Watershed: The 1967 Riots*. Hong Kong: Hong Kong University Press, 2009.

Bibliography

Chew, Matthew. "Contemporary Re-Emergence of the Qipao: Political Nationalism, Cultural Production and Popular Consumption of a Traditional Chinese Dress." *China Quarterly* 189 (2007): 144–61.

China Airlines, Ltd. *Financial Statements for the Years Ended December 31, 2019 and 2018 and Independent Auditors' Report.*

Chu, Cecilia Louise. "Speculative Modern: Urban Forms and the Politics of Property in Colonial Hong Kong." PhD diss., University of California, Berkeley, 2012.

Chung, Henry, Noriko Kanazawa, and Freddie Wong. *Good Bye Kai Tak* / 再見啓德 / さよか ら啓德. Hong Kong: 3-D Intellectual, 1999.

Chung, Stephanie Po-yin. *Chinese Business Groups in Hong Kong and Political Changes in South China, 1900–1920s.* London: Palgrave Macmillan, 1998.

Clayton, David W. "Hong Kong as an International Financial Centre: Emergence and Development, 1945–1965." In *Imagining Britain's Economic Future, c. 1800–1975*, edited by David Thackeray, Richard Toye, and Andrew Thompson, 231–51. Cham, Switzerland: Springer International Publishing, 2018.

Clifford, Mark. "Mainland Bounty." *Far Eastern Economic Review* 157, no. 4 (January 1994): 40.

Cochran, Sherman. "Capitalists Choosing Communist China: The Liu Family of Shanghai, 1948–1956." In *Dilemmas of Victory: The Early Years of the People's Republic of China*, edited by Jeremy Brown and Paul G. Pickowicz, 359–86. Cambridge, MA: Harvard University Press, 2007.

Cohen, Jim. "Divergent Paths, United States and France: Capital Markets, the State, and Differentiation in Transportation Systems, 1840–1940." *Enterprise & Society* 10, no. 3 (2009): 449–97.

Collier, Deirdre, Nandini Chandar, and Paul Miranti. "Marketing Shareholder Democracy in the Regions: Bell Telephone Securities, 1921–1935." *Enterprise & Society* 18, no. 2 (2017): 400–46.

Craik, Jennifer. "Is Australian Fashion and Dress Distinctively Australian?," *Fashion Theory* 13, no. 4 (2009): 409–41.

———. "The Cultural Politics of the Uniform." *Fashion Theory* 7, no. 2 (2015): 127–47.

Crane, Diana. *Fashion and Its Social Agendas: Class, Gender, and Identity in Clothing.* Chicago: University of Chicago Press, 2000.

Cwerner, Saulo, Sven Kesselring, and John Urry, eds. *Aeromobilities.* Abingdon, UK: Routledge, 2009.

Darwin, John. "Hong Kong in British Decolonisation." In *Hong Kong's Transitions, 1842–1997*, edited by Judith M. Brown and Rosemary Foot, 16–32. London: Macmillan, 1997.

———. *The Empire Project: The Rise and Fall of the British World-System, 1830–1970.* Cambridge: Cambridge University Press, 2009.

———. *Unlocking the World: Port Cities and Globalization in the Age of Steam, 1830–1930.* London: Allen Lane, 2020.

Da Silva Lopes, Teresa, and Mark Casson. "Entrepreneurship and the Development of Global Brands." *Business History Review* 81, no. 4 (2007): 651–80.

Bibliography

Davies, R. E. G. *A History of the World's Airlines*. London, New York: Oxford University Press, 1964.

———. *Airlines of Asia since 1920*. London: Putnam, 1997.

Decker, Stephanie. "Africanization in British Multinationals in Ghana and Nigeria, 1945–1970." *Business History Review* 92, no. 4 (2018): 691–718.

Derthick, Martha, and Paul J. Quirk. *The Politics of Deregulation*. Washington, DC: Brookings Institution, 1985.

Dobson, Alan P. *Anglo-American Relations in the Twentieth Century: Of Friendship, Conflict and the Rise and Decline of Superpowers*. New York: Routledge, 1995.

———. *Flying in the Face of Competition: The Policies and Diplomacy of Airline Regulatory Reform in Britain, the USA and the European Community, 1968–94*. Aldershot, UK: Ashgate Publishing, 1995.

———. *Peaceful Air Warfare: The United States, Britain, and the Politics of International Aviation*. Oxford: Oxford University Press, 1991.

———. "The Other Air Battle: The American Pursuit of Post-War Civil Aviation Rights." *Historical Journal* 28, no. 2 (1985): 429–39.

Doh, Jonathan P., Hildy Teegen, and Ram Mudambi. "Balancing Private and State Ownership in Emerging Markets' Telecommunications Infrastructure: Country, Industry, and Firm Influences." *Journal of International Business Studies* 35, no. 3 (2004): 233–50.

Donald, David C. *A Financial Centre for Two Empires: Hong Kong's Corporate, Securities and Tax Laws in Its Transition from Britain to China*. Cambridge: Cambridge University Press, 2014.

Duara, Prasenjit. *Decolonization: Perspectives from Now and Then*. New York: Routledge, 2004.

———. "Hong Kong as a Global Frontier: Interface of China, Asia, and the World." In *Hong Kong in the Cold War*, edited by Priscilla Roberts and John M. Carroll, 211–30. Hong Kong: Hong Kong University Press, 2016.

Dunnaway, Cliff. *Hong Kong High: An Illustrated History of Aviation in Hong Kong*. Hong Kong: Airphoto International Ltd, 2013.

Eather, Charles Edward James. *Airport of the Nine Dragons: Kai Tak, Kowloon*. Surfers Paradise, Australia: ChingChic Publishers, 1996.

Edgerton, David. *England and the Aeroplane: Militarism, Modernity and Machines*. New York: Penguin, 2013.

———. "From Innovation to Use: Ten Eclectic Theses on the Historiography of Technology." *History and Technology* 16, no.2 (1999): 111–36.

———. *Science, Technology and the British Industrial "Decline," 1870–1970*. Cambridge: Cambridge University Press, 1996.

———. "The Decline of Declinism." *Business History Review* 71, no. 2 (1997): 201–6.

———. *The Rise and Fall of the British Nation: A Twentieth-Century History*. London: Allen Lane, 2018.

Edwards, Louise. "Policing the Modern Woman in Republican China." *Modern China* 26, no. 2 (2000): 115–47.

Bibliography

Engel, Jeffrey A. *Cold War at 30,000 Feet: The Anglo-American Fight for Aviation Supremacy.* Cambridge, MA: Harvard University Press, 2007.

England, Joe, and John Rear. *Industrial Relations and Law in Hong Kong: An Extensively Rewritten Version of Chinese Labour Under British Rule.* Hong Kong; New York: Oxford University Press, 1981.

Fellows, James. "Britain, European Economic Community Enlargement, and 'Decolonisation' in Hong Kong, 1967–1973." *International History Review* 41, no. 4 (2019): 753–74.

Feng Xiaoyun, Anthony Yeh, Michael Enright, and Chang Ka Mun. *Creating Hong Kong's New Advantages in the Greater Bay Area—Identifying New Pathways to Growth and Opportunity.* Hong Kong: 2022 Foundation, 2021.

Finnane, Antonia. *Changing Clothes in China: Fashion, History, Nation.* New York: Columbia University Press, 2008.

———. "What Should Chinese Women Wear? A National Problem." *Modern China* 22, no. 2 (1996): 99–131.

Fitzgerald, Robert. *The Rise of the Global Company: Multinationals and the Making of the Modern World.* Cambridge: Cambridge University Press, 2015.

Foss, Richard. *Food in the Air and Space: The Surprising History of Food and Drink in the Skies.* Lanham, MD: Rowman & Littlefield, 2015.

Friedman, Milton, and Rose Friedman. *Free to Choose: A Personal Statement, the Classic Inquiry into the Relationship between Freedom and Economics.* New York: Harcourt, 1980.

Fu, Poshek. *Passivity, Resistance and Collaboration: Intellectual Choices in Occupied Shanghai, 1937–1945.* Stanford, CA: Stanford University Press, 1993.

Gallagher, John, and Ronald Robinson. "The Imperialism of Free Trade." *Economic History Review* 6, no. 1 (1953): 1–15.

GAO (US General Accounting Office). *Airline Deregulation: Boon or Bust?*, Washington, DC: General Accounting Office, 1981.

Gao, James Zheng. *The Communist Takeover of Hangzhou: The Transformation of City and Cadre, 1949–1954.* Honolulu: University of Hawai'i Press, 2004.

Gardner, John. "The Wu-fan Campaign in Shanghai: A Study in the Consolidation of Urban Control." In *Chinese Communist Politics in Action*, edited by A. Doak Barnett, 477–539. Seattle: University of Washington Press, 1969.

Gaudry, Marc, and Robert Mayes, eds. *Taking Stock of Air Liberalization.* Boston: Kluwer, 1999.

Gehlen, Boris, Christian Marx, and Alfred Reckendrees. "Ambivalences of Nationality— Economic Nationalism, Nationality of the Company, Nationalism as Strategy: An Introduction." *Journal of Modern European History* 18, no. 1 (2020): 16–27.

Goedhuis, D. "Sovereignty and Freedom in the Air Space." *Transactions of the Grotius Society* 41 (1955): 137–52.

Goldin, Claudia Dale. *Understanding the Gender Gap: An Economic History of American Women.* New York: Oxford University Press, 1990.

Goodstadt, Leo F. "Cowperthwaite." In *Dictionary of Hong Kong Biography*, edited by May Holdsworth and Christopher Munn, 108–10. Hong Kong: Hong Kong University Press, 2012.

———. "Fiscal Freedom and the Making of Hong Kong's Capitalist Society." In *Negotiating Autonomy in Greater China: Hong Kong and Its Sovereign before and after 1997*, edited by Ray Yep, 81–109. Copenhagen, Denmark: NIAS Press, 2013.

———. "Trench, Sir David Clive Crosbie." In *Dictionary of Hong Kong Biography*, edited by May Holdsworth and Christopher Munn, 435–36. Hong Kong: Hong Kong University Press, 2012.

———. *Uneasy Partners: The Conflict between Public Interest and Private Profit in Hong Kong*. Hong Kong: Hong Kong University Press, 2005.

Government of Hong Kong. *Papers on Development of Kai Tak Airport*. Hong Kong: Hong Kong Government Printers, 1954.

Graham, Brian. "The Regulation of Deregulation: A Comment on the Liberalization of the UK's Scheduled Airline Industry." *Journal of Transport Geography* 1, no. 2 (1993): 125–31.

Haise, Carrie Leigh, and Margaret Rucker. "The Flight Attendant Uniform: Effects of Selected Variables on Flight Attendant Image, Uniform Preference and Employee Satisfaction." *Social Behavior and Personality* 31, no. 6 (2003): 565–75.

Hamashita, Takeshi. "Tribute and Treaties: Maritime Asia and Treaty Port Networks in the Era of Negotiation, 1800–1900." In *The Resurgence of East Asia: 500, 150 and 50 Year Perspectives*, edited by Giovanni Arrighi, Takeshi Hamashita, and Mark Selden, 17–50. New York: Routledge, 2003.

Hannerz, Ulf. "Cosmopolitanism." In *A Companion to the Anthropology of Politics*, edited by David Nugent and Joan Vincent, 69–85. Malden, MA: Blackwell Publishing, 2007.

Harrison, Henrietta. *The Making of the Republican Citizen: Political Ceremonies and Symbols in China, 1911–1929*. Oxford: Oxford University Press, 2000.

Heracleous, Loizos, Jochen Wirtz, and Nitin Pangarkar. *Flying High in a Competitive Industry: Cost-Effective Service Excellence at Singapore Airlines*. Singapore: McGraw Hill, 2006.

Hesse-Biber, Sharlene Nagy, and Gregg Lee Carter. *Working Women in America: Split Dreams*. New York: Oxford University Press, 2000.

Hickson, Ken. *Mr. SIA: Fly Past*. Singapore: World Scientific Publishing, 2015.

Hisano, Ai. *Visualizing Taste: How Business Changed the Look of What You Eat*. Cambridge, MA: Harvard University Press, 2019.

Hong Kong Annual Digest of Statistics, 1978 Edition. Hong Kong: Government Printer, 1978.

Hong Kong Annual Digest of Statistics, 1981 Edition. Hong Kong: Government Printer, 1981.

Hong Kong Annual Digest of Statistics, 1985 Edition. Hong Kong: Government Printer, 1985.

Hong Kong Annual Digest of Statistics, 1990 Edition. Hong Kong: Government Printer, 1990.

Hong Kong Annual Digest of Statistics, 1994 Edition. Hong Kong: Government Printer, 1994.

Hong Kong Annual Digest of Statistics, 1998 Edition. Hong Kong: Government Printer, 1998.

Hong Kong Annual Digest of Statistics, 2000 Edition. Hong Kong: Government Printer, 2000.

Hong Kong By-Census 1976: A Graphic Guide. Hong Kong: Government Printer, 1976.

Hong Kong Civil Aviation Department. "Speech Delivered by Director of Civil Aviation, Mr. R A Siegel on the Light-out Ceremony of Kai Tak Airport," press release, July 5, 1998. https://www.cad.gov.hk/english/pressrelease_1998.html.

Hong Kong International Airport. "HKIA 2020 Passenger Volume Drops under Pandemic, Cargo Operations Remain Resilient," news release, January 15, 2021. https://www.hongkongairport.com/en/media-centre/press-release/2021/pr_1510.

Hong Kong Memory. "Kai Tak, An Old Neighbor." Last accessed December 19, 2021. http://www.hkmemory.org/kaitak/en/sub.html#&slider1=48.

Hong Kong Monthly Digest of Statistics, November 1991. Hong Kong: Government Printer, 1991.

Hong Kong Statistics, 1947–1967. Hong Kong: Government Printer, 1969.

Hope, Richard I. "Developing Airways in China." *Chinese Economic Journal* 6, no. 1 (1930): 104–16.

House of Commons. *Hansard Parliamentary Debates*, Commons, 5th series (1909–81), vol. 443, col. 862–3, October 29, 1947.

Howe, Stephen. "When (If Ever) Did Empire End? 'Internal Decolonisation' in British Culture since the 1950s." In *The British Empire in the 1950s: Retreat or Revival?*, edited by Martin Lynn, 214–37. Basingstoke, UK: Palgrave Macmillan, 2006.

Hsiao, Gene T. *The Foreign Trade of China: Policy, Law and Practice.* Berkeley: University of California Press, 1977.

Husain, Aiyaz. *Mapping the End of Empire: American and British Strategic Visions in the Postwar World.* Cambridge, MA: Harvard University Press, 2014.

Hynes, Geraldine E., and Marisa Puckett. "Feminine Leadership in Commercial Aviation: Success Stories of Women Pilots and Captains." *Journal of Aviation Management and Education* 1 (2011): 1–6.

Ikeya, Chie. "The Modern Burmese Woman and the Politics of Fashion in Colonial Burma." *Journal of Asian Studies* 67, no. 4 (2008): 1277–308.

International Civil Aviation Conference. *Proceedings of the International Civil Aviation Conference, Chicago, Illinois, November 1–December 7, 1944.* Washington, DC: US Govt. Printing Office, Department of State, 1948.

Jackson, Isabella. *Shaping Modern Shanghai: Colonialism in China's Global City.* Cambridge: Cambridge University Press, 2018.

JAL Group News. "Planned Integration of Japan Asia Airways with JAL." November 1, 2007. https://press.jal.co.jp/en/release/200711/003053.html.

Jensen, Casper Bruun, and Atsuro Morita. "Introduction: Infrastructures as Ontological Experiments." *Ethnos* 82, no. 4 (2017): 615–26.

Bibliography

Johnson, Chalmers A. *Japan, Who Governs?: The Rise of the Developmental State*. New York: W. W. Norton, 1995.

Jones, Geoffrey. *Beauty Imagined: A History of the Global Beauty Industry*. Oxford: Oxford University Press, 2010.

———. *British Multinational Banking, 1830–1990*. Oxford: Oxford University Press, 1995.

———. "Globalization." In *The Oxford Handbook of Business History*, edited by Geoffrey Jones and Jonathan Zeitlin, 141–68. Oxford: Oxford University Press, 2008.

———. *Merchants to Multinationals: British Trading Companies in the 19th and 20th Centuries*. Oxford: Oxford University Press, 2000.

Jones, Geoffrey, and Christina Lubinski. "Managing Political Risk in Global Business: Beiersdorf 1914–1990." *Enterprise and Society* 13, no. 1 (2012): 85–119.

Kahn, Alfred E. *Lessons from Deregulation: Telecommunications and Airlines after the Crunch*. Washington, DC: Brookings Institution Press, 2004.

———. *The Economics of Regulation: Principles and Institutions*. Cambridge, MA: MIT Press, 1971 (1988).

Kaufman, Victor S. "The United States, Britain, and the CAT Controversy." *Journal of Contemporary History* 40, no. 1 (2005): 95–113.

Keung, John. *Government Intervention and Housing Policy in Hong Kong: A Structural Analysis*. Cardiff: Department of Town Planning, University of Wales Institute of Science and Technology, 1981.

King, Ambrose Yeo-chi. *China's Great Transformation: Selected Essays on Confucianism, Modernization, and Democracy*. Hong Kong: The Chinese University Press, 2018.

Kirby, William C. *Germany and Republican China*. Stanford, CA: Stanford University Press, 1984.

———. "Traditions of Centrality, Authority, and Management in Modern China's Foreign Relations." In *Chinese Foreign Policy: Theory and Practice*, edited by Thomas W. Robinson and David L. Shambaugh, 13–29. Oxford: Oxford University Press, 1994.

Klein, Christina. *Cold War Orientalism: Asia in the Middlebrow Imagination, 1945–1961*. Berkeley: University of California Press, 2003.

Köll, Elisabeth. *Railroads and the Transformation of China*. Cambridge, MA: Harvard University Press, 2019.

Koo, Shou-eng. "The Role of Export Expansion in Hong Kong's Economic Growth." *Asian Survey* 8, no. 6 (1968): 499–515.

Krause, Richard C. *Class Conflict in Chinese Socialism*. New York: Columbia University Press, 1981.

Ku, Agnes S. "Immigration Policies, Discourse, and the Politics of Local Belonging in Hong Kong (1950–1980)." *Modern China* 30, no. 3 (2004): 326–60.

Ku, Agnes S., and Pun Ngai. "Introduction: Rethinking Citizenship in Hong Kong." In *Rethinking Citizenship in Hong Kong: Community, Nation, and the Global City*, edited by Agnes S. Ku and Pun Ngai, 1–17. London: Routledge, 2004.

Kuo Huei-ying. "Chinese Bourgeois Nationalism in Hong Kong and Singapore in the 1930s." *Journal of Contemporary Asia* 36, no. 3 (2006): 385–405.

Bibliography

Kurosawa, Takafumi, Neil Forbes, and Ben Wubs. "Political Risks and Nationalism." In *The Routledge Companion to the Makers of Global Business*, edited by Teresa da Silva Lopes, Christina Lubinski, and Heidi Tworek, 485–501. Abingdon, UK: Routledge, 2020.

Kwong Chi-man and Tsoi Yiu-lun. *Eastern Fortress: A Military History of Hong Kong, 1840–1970.* Hong Kong: Hong Kong University Press, 2014.

Kwong, Kai-sun. *Towards Open Skies and Uncongested Airports: An Opportunity for Hong Kong.* Hong Kong: Chinese University Press, 1988.

Larkin, Brian. "The Politics and Poetics of Infrastructure." *Annual Review of Anthropology* 42, no. 1 (2013): 327–43.

Lau Chi-pang 劉智鵬, Wong Kwan Kin 黃君健, and Chin Ho Yin 錢浩賢. *Tiankongxia de chuanqi: cong Qide dao Chila Jiao* 天空下的傳奇：從啓德到赤鱲角 (Legend under the Skies: From Kaitak to Chek Lap Kok). Hong Kong: Sanlian shudian (Xianggang) youxiangongsi, 2014.

Law Wing-sang. *Collaborative Colonial Power: The Making of the Hong Kong Chinese.* Hong Kong: Hong Kong University Press, 2009.

Le, Thuong T. "Reforming China's Airline Industry: From State-Owned Monopoly to Market Dynamism." *Transportation Journal* 37, no. 2 (1997): 45–62.

Leary, William M. *The Dragon's Wings: The China National Aviation Corporation and the Development of Commercial Aviation in China.* Athens: University of Georgia Press, 1976.

Lee, Danny. "Cathay Pacific's London Pilot Base Facing Shutdown, with 100 Jobs under Hong Kong Carrier at Risk." *South China Morning Post*, July 20, 2021. https://www.scmp.com/news/hong-kong/transport/article/3141859/cathay-pacifics-london-pilot-base-facing-shutdown-100-jobs.

———. "Cathay Pacific to Close Vancouver Base in June as Part of 'Ongoing Business Review' Putting 147 Jobs at Risk." *South China Morning Post*, March 6, 2020. https://www.scmp.com/news/hong-kong/hong-kong-economy/article/3073843/cathay-pacific-close-vancouver-base-june-putting.

———. "Hong Kong's Tough Pandemic Measures May Affect Cathay as Europe Pressures Airlines to Increase Flights or Lose Prized Runway Slots." *South China Morning Post*, August 1, 2021. https://www.scmp.com/news/hong-kong/transport/article/3143286/hong-kongs-tough-pandemic-measures-may-affect-cathay.

Lee, Grace O. M., and Ahmed Shafiqul Huque. "Transition and the Localization of the Civil Service in Hong Kong." *International Review of Administrative Sciences* 61, no. 1 (1995): 107–20.

Lee, Katon K. C. "Suit Up: Western Fashion, Chinese Society and Cosmopolitanism in Colonial Hong Kong, 1910–1980." PhD diss., University of Bristol, 2020.

Lee Keun, and Xuehua Jin. "The Origins of Business Groups in China: An Empirical Testing of the Three Paths and the Three Theories." *Business History* 51, no. 1 (January 2009): 77–79.

Leighton, Christopher R. "Capitalists, Cadres, and Culture in 1950s China." PhD diss., Harvard University, 2010.

Levine, Derek A. *The Dragon Takes Flight: China's Aviation Policy, Achievements, and International Implications.* Leiden, NL: Brill, 2015.

Bibliography

Li Kui-Wai. *Capitalist Development and Economism in East Asia: The Rise of Hong Kong, Singapore, Taiwan and South Korea.* London: Routledge, 2002.

———. *Economic Freedom: Lessons of Hong Kong.* Singapore: World Scientific, 2012.

Lieberthal, Kenneth G. *Revolution and Tradition in Tientsin, 1949–1952.* Stanford, CA: Stanford University Press, 1980.

Lin Pei-yin. "Writing beyond Boudoirs: Sinophone Literature by Female Writers in Contemporary Taiwan." In *Sinophone Studies: A Critical Reader,* edited by Shih Shu-mei, Tsai Chien-Hsin, and Brian Bernards, 255–69. New York: Columbia University Press, 2012.

Little, Virginia. "Control of International Air Transport." *International Organization* 3, no. 1 (1949): 29–40.

Longhurst, Henry. *The Borneo Story: The History of the First 100 Years of Trading in the Far East by the Borneo Company Limited.* London: Mewman Neame Limited, 1956.

Louis, W. M. Roger, and Ronald Robinson. "The Imperialism of Decolonization." *Journal of Imperial and Commonwealth History* 22, no. 3 (1994): 462–511.

Low, Donald A., and John Lonsdale. "East Africa: Towards the New Order 1945–1963." In *Eclipse of Empire,* edited by Donald A. Low, 164–214. Cambridge: Cambridge University Press, 1991.

Lubinski, Christina, and R. Daniel Wadhwani. "Geopolitical Jockeying: Economic Nationalism and Multinational Strategy in Historical Perspective." *Strategic Management Journal* 41, no. 3 (2019): 400–21.

Lynn, Martin. *The British Empire in the 1950s: Retreat or Revival?* Basingstoke, UK: Palgrave Macmillan, 2006.

Lyth, Peter. "Chosen Instruments: The Evolution of British Airways." In *Flying the Flag: European Commercial Air Transport since 1945,* edited by Hans-Liudger Dienel and Peter Lyth, 50–86. New York: St. Martin's Press, 1998.

———. "The Empire's Airway: British Civil Aviation from 1919 to 1939." *Revue Belge De Philologie et D'histoire* 78, no. 3 (2000): 865–87.

Ma, Ronald A., and Edward F. Szczepanik. *The National Income of Hong Kong, 1947–1950.* Hong Kong: Hong Kong University Press, 1955.

Mark, Chi-Kwan. "Lack of Means or Loss of Will? The United Kingdom and the Decolonization of Hong Kong, 1957–1967." *International History Review* 31, no. 1 (2009): 45–71.

———. "Vietnam War Tourists: US Naval Visits to Hong Kong and British-American-Chinese Relations, 1965–1968." *Cold War History* 10, no. 1 (2010): 1–28.

Matejova, Miriam, and Don Munton. "Western Intelligence Cooperation on Vietnam during the Early Cold War Era." *Journal of Intelligence History* 15, no. 2 (2016): 139–55.

McCarthy, Faye, Lucy Budd, and Stephen Ison. "Gender on the Flightdeck: Experiences of Women Commercial Airline Pilots in the UK." *Journal of Air Transport Management* 47 (2015): 32–38.

Melzer, Jürgen P. *Wings for the Rising Sun: A Transnational History of Japanese Aviation.* Cambridge, MA: Harvard University Asia Center, 2020.

Meyer, David R. *Hong Kong as a Global Metropolis.* Cambridge: Cambridge University Press, 2000.

Mitchell, Jim, Alexandra Kristovics, and Leo Vermeulen. "Gender Issues in Aviation: Pilot Perceptions and Employment Relations." *International Journal of Employment Studies* 14, no. 1 (2006): 35–59.

Mitchell, William, and Thomas Fazi. "We Have a (Central) Plan: The Case of Renationalisation." In *Reclaiming the State: A Progressive Vision of Sovereignty for a Post-Neoliberal World*, 248–62. London: Pluto Press, 2017.

Namba, Tomoko. "School Uniform Reforms in Modern Japan." In *Fashion, Identity, and Power in Modern Asia*, edited by Kyunghee Pyun and Aida Yuen Wong, 91–113. Cham, Switzerland: Palgrave Macmillan, 2018.

Ng, James 吳邦謀. *Xianggang hangkong 125 nian* 香港航空125年 (125 Years of Hong Kong Aviation). Hong Kong: Zhonghua Shuju, 2015.

———. *Zaikan Qide: cong Rizhan shiqi shuoqi* 再看啓德: 從日佔時期説起 (Kaitak reexamined: Starting from the period of Japanese occupation). Hong Kong: ZKOOB Limited, 2009.

Ng, James 吳詹仕, and Ho Yiu-sang 何耀生. *Cong Qide chufa* 從啓德出發 (Taking off from Kaitak). Hong Kong: et press, 2007.

Ng, Sandy. "Gendered by Design: Qipao and Society, 1911–1949." *Costume—Journal of the Costume Society* 49, no. 1 (2015): 55–74.

Ng Wing-leung 伍永樑. *Qide: zuihou de guangjing* 啓德: 最後的光景 (*Kai Tak: The Final Days*). Hong Kong: Softrepublic Limited, 2008.

Ngo, Tak-Wing. "Industrial History and the Artifice of *Laissez-faire* Colonialism." In *Hong Kong's History: State and Society under Colonial Rule*, edited by Tak-Wing Ngo, 119–40. New York: Routledge, 1999.

Obendorf, Simon. "Consuls, Consorts or Courtesans? 'Singapore Girls' between the Nation and the World." In *Women and the Politics of Representation in Southeast Asia: Engendering Discourse in Singapore and Malaysia,* edited by Adeline Koh and Yu-Mei Balasingamchow, 33–59. New York: Routledge, 2015.

Obstfeld, Maurice. "The Global Capital Market: Benefactor or Menace?" *Journal of Economic Perspectives* 12, no. 4 (1998): 9–30.

Ong, Aihwa. *Flexible Citizenship: The Cultural Logics of Transnationality.* Durham, NC: Duke University Press, 1999.

Osterhammel, Jürgen. *The Transformation of the World: A Global History of the Nineteenth Century.* Princeton, NJ: Princeton University Press, 2014.

Otmazgin, Nissim. "A New Cultural Geography of East Asia: Imagining a 'Region' through Popular Culture." *Asia-Pacific Journal* 14, no. 7 (2016): 1–12

Oudshoorn, Nelly, and Trevor Pinch. *How Users Matter: The Co-construction of Users and Technology.* Cambridge, MA: MIT Press, 2003.

Pearson, James. "10 to 6: How Cathay's North American Operations Have Shrunk," *Simple Flying,* July 22, 2021. https://simpleflying.com/10-to-6-how-cathays-north -american-operations-have-shrunk/.

Pearson, Margaret M. "The Business of Governing Business in China: Institutions and Norms of the Emerging Regulatory State." *World Politics* 57, no. 2 (2005): 296–322.

Bibliography

Peters, Ed. "Remembering Kai Tak: Hong Kong Airport That Closed 20 Years Ago Is Gone but Not Forgotten." *South China Morning Post*, July 1, 2018. https://www.scmp.com/magazines/post-magazine/long-reads/article/2153099/remembering-kai-tak-hong-kong-airport-closed-20.

Pigott, Peter. *Kai Tak: A History of Aviation in Hong Kong*. Hong Kong: Government Printer, 1989.

Pirie, Gordon. *Cultures and Caricatures of British Imperial Aviation: Passengers, Pilots, Publicity*. Manchester, UK: Manchester University Press, 2012.

Polsky, Anthony. "Hong Kong, A Valuable Pawn." *Far Eastern Economic Review*, August 29, 1968, 411–12.

Pyun, Kyunghee. "Hybrid Dandyism: European Woolen Fabric in East Asia." In *Fashion, Identity, and Power in Modern Asia*, edited by Kyunghee Pyun and Aida Yuen Wong, 285–306. Cham, Switzerland: Palgrave Macmillan, 2018.

Report of the Salaries and Wages Committee, June 1955. Hong Kong: University of Hong Kong, 1955.

Richards, Isabella. "Cathay Confirms It's Closing Australian Base." *Australian Aviation*, June 4, 2021. https://australianaviation.com.au/2021/06/cathay-confirms-its-closing-australian-base/.

Rietsema, Kees Willem. "A Case Study of Gender in Corporate Aviation." PhD diss., Capella University, 2003.

Rimmer, Peter J. *Asian-Pacific Rim Logistics: Global Context and Local Policies*. Cheltenham, UK: Edward Elgar Publishing, 2014.

———. "Australia through the Prism of Qantas: Distance Makes a Comeback." *Oteman Journal of Australian Studies* 31 (2005): 135–57.

Robbins, Bruce. "Comparative Cosmopolitanisms." In *Cosmopolitics: Thinking and Feeling beyond the Nation*, edited by Pheng Cheah and Bruce Robbins, 246–64. Minneapolis: University of Minnesota Press, 1998.

Roces, Mina. "Dress, Status, and Identity in the Philippines: Pineapple Fiber Cloth and Ilustrado Fashion." *Fashion Theory* 17, no. 3 (2013): 341–72.

———. "Gender, Nation and the Politics of Dress in Twentieth-Century Philippines." *Gender & History* 17, no. 2 (2005): 354–77.

San Francisco Airport Commission. *Famous Firsts: The John T. McCoy Pan Am Watercolors*. San Francisco: San Francisco Commission Aviation Library, 2005.

Schenk, Catherine R. "Negotiating Positive Non-interventionism: Regulating Hong Kong's Finance Companies, 1976–1986." *China Quarterly* 230 (2017): 348–70.

———. "The Empire Strikes Back: Hong Kong and the Decline of Sterling in the 1960s." *Economic History Review* 57, no. 3 (2004): 551–80.

Schularick, Moritz. "A Tale of Two 'Globalizations': Capital Flows from Rich to Poor in Two Eras of Global Finance." *International Journal of Finance and Economics* 11, no. 4 (2006): 339–54.

Sekine Hiroshi 関根寛. *Keitoku kaisō: Honkon kokusai kūkō no rekishi to miryoku o tsumuida memoriaru sutōrī* 啓徳懐想: 香港国際空港の歴史と魅力を紡いだメモリアル・ストーリー (Reminiscing about Kaitak: Spinning the memorable story of the history

Bibliography 325

and charm of the Hong Kong International Airport). 1st ed. Tokyo: Tokimeki Publishing, 2008.

Silberstein, Rachel. "Fashioning the Foreign: Using British Woolens in Nineteenth-Century China." In *Fashion, Identity, and Power in Modern Asia*, edited by Kyunghee Pyun and Aida Yuen Wong, 231–58. Cham, Switzerland: Palgrave Macmillan, 2018.

Sinha, Dipendra. *Deregulation and Liberalisation of the Airline Industry: Asia, Europe, North America and Oceania*. Aldershot, UK: Ashgate, 2001.

Sinn, Elizabeth. *Pacific Crossing: California Gold, Chinese Migration and the Making of Hong Kong*. Hong Kong: Hong Kong University Press, 2012.

Smith, Andrew. "The Winds of Change and the End of the Comprador System in the Hongkong and Shanghai Banking Corporation." *Business History* 58, no. 2 (2016): 179–206.

Southern, R. Neil. "Historical Perspective of the Logistics and Supply Chain Management Discipline." *Transportation Journal* 50, no. 1 (2011): 53–64.

Stallings, Barbara. "The Globalization of Capital Flows: Who Benefits?" *Annals of the American Academy of Political and Social Science* 610, no. 1 (2007): 202–16.

Stockwell, Sarah. "Imperial Liberalism and Institution Building at the End of Empire in Africa." *Journal of Imperial and Commonwealth History* 46, no. 5 (2018): 1009–33.

Sugihara, Kaoru. *Japan, China, and the Growth of the Asian International Economy, 1850–1949*. Oxford: Oxford University Press, 2005.

Tagliacozzo, Eric, Helen F. Siu, and Peter C. Perdue, eds. *Asia Inside Out: Changing Times*. Cambridge, MA: Harvard University Press, 2015.

———. *Asia Inside Out: Connected Places*. Cambridge, MA: Harvard University Press, 2015.

Tarlo, Emma. *Clothing Matters: Dress and Identity in India*. London: Hurst, 1996.

———. "The Problem of What to Wear: The Politics of Khadi in Late Colonial India." *South Asia Research* 11, no. 2 (1991): 134–57.

Taylor, Jean Gelman. "Costume and Gender in Colonial Java, 1800–1940." In *Outward Appearances: Dressing State and Society in Indonesia*, edited by Henk Schulte Nordholt, 85–116. Leiden, Netherlands: KITLV Press, 1997.

Taylor, Peter J., and Derudder, B. *World City Network*. 2nd edition. London: Routledge, 2016.

Tsang, Steve. *A Modern History of Hong Kong*. Hong Kong: Hong Kong University Press, 2004.

———. *Governing Hong Kong: Administrative Officers from the Nineteenth Century to the Handover to China, 1862–1997*. Hong Kong: Hong Kong University Press, 2007.

Tsurumi, E. Patricia. *Factory Girls: Women in the Thread Mills of Meiji Japan*. Princeton, NJ: Princeton University Press, 1990.

Urata, Shujiro, Chia Siow Yue, and Fukunari Kimura, eds. *Multinationals and Economic Growth in East Asia: Foreign Direct Investment, Corporate Strategies and National Economic Development*. London: Routledge, 2006.

Ure, Gavin. *Governors, Politics and the Colonial Office: Public Policy in Hong Kong, 1918–58*. Hong Kong: Hong Kong University Press, 2012.

Vahtra, Peeter, Kari Liuhto, and Harri Lorentz. "Privatisation or Re-nationalisation in Russia? Strengthening Strategic Government Policies within the Economy." *Journal for East European Management Studies* 12, no. 4 (2007): 273–96.

Van Hook, James C. "From Socialization to Co-Determination: The US, Britain, Germany, and Public Ownership in the Ruhr, 1945–1951." *Historical Journal* 45, no. 1 (2002): 153–78.

Van Vleck, Jenifer. *Empire of the Air: Aviation and the American Ascendancy.* Cambridge, MA: Harvard University Press, 2013.

Varg, Paul A. "Myth of the China Market, 1890–1914." *American Historical Review* 73, no. 3 (1968): 742–58.

Vogel, Ezra F. *Canton under Communism: Programs and Politics in a Provincial Capital, 1949–1968.* Cambridge, MA: Harvard University Press, 1969.

———. *The Four Little Dragons: The Spread of Industrialization in East Asia.* Cambridge, MA: Harvard University Press, 1991.

Wang, David. "Wenxue xinglu yu shijie xiangxiang" "文學行路與世界想像" (Literary trajectories and world imagination), *United Daily News* 聯合報, July 8 and 9, 2006.

Wang Gungwu. "Hong Kong's Twentieth Century: The Global Setting." In *Hong Kong in the Cold War,* edited by Pricilla Roberts and John M. Carroll, 1–14. Hong Kong: Hong Kong University Press, 2017.

Welsh, Frank. *A History of Hong Kong.* Revised ed. London: Harper Collins, 1997.

White, Lynn T. *Careers in Shanghai: The Social Guidance of Personal Energies in a Developing Chinese City, 1949–1966.* Berkeley: University of California Press, 1978.

White, Nicholas J. *British Business in Post-Colonial Malaysia, 1957–70.* Abingdon, UK: Routledge, 2004.

———. "The Business and the Politics of Decolonization: The British Experience in the Twentieth Century." *Economic History Review* 53, no. 3 (2000): 544–64.

Wickramasinghe, Nira. *Dressing the Colonised Body: Politics, Clothing and Identity in Colonial Sri Lanka.* Hyderabad: Orient Longman, 2003.

Wilkins, Mira. "Foreword." In *The Routledge Companion to the Makers of Global Business*, edited by Teresa da Silva Lopes, Christina Lubinski, and Heidi J. S. Tworek, xiv–xviii. New York: Routledge, 2019.

———. "Role of Private Business in the International Diffusion of Technology." *Journal of Economic History* 34, no. 1 (1974): 166–88.

———. "The History of Multinationals: A 2015 View." *Business History Review* 89, no. 3 (2015): 405–14.

Williamson, Jeffrey G. "Globalization, Convergence, and History." *Journal of Economic History* 56, no. 2 (1996): 277–306.

Wong, Joanne. "Cathay Dragon Traffic Rights Are up in the Air: Govt." *RTHK*, November 9, 2020. https://news.rthk.hk/rthk/en/component/k2/1559047-20201109.htm.

Woods, Randall Bennett. *A Changing of the Guard: Anglo-American Relations, 1941–1946.* Chapel Hill: University of North Carolina Press, 1990.

Bibliography

Wu, Yu-Shan. "Taiwan's Developmental State: After the Economic and Political Turmoil." *Asian Survey* 47, no. 6 (2007): 977–1000.

Yano, Christine. *Airborne Dreams: "Nisei" Stewardesses and Pan American World Airways.* Durham, NC: Duke University Press, 2011.

Yoon, Sang Woo. "Transformations of the Developmental State into the Post-Developmental State: Experiences of South Korea, Japan, and Taiwan." *Asia Review* 9, no. 2 (2020): 159–89.

Young, Gavin. *Beyond Lion Rock: The Story of Cathay Pacific Airways.* London: Hutchinson, 1988.

Zhang Rui, Ngo Thi Viet Ha, and Wang Jianping. "Optimizing Sleeves Pattern for Vietnamese Airlines Stewardess Uniform—Ao Dai." *Proceedings of the 2015 International Conference on Computational Science and Engineering* 17 (2015): 246–49.

Zheng Yangwen, Hong Liu, and Michael Szonyi, eds. *The Cold War in Asia: The Battle for Hearts and Minds.* Leiden, Netherlands: Brill, 2010.

Index

Page numbers for figures and tables are in italics.

Aeroplane and Commercial Aviation News, 152

Ah Tai, 117, *118*

Air Cambodge, 188

Air China, 294, 296, 302

aircraft movements, 148, 184; Cathay Pacific, 94; Cold War dynamics and, 93; decrease after Communist takeover of mainland, 55; growth in, 67, 74, 154, 157, 178; lackluster recovery of, 64; number of, 4–5, 50, *95*, 183, *292, 298*; wide-body, 184, 188

Air France, 40, 62, 65

Air India, 66, 70, 136n88, 169, 301

Air Lao, 188

airport development, 68–74, 155–56

Airport Progress Committee, 70

Air Vietnam, 63–65, 188

ANA (Australian National Airways), 81–84, 88, *92*, 243–45

Anglo-Chinese air agreement (1947), 50

Around the World in Eighty Days (Verne, 1873), 41

Atkins, H. E., 176

ATLA (Air Transport Licensing Authority), 163, 209, 213, 214, 216–17, 223;

Dragonair and, 233, 254; Laker Airways and, 206–7, 223

Australia, 20, 46, 62, 76, 83, 97, 106, 195, 196, 307; BOAC operations in, 169, 170; Cathay Pacific pilots from, 264; Cathay Pacific routes expanded to, 198–203, *199, 201;* Imperial Airways route to, 24, 27, 31, 32, 40

aviation, commercial: capital investment in, 185–86, 292; centrality of Hong Kong, 10, 17, 295; dominance of North American airlines, 43, 149; global circuitries and, 303–9; military use of, 10, 49, 57; political history of, 22; remapping of global and regional networks, 21; as "space of flows," 14, 194, 287

Azuma Taruko, 111

baggage handling, 69, 178, 182

Balmain, Pierre, 133n81, 136n88, 136n90, 139, 202

Bangkok, 40, 62, 63, 64, 76, 89, 94, 117, *159, 199*

Baxter, Peter, 267, 269–70

Index

Beijing, 226–31, 233–34, 248, 249, 255–56, 280, 281, 294, 300
Bell, William H., 26
Bennett, J. C. Sterndale, 43
Berle, Adolf A., 44
Bermuda Agreement, 45, 47
Bixby, H. M., 38
Black, Robert, 70
Bluck, Duncan, 200, 250
BOAC (British Overseas Airways Corporation), 43, 44–45, 47, 48, 99, 235, 244, 290; Britannia aircraft, 69; cargo traffic and, 152; Cathay Pacific as rival, 101; Cathay Pacific shares owned by, 245–46, 282; Comet 4 plane, 71; competition to be Hong Kong's airline and, 85, 86–93; competitors for control of civil aviation, 80; foreign traffic rights restricted in interests of, 161; Hong Kong Airways as subsidiary of, 77–79; Hong Kong/Australian franchise and, 83; lament over lack of UK government support, 63; leading position in Hong Kong, 66; local operators and, 100; Malayan Airways funded by, 301; number of weekly flights, 62, 63, 64; passenger revenues in Hong Kong, 169; pilots, 264; restrictive landing rights for benefit of, 177; share of traffic in Hong Kong, 68; shares of Malaysian Airways owned by, 189; Speedbird emblem, 105; traffic rights in Hong Kong, 97, 170
Board of Trade, 162, 163, 166–67, 169; financing of infrastructure upgrade and, 172, 174, 180; value of carriers' rights and, 170
Boeing 747s. See "Jumbo jets" (Boeing 747s)
Bombay, 66, 159
Borneo Company Limited, 84–85, 92, 245
Bowden, Herbert, 157
Braathens SAFE Airtransport A/S, 62
Britain/British Empire, 22, 23, 27, 32, 41, 301; attempted postwar reestablish-

ment of, 47; CAA (Civil Aviation Authority), 212–19, 223, 224; Cathay Pacific pilots from, 264; civil aviation as part of British strategy, 43; Colonial Office, 72, 78, 83, 88, 99, 100; Commonwealth Office, 163, 165–66; decolonization and, 61, 68, 97, 99; importance of Hong Kong to, 51, 61, 97; local interests of Hong Kong and, 197; maximized value of colonial holdings, 165–71; refashioned into the Commonwealth, 58; US-British competition, 44–46, 49, 57, 63, 81. See also Sino-British treaties and agreements
British Aerospace Flying College (Prestwick, Scotland), 267, 268, 269, 277
British Airports Authority, 186
British Airways, 8, 200, 203, 210, 211, 223, 229; Cathay Pacific shares owned by, 246; divergence of interests with Cathay Pacific, 247; London–Beijing route and, 228; monopoly over Hong Kong–London route, 212–15, 220, 222, 235; privatization and, 249
British Caledonian Airways, 136n88, 212–14, 216, 219, 220, 222, 224, 225
British Commonwealth, 64, 96, 97, 113, 239, 278, 283, 284, 290, 301; Cathay Pacific pilots from, 122, 123, 263; leading role in civil aviation in Hong Kong, 65
British Ministry of Civil Aviation. See Ministry of Civil Aviation
Broadbent Report, 69
Browne, H.J.C. (John), 175
Brunei, 94, 159, 199, 277n170
Bryan, Paul, 215
Burma, 27, 63, 75, 170
Butterfield & Swire, 82, 84, 91
Butters, H. R., 31

C-47 planes, 75, 76
CAAC (Civil Aviation Administration of China), 228, 229, 230–34, 259

Index

Calcutta, 63, 65, 89, 94, 108, 109, 121, *159*
Caldecott, Andrew, 31, 36
Canada, 46, 64, 196, 203–9, 307
Canadian Pacific Airlines, 62–63, 64, 70, 71, 204, 205, 209
Canton, 36, 40, 47, 49–51 77, 86; captured by Communist forces (1949), 53; Hong Kong Airways routes and, 79–80, 85; Japanese capture of, 42
Cantonese language, 273, 278
cargo traffic, 152, 164, 179, 184; asymmetrical distribution of, 152, 183, 187; post-handover growth in, 297, *298*
Cartier, Carolyn, 304
CAT (Civil Air Transport), of Nationalist China, 63, 65–66
Cathay Pacific Airways, 1, 8, 65, 80, 175, 269; bilateral air service agreements and, 162, 253; BOAC agency and, 88; BOAC as rival, 101; as British Commonwealth concern, 85; cadet program for local recruitment, 272–73, 274, 277–80; cosmopolitan branding of, 103–4; COVID-19 pandemic effects on, 306; customers, 67; Dragonair incorporated by, 296–97, 303, 307; Dragonair rivalry with, 255, 270–72, 274–75, 280; established as British airline in Hong Kong, 74–85; "geopolitical jockeying" of, 242–43, 300–301; Hong Kong authorities' identification with, 197; as "Hong Kong's airline," 220; as iconic Hong Kong carrier, 18; incorporation of (1946), 76; investor base, 19; Kantai service for Japanese clientele, 129, 131; as leader in Singapore market, 66–67; as "local" British airline, 97; local Hong Kong identity and, 246–54; logo designs of, 106–7, *107*; mainland Chinese investments in, 254–63; maintenance crews, 141; majority stake sold to Swire, 81; merger with Hong Kong Airways, 90–92, *92*; number of weekly flights, 62, 64; passenger

counts, *96*, 202, 204, 222, 232; political positioning and continuing success of, 299–303; politics in competition to be local airline, 270–86; PRC regime and, 238; private ownership of, 242; recast identity of, 18; as regional player in Southeast Asia, 194; sovereignty transfer to PRC and, 283–85; staffing policy, 103. *See also* flight attendants, Cathay Pacific; pilots, Cathay Pacific
Cathay Pacific Airways advertising, 102, 107–9, 141, 143n105, 193; for Hong Kong–London route, *221*; "many faces of the Orient" ad, *142*; for nonstop service to Vancouver, 207–9, *208*
Cathay Pacific branding, 116, 140–41, *142*, 143–47; brand refreshment in changing business environment, 133; corporate name change (2021), 308; staffing profile, 109–23; uniform of female cabin crew and, 104–5, 123, *124*, 125–29, *130*, 131–33, *132*, *135*, *139*, 140
Cathay Pacific routes: expansion to Australia, 198–203, *199*, *201*, 234; into London as heart of imperial network, 210–25, *221*; mapping Pacific routes of, 105–9, *108*; opening of routes into PRC, 225–37, 300; Southeast Asia regional pattern, 93, *94*; trans-Pacific route to North America, 203–10, *208*, 234, 295–96, 307
Cathy Pacific planes: *Betsy* (DC-3), 106, *107*; Boeing 707s, 200, 202, 207, 210; Convair 880-22M fleet, 200; DC-3 planes, 70; DC-6B planes, 70; DC-10s, 214; *Electra*, 109–10, 111, 202; "Jumbo jets" (Boeing 747s), 214, 266; Lockheed L-1011 Super TriStar, 207, 266; *Niki* (DC-3), 106, *107*
Cathay Pacific shareholders, 245–54, *251*, 286, 303; CNAC, 263; HSBC, 215, 244, *246*, 247, *251*, 252, *253*, 255–57, 259; Jardine Matheson, *92*; Swire, *92*, *246*, 247, *253*, 255, 260, 263, 281, 299–302

332   *Index*

Central Air Transport Corporation, 55, 85. *See also* Eurasia

Chak, Rosa, 272, 275

Chan, Anson, 2

Chao Kuang-pui, 254

Chek Lap Kok Airport, 4, 6, 7, 262; Air Force One at, 294; Cathay Pacific *Polar One* route and, 296; inaugural day of, 294, *295*

Cheng, Alice, 110

Cheng, Diane, 119

Cheng, Josephine, 111–12, 114, 115

cheongsam (*qipao*), uniform design and, 126, 128, 129, 131–33, 137, 146

Cheuk, Katherine, 109

Cheung, Oswald, 215

Cheung Kong (Holdings) Limited, 252, *253*

China, mainland, 7, 19, 22; Civil War in, 51, 53; Communist takeover (1949), 11, 17, 23, 57–58, 60, 86, 98–99, 156, 303; Hong Kong as gateway to, 20; investors in Cathay Pacific from, 254–63; markets of, 9; reconstituted ties to, 13, 225–34; Reform Era, 194, 293; Republican (Nationalist) government in, 25. *See also* PRC (People's Republic of China); Sino-British treaties and agreements

China, People's Republic of. *See* PRC

China Airlines, 226, 302

China Clipper, 37

China Navigation Co. Ltd., 82, 83, 84, *92*, 245

Chongqing (Chungking), 1, 40, 49, 50

Chu Chang-sing, 38

CITIC (China International Trust and Investment Corporation), 255–59, 261, 263, 265, 282, 296

Civil Aeronautics Board, US, 46

Civil Aviation Acts, 224

Civil Aviation Department, Hong Kong, 42, 188, 196, 202, 268

Clemmow, Captain Gerry, 274

Clinton, Bill, 294

clippers ("flying boats"), 17, 34, 35, 37, 38, 288

CNAC (China National Aviation Corporation), 21, 33, 38–40, 49, 50, 55, 74–75; collapse of, 302; "flying boat" of, 35–36; Hong Kong as junction for, 41; Pan Am's investment in, 34; Shanghai route and, 85; trial runs from Hong Kong to Shanghai, 25

CNAC, reincarnated under PRC, 282; Air China and, 296; Cathay Pacific shareholding percentage with, 259–61, 263

Cold War, 7, 9, 12, 13; Cathay Pacific's regional space and, 103; commercial air flow and, 93; decolonization and, 61–62; economic growth and, 293; escalation of, 67; Hong Kong as foothold of "free world" in China, 11, 17; Hong Kong exports to Western nations and Japan, 149; industrialization during, 18; shifting dynamics of, 183. *See also* geopolitics

colonialism, 191, 243, 248

Concorde (aircraft), 157

Convention for the Extension of Hong Kong Territory (1898), 72

Convention on International Civil Aviation (Chicago, 1944), 44, 45, 162, 174

Convention of Peking (1860), 72

Convention on the Territorial Sea and the Contiguous Zone (1958), 72

Cortazzi, Hugh, 211–12

cosmopolitanism, 15, 103–4, 109, 122, 133, 140, 144, 145

COVID-19 pandemic, 306, 309

Cowden, William, 177–78

Cowperthwaite, John, 148, 163–64, 189; as advocate for Hong Kong autonomy, 173–74; on British responsibility to Hong Kong, 175, 180; career background of, 173

Craik, Jennifer, 131
Critchley, Alfred Cecil, 81
CTS (China Travel Service), 259, 282

decolonization, 61, 68, 99, 234, 301; economic and political, 284; ethnic identities and, 146; Hong Kong dissociated from metropolitan interests, 190, 191; shrinking British footprint and, 97; successor institutions and, 281
Deep Bay, 48, 49, 51, 99; abandonment of airport project at, 54, 55, 68; airport construction at, 52; plans for, 52–53
De Kantzow, Syd, 75, 82, 83, 84
deregulation, 193, 194, 223, 236; liberalization of air traffic rights and, 235; limitations of, 224
Donald, Alan, 232
Dorado (Imperial Airways liner), 31–32, 39
Dragonair, 1, 233, 249, 250, 252, 254–55, 260–63; Cathay Pacific purchase of, 296–97, 303, 307; as challenger to Cathay Pacific, 255, 270–72, 274, 280; mainland Chinese investors and, 281; partnership with Cathay Pacific, 257–58; pilots, 278
Dunnaway, Cliff, 242
Dupuy, M., 40

East Pakistan, 63
Eddington, Rod, 274
Edgerton, David, 58
EEC (European Economic Community), 167, 168
English language: in Cathay Pacific advertisements, *201*, 208–9, *221;* Cathay Pacific flight attendants and, 102, 111, 114–18, 120–22, 143, 263; Cathay Pacific recruitment and, 267, 268, 273, 277, 278; Hong Kong newspapers in, 24, 35, 92, 121, 129, 181, 200, 217, 255, 259; Hong Kong pilots and, 268
entrepreneurs, 20, 22, 42

ethnicity, fashioning of, 127–28, 131–32, *132*
Eurasia (Sino-German venture), 39, 40, 55

"Far East," 107
Far Eastern Aviation, 38, 80, 82
Far East Regional Air Services, 91
Farrell, Roy, 75, 83
"first" flights, 41
flight attendants, Cathay Pacific, 103, 109–10, 270; hiring and retention policy for, 116, 117; ideologies of Asian femininity and, 121–22, 264; "Kathy the air hostess" cartoon, 117, *118*; language skills of, 114, 116, 117–18, 121, 141, 143; male, 128; names given, 109, 110, 111; portrayed as having glamorous lives, 112–13, 116; staff attrition through matrimony, 113–14, 116; training program for, 115; Western beautification standards and, 118–20. *See also* uniforms, of Cathay Pacific flight attendants
Foreign and Commonwealth Office, British, 167–68, 175, 177, 180, 210–12, 215, 230–31
Foreign Office, British, 53
France, 25, 35, 50
"freedoms of the air," 45–46, 169

Gallagher, John, 190
Gatwick Airport (London), 214, 218, 220
geopolitics, 7, 9, 10, 176, 286; airlines' cessation of operations and, 188; business choices and, 284; commercial interests and, 22; corporations and, 238; development of air routes and, 100; economic opportunities and, 151; "freedoms of the air" and, 45–46; "geopolitical jockeying," 239–43, 300–301; hopes engendered and dashed by, 98; reshaped by air travel, 14; risks for Hong Kong after victory in Civil War, 58; shifts in, 12, 18, 19. *See also* Cold War

334   *Index*

Germany, 307
Gillespie, Ronald D., 50
globalization, 6, 7, 10, 11–13, 19, 305
Grantham, Alexander, 52–54
Guam, 36, 46, 206, 207

Hainan, 50, 65
Hanoi, 29, 33, 40, 62, 63, 65, 89
helicopters, 183
Heracleous, Loizos, 242
Heron-Webber, Stephanie, 276
Hickson, Ken, 242
Ho, Cherry, 131
Hole, G. F., 31
Holyman, Ivan, 82
Hong Kong: boundaries of, 72, *73*, 74;
as British imperial outpost, 22, 42, 57;
British return (1945), 43, 289; cessation
of aerial connections with mainland
China, 54–55, 86, 303; as Cold War
contact zone, 61–62; commercial
aviation in regional and global history,
10–16; conflicting interests with
London, 9, 161–65, 191, 212; "flexible
citizenship" and, 285; growth as
metropolis, 6–7, 20; Japanese occupation in World War II, 42–43; Legislative
Council, 50, 52, 154, 157, 158, 165, 175,
179, 180, 184, 211, 215; in "Little
Dragons" group, 190, 240–41; local
identity of Chinese migrants in, 145;
proximity to mainland China, 9,
289–90; riots (1967), 160, 175, 179; as
shipping center, 12–13, 24, 49, 56, 288;
sovereignty transfer to PRC, 6, 11, 239,
256, 283; as Special Administrative
Region of PRC, 233, 239, 253, 260,
301; tourism industry in, 112, *113*,
134, 138, 153, 156; transnational qualities
of, 143; wired into global circuitries,
303–9
Hong Kong, as aviation hub, 1, 5, 6, 102,
289, 299; adaptation to Communist
takeover of mainland, 60; beginning
as linchpin (1937), 40; British colonial

authorities and, 26–27; British hub for
Chinese traffic, 42–59; COVID-19
pandemic and, 309; making of prewar
hub, 23–27, *28*, 29–42; overcoming of
obstacles and, 57; post-handover
networking, 294–97, *295*, *298*, *299*;
reconfigured air space in Cold War,
62–68; reopening of mainland market
and, 305; in Southeast Asia, 66;
transition from seaport, 25, 38; World
War II allied planning and, 98
"Hong Kong, A Valuable Pawn" (*Far
Eastern Economic Review* article, 1968),
164–65
Hong Kong, economic growth/
development of, 8, 10, 13–14, 16, 102;
air cargo and, *150*, *298*; autonomy of
the colony, 195–98; business history of
aviation and, 19; costs of, 156–65;
economic opportunities in politically
liminal space, 288–94, *292*; export-led,
149, *150*; financial autonomy of the
colony, 171–75; globalization and,
305–6; ties to British pound sterling,
160
Hong Kong Air Transport Licensing
Regulations, 163
Hong Kong Airways Ltd., 63, 70, 229;
competition to be Hong Kong's airline
and, 85–90, 93; merger with Cathay
Pacific, 90–92, *92*; routes into mainland
cities, 79–80; as subsidiary of BOAC,
77–79
Hongkong Clipper, 38–39, *39*, 40
Hong Kong International Airport, 185,
192, *298*
Hong Kong Management Association,
216
Hong Kong Rotary Club, 24, 32
Honolulu, 36, 46, 206
HSBC (Hongkong and Shanghai
Banking Corporation), 215, 244, 245,
246, 246–48, 250, *251*, *253*, 255–56, 259,
283
Hua Guofeng, 227

Hutchison Whampoa Limited, 252, *253*
Hysan Development Company Limited, 252, *253*, 259

IATA (International Air Transport Association), 307
Imperial Airways, 22, 26, 39, 56, 57, 289; British colonial government and, 27; entry into Hong Kong, 30–32; founding of (1924), 23–24; Hong Kong as junction for, 41; "horseshoe" service, *28*; mail delivery as primary task of, 24, 30, 31–32; rivalry with Pan Am, 33; routes of, 43; Swire and, 25
imperialism, 61, 190, 289
India, 24, 43, 63, 170, 196
Indo-China, French, 37, 64, 79, 106
Indonesia, 63, 170, 196
industrialization, 146, 152, 160, 190
infrastructure, aviation, 47, 57, 59, 68; baggage containerization and, 182; colonial government and, 18; economic expansion and, 148; funding of, 185–87, 195; geopolitics and, 60; Hong Kong as global linchpin of, 40; investment in, 61, 185; in mainland China, 282; reduced demand for development of, 99; reduced scale of construction after Communist victory, 58; runway extension, 171–72, 188; technology and, 287; World War II expansion of, 43
International Civil Aviation Organization, 67, 164
Ip, Jack, 275

Jackson, Isabella, 191
Jandacot Flying College (Australia), 268
Japan, 17, 63, 78, 87, 100, *113*, 294; air service agreements and, 196; increase in visitors to Hong Kong, 138; liberalization of outbound travel (1964), 121
Japan Air Lines, 66, 136n88, 154, 170, 182, 204, 301
Japan Asia Airways, 226

Jardine Matheson company, 13, 77, 86, 88, 89, 93; Cathay Pacific shareholding percentage with, *92*; resurfacing in 1970s, 233; Swire rivalry with, 90, 91, 288
Jeaffreson, D. G., 210, 211
Jeffries, C. W., 31
Jiang Zemin, 294
"Jumbo jets" (Boeing 747s), 175–83, 207, 211, 235, 266, 267, 296; Cathay Pacific trans-Pacific route and, 205; on Hong Kong–London route, 214–15, 218; increase in passenger count and, 183

Kadoorie, Lawrence, 46–47
Kai Tak Airport, 2, 35, 275–76, 295; accelerated traffic growth and, 151–56; annual number of passengers, *150*; beauty course for flight attendants at, 119, 120; capacity of, 179; China as predominant user of, 50; conflict over cost of expansion project, 156–65; cosmopolitanism and, 144; critiqued for drawbacks, 5–6, 47–49; enlarged by Japanese occupation forces, 42–43, 47; evacuated in World War II, 42; expansion/extension of, 55; funding of expansion for, 172; growth in volume of traffic, 52; infrastructure development at, 69–71; "Jumbo jet" (Boeing 747) arrival (1970), 182–83; laments over closure of, 4; last flights, 1, *3*; military use of, 49; modernization of, 51; number of airlines with service to, 50; origins of, 42; passenger counts at, *95*, *96*; planning for Jumbo Jets, 175–81, 194; runway extension, 185, 188; runway lengths, 47, 153, *159*, 160; technological upgrading of, 93; terminal, 155, *155*
King, Ambrose, 279
KLM (Royal Dutch Airlines), 40, 165–69
Knollys, Lord, 43
Korea, South, 17, 63, 66, 100, 196, 294, 301
Korean Air Lines, 301

Index

Korean National Airlines, 66, 301
Korean War, 66
Kowloon Bay, 38, 42, 72
Kowloon City, 1, 4, 72
Kuala Lumpur, 32, 153, *159, 199*
Kuching, 94
Kunming, 50
Kwan, Bancho, 271

Labuan, 62, 64, 67, 89
Laird, E. O., 180
Lai Wah of Kowloon, 134, 136, 145
Laker, Freddie, 224
Laker Airways, 205–7, 212–14, 218, 219, 223–24
Lane Crawford (store), 119
Lantau Island, 4
Lau, Eddie, 297
Lau Chi-pang, 242
Lee, H.C., 259
Levine, Derek, 242
Lewis, Mary, 110
Liang Bun, Dennis, 269
Li Ka-shing, 252, 259
Lock, Captain J. H., 32–33
Los Angeles, 46, 66, 68, 204, 206, 207, 295

Macao, 29, 30, 33, 76, 78, 83, 258; *China Clipper* arrival in (1937), 37–38; inability to compete with Hong Kong as hub, 41; as official Pan Am terminus, 36–37; as proposed "Oriental terminus," 34
MacLehose, Murray, 196–97, 203–4, 217, 230; on Cathay Pacific route to London, 219; on Sino-British negotiations over air routes, 231–32
Malaya, British, 17, 27, 106, 189
Malayan Airways, 82, 90, 127, 301
Malaysia, 121, 170, 212
Malaysia Airlines, 301
Malaysian Airways, 189, 249
Manila, 33, 47, 63, 83, 89, 153, *159, 199*; Bermuda Agreement and, 46; Cathay

Pacific routes and, 94; Pan Am clippers in, 36, 37, 39, 40; US control of, 41
Mao Zedong, 227
Mayuzumi Keiko, 131
McLaren, R.J.T., 210
Middle East, 27
Midway Island, 36, 46
Miles, H.M.P. (Michael), 221, 249, 256
Miller, Steve, 278
Ministry of Aviation, British, 156, 177
Ministry of Civil Aviation, 69, 78, 80, 88
modernity, 7, 14, 15, 146
Mori, Hanae, 136n88
Mulley, Fred, 166

Nagahata Tomoko, 131
Nanjing, 47, 50, 51, 81, 230
Nelson, E., 31
Netherlands, 25, 35, 165, 167, 168, 196
New Territories, 53, 72
New Zealand, 27, 46, 170, 203, 264, 307
Ng, James, 242
Ng Siu-hoi, Jennifer, 278
Niugini (Papua New Guinea), 196
North America, 13, 17, 21, 34, 66, 208, 294
North Borneo, 65
Northwest Airlines, 46, 87, 89, 154, 182
Norway, 62
Nott, John, 213, 216, 223

oil crisis, global (1973), 183
Okinawa, 87, 89
Ong Ee-lim, 32
"open skies" policy, 44, 165–66
"Orient, the," 103, 107–9, 131–33, 143, 263
Oriental Airways Express Company, 77
Orient Travel (monthly news magazine), 116, 119

P&O (Peninsular & Oriental Steam Navigation Company), 84, *92*, 245–46

Index

Pan American (Pan Am), 8, 25, 40, 70, 204, 302; advertising for, 143n105; all-male clipper crews, 110; Asian network of, 21–22; bases for trans-Pacific service, 29; BOAC competition with, 66; British colonial resistance to, 27; clippers ("flying boats"), 35, 110; DC-8 plane, 68; expanded routes of, 23, 24; founding of (1927), 23; Hong Kong as junction for, 41; logo designs of, 105–6, *106*; Manila route dominated by, 86; potential funding of Kai Tak expansion and, 173; rivalry with Imperial Airways, 33; routes of, 43; share of traffic in Hong Kong, 68; uniforms of female cabin crew, 138n95; US control of Manila and, 41; World Airways trunk route, 64–65

Pangarkar, Nitin, 242

Pao, Sir Yue-kong, 252, 254, 255, 258

passenger counts, 5, *5*, 52, 93, 183, 204, *298*; Cathay Pacific as leader in, 202, 204, 222; on Cathay Pacific Shanghai route, 232; at Kai Tak Airport, *95, 96, 150*

"passenger flow," 178

Patten, Chris, 262

Pearl River Delta, 22

Penang, 27, 32, 41

Philippine Air Lines, 63, 65, 66, 301; logo design of, 105–6; as "Route of the Orient Star," 105

Philippines, 17, 36, 50, 66, 100, 294; in Cathay Pacific route map, 106; as gateway to United States, 64; Pan Am 1930s flights to, 23, 26

Phnom Penh, 94, *159*

pilots, Cathay Pacific: Caucasian pilots, 74–75, 122, 128, 141, *142*, 276; female pilots, 274–76; local Chinese presence in cockpit, 18–19, 263–70, 275, 276–79

Pirie, Gordon, 242

Polo, Marco, 103, 207, 308

pooling arrangements, 170

Portugal, 29, 41

PRC (People's Republic of China), 8, 60, 217, 225; Cathay Pacific routes into, 225–37; Hong Kong exports to, 291, *292*; Hong Kong sovereignty transfer to, 6, 11, 239, 256, 283–85; June Fourth crisis (1989), 257; reemergence in 1970s after Cultural Revolution, 9; reopened airways of, 236. *See also* China, mainland; Sino-British treaties and agreements

privatization, 8, 18, 223, 235, 241, 305

Qantas (Queensland and Northern Territory Aerial Service), 24, 70, 83, 88, 169; Boeing 707 and, 244; Cathay Pacific service to Australia and, 200, 202, 203; competition with Cathay Pacific, 95; "Kangaroo service" with UK, 97; pilots, 264; Qantas Empire Airways, 62, 64; uniforms of female cabin crew, 136n88

Qian Qichen, 262

RAF (Royal Air Force), 31, 48, 70, 268

Rangoon, 63, 77, 89, *159*

Roberts, C.C., 88

Robinson, Ronald, 190

Rodgers, William, 178

Ross, George, 154, 158, 171–72

Roy Farrell Export-Import Company, 75, 76

Royle, Anthony, 176

Saigon, 31, 32, 47, 62, 63, 65, 89, *159, 199*

San Francisco, 206

Sato Sachiko, 111, 112

Seattle, 206

Seoul, 66, 89, 92, 108, *159, 199*

Shanghai, 25, 36, 37, 50, 75, 126, 191, 228; Cathay Pacific routes into, 225, 229; Hong Kong Airways routes and, 79

Shek Kong, 54

Shepherd, Malcolm, 177

Shull, Rudella, 125, 126

Siegel, Richard, 1, 2

338   *Index*

Sikorsky planes, 38

Singapore, 17, 24, 30, 47, 63–65, 76, 89, 108, *159*, *199;* air service agreements and, 196; Paya Lebar airport, 176

Singapore Airlines, 136n88, 139, 143n105, 204, 242, 249, 264n124, 277n170, 301

"Singapore Girl," 133n81, 139, 143n105

Sino-British treaties and agreements, 72, 73, 79, 263; CMUs (confidential memoranda of understanding), 228, 230; Joint Declaration (1984), 226–33, 248, 252, 253, 260, 279, 294

Skyways, 80, 81

Slough, David, 272

Smith, N. L., 38

So, Eleanor, 112

Southeast Asia, 63, 79, 85, 93, *94*, 140, 234, 287; colonial regimes in, 22; network connections to Europe and North America, 294

Sun Fo, 35

Sun Yat-sen, 35

Sutch, Peter, 248, 251, 254, 257, 259, 263

Swire, Adrian, 211

Swire, John Kidston (Jock), 56, 82, 83

Swire company, 13, 19, 81, 175, 215, 233, 248, 257; Cathay Pacific shareholding percentage with, 82, 84, *92*, 245, *246*, 247, *253*, 255, 256, 259, 260, 263, 281, 299; as Commonwealth conglomerate, 243; development of commercial aviation and, 56; Jardine Matheson rivalry with, 90, 91, 288; perceived Britishness of, 281; setbacks to expansion, 88; as shipping company, 25

Sydney, 75, 76, 94, 199, *199*, *201*

Taipei, 63, 64, 86–87, 89, 92, *159*, *199*

Taiwan, 60, 78, 100, 225, 294; in Cathay Pacific route map, 106; Hong Kong–Taiwan flights as "regional air traffic," 226; as Nationalist base, 55

Takeuchi Suma, 111

Tam, Rose, 112

Tata Air Lines, 301

technology, 7, 17, 58, 137, 236, 287; common platform, 40; consumption of, 14; evolution of, 56; improvements in commercial aviation and, 148; legacy of World War II and, 144; rapid developments in, 144

Thailand (Siam), 50, 196

Thatcher, Chris, 271

Thatcher, Margaret, 223, 224, 232

Tianjin, 230

Tokyo, 47, 64, 87, 109, *159*, *199*, 205, 206

Tongtaam, Somruedee, 121

Tourane (Da Nang), Vietnam, 31, 32

tourism, 9, 134, 138, 153, 156; airport expansion and, 159; incoming visitors to Hong Kong by nationality, 112, *113*

transit passengers, 183

transnationalism, 8, 15–16

Trench, David, 156, 157, 158, 160, 161

Trippe, Juan, 29

trunk routes, 30, 40, 51, 66, 174; BOAC monopoly over, 88; defined, 27n29; global circuitries and, 304; Hong Kong as hub for, 64; of Pan Am World Airways, 64–65

Tsang, Donald, 2

Tsang, Steve, 279

Tung, Jenny, 120

Tung Chee-hwa, 294

TWA (Trans World Airlines), 153–54, 301

uniforms, of Cathay Pacific flight attendants, 104–5, 123, 146–47; Chinese-inspired design (1962–1969), 125–29, *130*, 134, 145, 146; conservative military-inspired design (1940s–1950s), 123, *124*, 125; mixed with national costumes, 129, 131–33, *132*; post-handover Lau design (1998), 297; pragmatic design (1969–1974), 134, *135*, 136–38; Tung Hoi (Eastern Seas) ensemble and French couture (1974–1998), 138–40, *139*, 146, 202

Union of Burma Airways, 188

Index

United Arab Emirates, 224
United Kingdom (UK), 43, 47, 63, 65, 169; Air Navigation Order, 163; air service agreements with foreign countries, 189, 196; BOAC trunk routes from, 64, 78; British Airports Authority, 186; Cathay Pacific charter to, 76; "colonial formula" and, 174; Hong Kong exports to, 149, *292*; Qantas "Kangaroo service" to, 97; US-British competition in aviation, 87, 161. *See also* Britain/British Empire; Sino-British treaties and agreements
United Nations (UN), 217
United States, 63, 291; emergence as global leader, 43, 56; expansion in Far East, 6; Hong Kong exports to, *292*; retreat from Southeast Asia, 183; US-British competition, 44–46, 49, 57, 63, 81
United States Aircraft Export Corporation, 24

Vancouver (Canada), 203, 206, 207, 209, 295
Verne, Jules, 41
Vientiane, 94, *159*
Vietnam, 29, 31, 65

Wake Island, 36, 46
Wales, Captain Alec, 264
Walsh, Willie, 307
Wilson, David, 267
Wirtz, Jochen, 242
Wong, Wilfred, 185
Wong Kwok Ho, 269
World War II, 42–43, 44, 173, 309; Cathay Pacific pilots as veterans of, 122; commercial aviation and, 74, 75; female employees' entry into cabin crew during, 110; vulnerability of colonial powers exposed by, 61
Wu, Candy, 275–76

Harvard East Asian Monographs
(most recent titles)

438. Lawrence C. Reardon, *A Third Way: The Origins of China's Current Economic Development Strategy*
439. Eyck Freymann, *One Belt One Road: Chinese Power Meets the World*
440. Yung Chul Park, Joon Kyung Kim, and Hail Park, *Financial Liberalization and Economic Development in Korea, 1980–2020*
441. Steven B. Miles, *Opportunity in Crisis: Cantonese Migrants and the State in Late Qing China*
442. Grace Huang, *Chiang Kai-shek's Politics of Shame: Leadership, Legacy, and National Identity*
443. Adam Lyons, *Karma and Punishment: Prison Chaplaincy in Japan*
444. Craig A. Smith, *Chinese Asianism, 1894–1945*
445. Sachiko Kawai, *Uncertain Powers: Sen'yōmon and Landownership by Royal Women in Early Medieval Japan*
446. Juliane Noth, *Transmedial Landscapes and Modern Chinese Painting*
447. Susan Westhafer Furukawa, *The Afterlife of Toyotomi Hideyoshi: Historical Fiction and Popular Culture in Japan*
448. Nongji Zhang, *Legal Scholars and Scholarship in the People's Republic of China: The First Generation (1949–1992)*
449. Han Sang Kim, *Cine-Mobility: Twentieth-Century Transformations in Korea's Film and Transportation*
450. Brian Hurley, *Confluence and Conflict: Reading Transwar Japanese Literature and Thought*
451. Simon Avenell, *Asia and Postwar Japan: Deimperialization, Civic Activism, and National Identity*
452. Maura Dykstra, *Empire of Routine: The Administrative Revolution of the Eighteenth-Century Qing State*
453. Marnie S. Anderson, *In Close Association: Local Activist Networks in the Making of Japanese Modernity, 1868–1920*
454. John D. Wong, *Hong Kong Takes Flight: Commercial Aviation and the Making of a Global Hub, 1930s–1998*
455. Martin K. Whyte and Mary C. Brinton, compilers, *Remembering Ezra Vogel*
456. Lawrence Zhang, *Power for a Price: The Purchase of Official Appointments in Qing China*
457. J. Megan Greene, *Building a Nation at War: Transnational Knowledge Networks and the Development of China during and after World War II*